Climate Action

Published by **Sustainable Development International** in partnership with the **United Nations Environment Programme**

ISBN: 978-0-9554408-5-4
Published December 2007

Sustainable Development International
Henley Media Group
Trans-world House, 100 City Road
London EC1Y 2BP
Tel: +44 (0)207 871 0123
Fax: +44 (0)207 871 0101
www.climateactionprogramme.org

Division of Communications and Public Information
United Nations Environment Programme (UNEP)
PO Box 30552
Nairobi, Kenya
Tel: +254 20 762 3292
Fax: +254 20 762 3927
www.unep.org

Publisher:	Jane Henry
Commercial Director:	David Amos
Sponsorship (UK Office):	Keith Bailey, Michael Gray, John Fellows, Raj Power, David Ramos, Geraldine Roux
Editorial Managers:	Natalia Marshall, Matt Duncan
Editorial Assistants:	John Cummings, Elke Bode
Sub Editor:	Lesley Smeardon
Design:	Andy Crisp
Cover Design:	Tina Khanian
Print:	Westerham Press
Distribution Manager:	Diva Rodriguez

To purchase further copies of this publication, please visit www.climateactionprogramme.org

As part of Henley Media Group's commitment to corporate social responsibility we have chosen to aim for carbon neutrality. The carbon footprint of producing and distributing Climate Action, including the consumption of energy and water, and generation of waste and by-products, will be calculated and offset.

Printed on:
cyclus
let the paper talk
CyclusPrint 100gsm
CyclusPrint 350gsm
www.cyclus.dk

CLIMATE ACTION

The Publishers wish to thank all the individuals and organisations who have contributed to this book.

In particular we acknowledge the following for their help and advice in producing Climate Action:

Eric Falt, Director, Division of Communications and Public Information, UNEP

David Simpson, Editor, Division of Communications and Public Information, UNEP

Robert Bisset, Head of Communications, Division of Technology, Industry and Economics, UNEP

Cornis Van Der Lugt, Programme Officer Corporate Responsibility, Division of Technology, Industry and Economics, UNEP

Geoff Lean, Consultant

Peter Fries, Consultant

VISIT: WWW.CLIMATEACTIONPROGRAMME.ORG

KPMG

CONTENTS

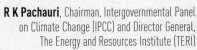

CONTENTS

CONTENTS

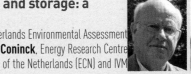

Foreword

ACHIM STEINER
UN UNDER-SECRETARY GENERAL AND
EXECUTIVE DIRECTOR
UN ENVIRONMENT PROGRAMME (UNEP)

CLIMATE ACTION

Climate Action is produced by Sustainable Development International in partnership with the United Nations Environment Programme to encourage and assist governments and business to lower greenhouse gas emissions. This book and supporting website feature a range of articles that encourage the sharing of best practice and the development of new technologies and initiatives and illustrates the opportunities for business and governments to reduce costs and increase profits while tackling climate change.
There are a number of steps that you, as business and government leaders, can take to reduce your carbon footprint; the second part of this book is dedicated to these actions. Some require little investment in time or money, while others require substantial time and capital. What they all require is a commitment to succeed.

The year 2007 may well prove to be a defining moment in the international effort to combat climate change.

Across the world, countries, companies, cities and citizens are rising to the challenge of climate change and at a scale and pace which is truly unprecedented. Some are committing to big cuts in greenhouse gas (GHG) emissions with clear targets and timetables, while others are going that extra mile, pledging to be carbon neutral. Nations spearheading this include Costa Rica, Norway, New Zealand and the Vatican City and many companies have successfully risen to the challenge. As Ban Ki-Moon, the UN Secretary General, writes in this special edition of Climate Action, the United Nations is also stepping up to the mark.

A NEW CARBON NEUTRAL NETWORK FOR THE UN

In October UN agencies agreed to move to carbon neutrality too. UNEP, which is putting the final touches to its strategy, will be one of the 'early movers', cutting emissions in offices and operations and using UN approved standards to offset the rest.
This worldwide interest is one that we cannot simply allow to slip by. At the request of the Environment Minister of Costa Rica, UNEP will shortly be establishing a carbon neutral network – an Internet based information exchange service bringing carbon cutting and carbon shedding countries, companies and cities into close association. A central focus will be on strategies, policies, creative market mechanisms and technologies being tried, tested and piloted en route to lower carbon or carbon-free societies.
The relevance of the network to developing countries will be paramount including access to information but also to financing and suggestions on climate friendly and development-empowering projects. These will include projects ranging from renewable energy and forestry related schemes, to modern energy efficient factories, offices, homes and urban transportation systems backed under mechanisms like the Clean Development Mechanism of the Kyoto Protocol.

> “Across the world, countries, companies, cities and citizens are rising to the challenge of climate change and at a scale and pace which is truly unprecedented.”

Over the coming months and during the Bali climate change convention meeting we will be looking to recruit more pioneers to the network. The aim is to formally launch the full system at UNEP's Governing Council/ Global Ministerial Environment Forum in February 2008. We will also link this undertaking with follow up to the Caring for Climate initiative with business leaders that has been undertaken by the UN Global Compact, UNEP and the World Business Council for Sustainable Development. Readers of Climate Action, and its associated website, with carbon cutting strategies in place or in gestation are most welcome too.

The fact that a global carbon neutral network is being called for reflects the momentum of an issue that has become the challenge for this generation.

THE KEY TO THE FUTURE

The demand for global action reflects the milestones we have passed and the barriers that have been hurdled in this 10th anniversary year of the Kyoto Protocol and the 20th year of the Brundtland Commission report, the report that first popularised the concept of sustainable development.

The momentum is in no small part due to the Intergovernmental Panel on Climate Change (IPCC), established by UNEP, and the World Meteorological Organisation and its series of fourth assessment reports. It has put the full stop behind the science; climate change is happening, it is 'unequivocal'.

The IPCC's 2,000 plus scientists have also outlined the likely impacts with greater clarity and certainty. Impacts that are likely within this generation, such as the melting of substantial numbers of Himalayan glaciers with huge ramifications for industrial, agricultural and domestic water supplies in Asia.

It is a veritable Pandora 's Box with the lid just beginning to open. The link between climate change and instability has this year been implicitly recognised with the IPCC winning the Nobel Peace Prize jointly with Al Gore, the former US Vice-President.

But the IPCC has also outlined the design of a key to another box, an escape box of opportunities rather than calamities. It estimates that overcoming serious climate change may cost perhaps as little as 0.1 per cent of global GDP a year for three decades.

In Bali, governments must begin forging that key. They must get down earnestly and urgently to agree the parameters for negotiating a new and decisive post-2012 treaty along with a timetable for when those negotiations end.

© Gilles Deniaud/Fotolia

The existing Kyoto Protocol has achieved a great deal but it was only ever going to be a first step. Nevertheless, it has spawned new and novel markets in carbon trading. Kyoto's Clean Development Mechanism (CDM), which allows developed countries to offset some of their emissions via cleaner and greener energy projects in developing countries, is set to channel US$100 billion from North to South. This in part is helping to drive new markets and new employment opportunities in new businesses.

> “The fact that a global carbon neutral network is being called for reflects the momentum of an issue that has become the challenge for this generation.”

Renewables may still only account for some two per cent of electricity generation. But according to a new study by the UNEP linked Sustainable Energy Finance Initiative, renewables now account for 18 per cent of new investment.

© Jürgen Rode/Fotolia

Studies in Germany indicate that more people will be employed in environmental technologies in 2020 than are employed in the car industry.

And it is not just in Germany. Two of the biggest and fastest growing wind power companies are in China and India. Brazil is the indisputable leader in ethanol fuels and is convinced it can boost exports without encroaching on the Amazon.

"Overcoming serious climate change may cost perhaps as little as 0.1 per cent of global GDP a year for three decades."

The CDM is also having other spin offs. One area of persistent division between the developed and developing world has been the singular failure of rich countries to meet the overseas aid commitments of the past. Few countries spend 0.7 per cent of their domestic GDP on this. The US$100 billion set to be delivered by CDM is in many ways helping to meet at least some of that unfulfilled promise.

Unfortunately, the benefits of the CDM are being spread far from evenly. The lion's share of projects are going to the big rapidly developing economies of Brazil, China and India. UNEP in partnership with the UN Development Programme (UNDP) last year launched a new initiative with the twin aims of building the capacity of smaller developing countries while assisting to climate proof their economies. One of the key areas being targeted is sub-Saharan Africa.

The fact that so much activity is underway is also due to the anticipation that countries will indeed agree a treaty that engages post-2012 and one that delivers much bigger cuts than Kyoto. These cuts will put the world on track to the up to 80 per cent emission reductions deemed necessary by scientists to avert dangerous climate change.

UNEP's role in Bali is to support Yvo de Boer, Executive Secretary of the UN Framework Convention on Climate Change and his team. It is they who are principally responsible for coordinating the negotiations. Part of that support involves looking at the wider landscape, economically, environmentally and socially.

It is about sensitising governments, but also companies, local authorities, civil society and citizens, to the threats and opportunities of climate change while also demonstrating creative and innovative solutions that can, in the jargon, be 'scaleable and replicable'.

THE WIDER LANDSCAPE

The wider landscape was set out in the IPCC reports but also in UNEP's *Global Environment Outlook-4 (GEO-4)*, launched in late October 2007. The peer reviewed work of some 1,300 scientists, GEO-4 makes it starkly clear that carbon dioxide emissions have risen by 30 per cent a year over the past 20 years and that unchecked climate change will impact on the wider development aims including the Millennium Development Goals.

GEO-4 also points out climate change is happening against a backdrop of declining ecosystem services in many areas as a result of impacts, such as overfishing, air pollution, land degradation and rapidly declining biodiversity.

Combating climate change however may help reverse some of these declines. Standing forests, home to some of the world's last biodiversity and important ecosystems for water supplies, land stability and livelihoods in many poorer parts of the globe, currently fall outside the

© Johanna Mühlbauer/Fotolia

> "The fact that so much activity is underway is also due to the anticipation that countries will indeed agree a treaty that engages post-2012 and one that delivers much bigger cuts than Kyoto."

carbon markets, including the CDM, despite the fact that they absorb significant amounts of atmospheric carbon pollution.
Conversely, deforestation currently accounts for an estimated 20 per cent of greenhouse gas emissions. Financial incentives for standing forests, perhaps as part of the CDM, become increasingly possible under a new post-2012 climate regime and need to be part of the Bali discussions.
Some of the wider economic and environmental benefits of fighting climate change also remain largely unnoticed. UNEP, at the request of governments, is currently involved in discussions on how to reduce global emissions of the highly toxic and poisonous heavy metal, mercury. The fastest rising source of new mercury into the environment, where it can often spread as far as the poles, is coming from increasing coal-fired power generation in the rapidly growing economies of Asia.
Meanwhile there is some evidence that old mercury, deposited in the sediments of lakes in places like North America, are re-emerging as a result of rising lake temperatures linked with climate change. The increasing melting of ice in the Arctic may also be re-introducing old mercury back into the environment, entering the food chain and thus humans.
Combating climate change may therefore also assist in reducing new emissions and the reemergence of old deposits of mercury with significant economic, health and environmental spin offs. There are many more examples, from reduced health hazardous air pollution in cities to cuts in the costs of natural but climate-fuelled disasters.

TRANSITION TO A LOW CARBON SOCIETY

UNEP's slogan at the climate convention talks will be 'Transition to a Low Carbon Society – Bali and Beyond' with the sub themes, echoing to the challenges and the solutions: 'Climate Proofing Economies; Financing a Low Carbon Transition and Technologies for a Low Carbon World'.
A carbon neutral network fits into this strategy and will be underlined again in New Zealand, host of World Environment Day 2008, with the theme 'Kick the CO_2 Habit'.

I would like to congratulate this *Climate Action* initiative of Sustainable Development International which is very much embedded in this new thinking; thinking that is no longer just about the threats but about self evident (and often not so self evident) cost effective solutions.
Climate change offers the chance to transform the way we do business on this planet. All that is really required now is the political commitment of governments to unleash human ingenuity. This is a commodity that is freely available, extraordinarily versatile, in limitless supply and with an already distinguished track record.

> "Climate change offers the chance to transform the way we do business on this planet. All that is really required now is the political commitment of governments to unleash human ingenuity."

Achim Steiner
Achim Steiner is UN Under-Secretary General and Executive Director of the United Nations Environment Programme (UNEP). He has worked both at the grassroots level and the highest levels of international policymaking to address the interface between environmental sustainability, social equity, and economic development.
His professional career has included assignments with governmental and non-governmental, as well as international organisations in different parts of the world. In 2001, he was appointed Director-General of the World Conservation Union (IUCN), widely regarded as one of the most influential and highly respected organisations in the field of conservation, environment and natural resources management.
Mr Steiner serves on a number of international advisory boards, including the China Council for International Cooperation on Environment and Development (CCICED), and Planet Green, an initiative of the Discovery Channel.

CLIMATE ACTION WISHES TO THANK OUR SUPPORTING ORGANISATIONS:

The **Carbon Trust** is a private company set up by the UK Government in response to the threat of climate change, to accelerate the transition to a low carbon economy. The Carbon Trust works with business and the public sector to create practical business-focused solutions through its external work in five complementary areas: insights, solutions, innovations, enterprises and investments. Together these help to explain, deliver, develop, create and finance low carbon enterprise.

Ceres is a US-based network of investors, environmental organisations and other public interest groups working with companies and investors to address environmental and social challenges such as global climate change. Its mission is to integrate sustainability into company practices and the capital markets to protect the health of the planet and its people. Ceres directs the Investor Network on Climate Risk, a network of more than 60 institutional investors managing more than $4 trillion in assets focused on the business impacts from climate change.

The **FTSE4Good** Index Series has been designed to measure the performance of companies that meet globally recognised corporate responsibility standards, and to facilitate investment in those companies. As the impact of climate change is having a significant and increasing influence on the global environment, society and on the economic value of companies, eligibility for inclusion in the FTSE4Good Index Series has been expanded to include climate change.

The **Investor Network on Climate Risk** is a network of institutional investors and financial institutions that promotes better understanding of the financial risks and investment opportunities posed by climate change. Much of INCR's focus is aimed at improving corporate disclosure and governance practices on climate change. The four-year-old network, coordinated by Ceres, includes more than 60 investor members with collective assets totalling more than US$4 trillion.

Investor Network on
CLIMATE RISK

The **Pew Center on Global Climate Change** is an independent, nonprofit, nonpartisan organisation dedicated to promoting practical and effective climate change policies in the US and internationally. The Pew Center produces expert analysis of climate science, economics, solutions, and policy issues; facilitates dialogue among policymakers, stakeholders, and experts; and contributes directly to the policymaking process.

CLIMATE ACTION WISHES TO THANK OUR PLATINUM SPONSORS:

InterCall provides the broadest choices of audio conference calls and web/video conferencing for business collaboration. Using InterCall, you can connect with your customers, teams and vendors without the significant carbon footprint (and expense) caused by commuting or air travel. Collaborate better, communicate faster and reduce environmental impact with InterCall!

Johnson Controls, Inc. integrates technologies, products and services to develop smart environments and create a more comfortable, safe and sustainable world. For buildings, we optimize energy use and improve comfort and security. We also provide innovative automotive interiors and batteries for vehicles, including hybrid-electric, along with systems engineering and service expertise.

KPMG is the global network of professional services firms who provide audit, tax and advisory services. With a strong record of environmental management KPMG operates from 22 offices across the UK with over 10,000 partners and staff. Its Carbon Advisory Group is providing insight and strategies to help organisations understand and manage the many business implications of climate change.

The **National Mining Association** (NMA) represents the interests of the US coal mining industry before government on legislative, regulatory and legal matters. NMA and its industry members are committed to the development of technologies and policies that address climate change, including reductions in greenhouse gas emissions from all sources.

Siemens AG, headquartered in Berlin and Munich, is a global powerhouse in electrical engineering and electronics. Its 475,000 employees provide innovative technologies and comprehensive know-how to benefit customers in some 190 countries. Founded 160 years ago, Siemens focuses on the areas of Information and Communications, Automation and Control, Power, Transportation, Medical, and Lighting.

SIEMENS

Vestas is the No. 1 in Modern Energy. The Group's core business comprises the development, manufacture, sale and maintenance of high-technology wind power systems which use wind power to generate electricity. Vestas has installed 33,500 wind turbines worldwide. Vestas turbines generate 50 million MWh a year, enough to provide electricity for every household in a country the size of Spain with its 45 million inhabitants.

Yara International helps provide food and energy for a growing world population and aims to play a defining role in making the planet a safe and healthy place to live. Yara has developed technologies that can dramatically reduce greenhouse gas emissions and has committed to cut down emissions by 25 percent before 2009.

Greenhouse gases cause both flooding and severe drought.

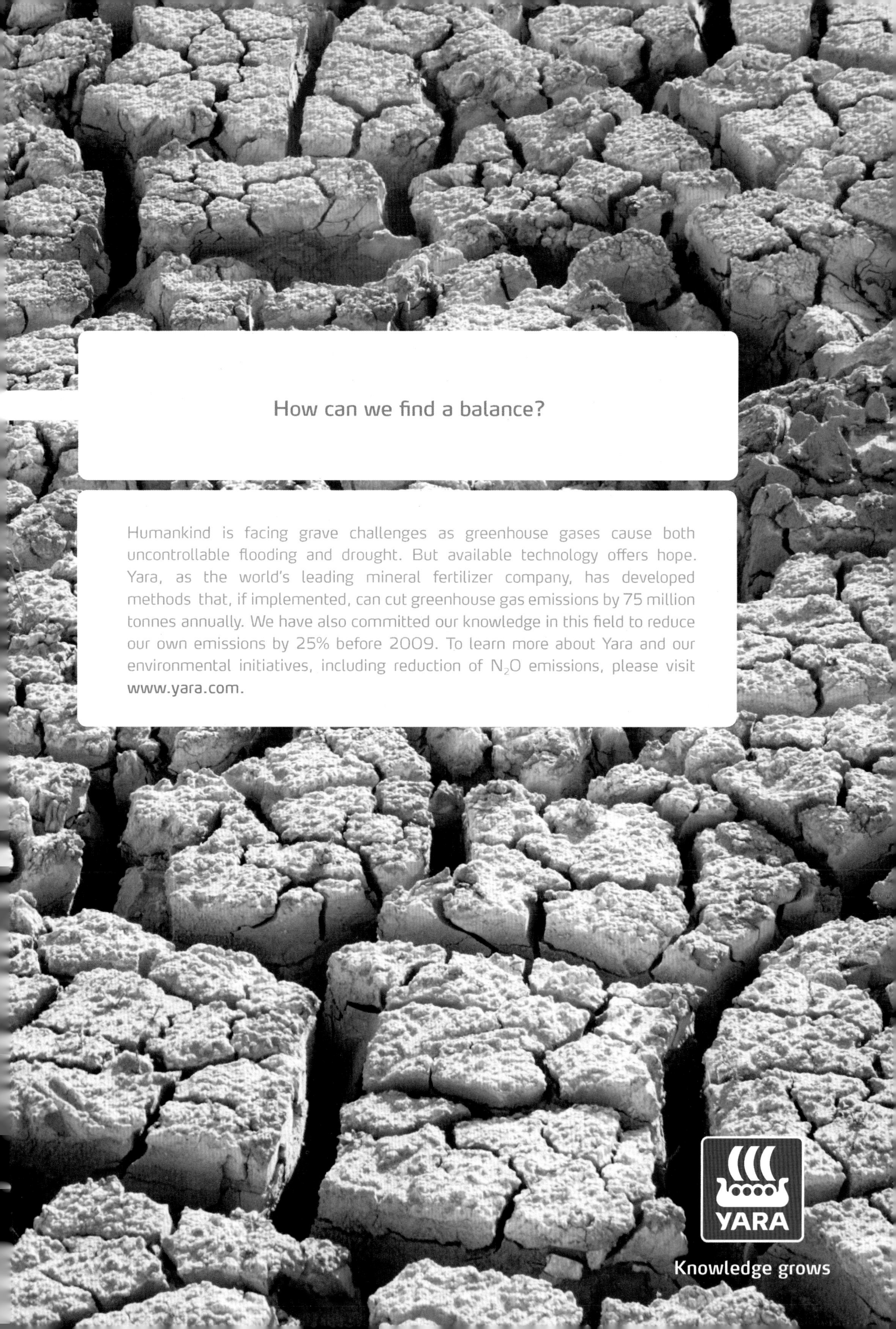

How can we find a balance?
Humankind is facing grave challenges as greenhouse gases cause both uncontrollable flooding and drought. But available technology offers hope. Yara, as the world's leading mineral fertilizer company, has developed methods that, if implemented, can cut greenhouse gas emissions by 75 million tonnes annually. We have also committed our knowledge in this field to reduce our own emissions by 25% before 2009. To learn more about Yara and our environmental initiatives, including reduction of N_2O emissions, please visit www.yara.com.
YARA
Knowledge grows

The need for climate action

BAN KI-MOON
SECRETARY-GENERAL OF THE UNITED NATIONS

The time to act on climate change is now. We can no longer consider the issues relating to climate change as primarily environmental. Quite simply, climate action has become a matter of strategic consequence, a core political issue for every government on Earth.

Another day, another weather disaster. In November it was Mexico's Tabasco state, swamped by floods, all crops ruined and more than two million people directly affected. Next month, who knows? Only one thing is for certain, climate change is posing challenges that we have yet to adequately address.
One day, we learn that the Arctic Ocean may be free of ice by 2050, that cyclones are forming ever further north in the Indian Ocean and that some Caribbean islands, long deemed by Lloyds of London to lie outside the Caribbean hurricane belt, are under increasing threat. The next we hear that world leaders meeting in New York have pledged their best efforts to tackle climate change.

THE TWO FACES OF THE CLIMATE ISSUE

These are the two faces of what has now become a defining issue of our era. On the one hand, we have worsening cases of extreme weather, accompanied by scientific evidence that humankind is the cause. On the other, clear signs that the world is waking up to the scale of the problem.

> “Certain words ran like a thread through our discussions: 'urgency', 'action' and 'now'.”

In September this year, I called a high level meeting on climate change at the United Nations. I wanted to spur conversation and get global policy makers together to make common cause in finding solutions to a common problem. In this, we succeeded beyond expectation. Certain words ran like a thread through our discussions: 'urgency', 'action' and 'now'.
It was the largest such meeting ever held, with more than 80 heads of state. I sensed something remarkable happening, something transformative – a sea change, whereby leaders showed themselves willing to put aside blame for the past, and pose to themselves more forward-looking questions. Questions such as: Where do we go from here? What can we do, together, in the future? President Michelle Bachelet of Chile put it bluntly, likening our fragile planet to an island in the universe: "We can destroy it," she said, "or save it."

“India, one of the world’s most rapidly advancing economies, is devoting two per cent of GDP annually to flood control and food security programmes, as well as mandating tough energy efficiency standards.”

THE BAD NEWS

Since that meeting, I have visited South America myself to see first hand some of the changes being inflicted on our warming world. It is a grim tale, with echoes across the globe. Unusually severe El Niño weather cost Peru 4.5 per cent of its GDP last year, while, two oceans away, fishermen in Mauritius blame warmer weather for driving tuna into deeper waters, diminishing their catch and making it harder to earn a living.

In Chile, I also saw how glaciers are rapidly melting, just as they are in all the world’s frozen regions. Such melting will mean severe shortages of water for half a billion people in South Asia. Much of northern China may become desert. Seeing the rapidly melting ice also made me recall the words of Micronesia’s delegate at the September climate meeting, when he worried openly about his country sinking beneath the rising seas. “How do we explain to our people, to future generations, that we have nothing for them?” he asked, meaning it literally.

THE GOOD NEWS

But there was good news on my trip, too. In Brazil, I learned that the country has reduced deforestation in the Amazon basin by 50 per cent. This is a trend we need to see reflected in all major forested regions of the tropics, with international support where necessary.

The good news does not stop there. India, one of the world’s most rapidly advancing economies, is devoting two per cent of GDP annually to flood control and food security programmes, as well as mandating tough energy efficiency standards. China, too is promoting cleaner and more resource-efficient development. Meanwhile, California, a crucible for technological revolutions, is mobilising, both politically and entrepreneurially, to fight climate change, and taking a lead that cities, states, businesses and communities are jostling to follow.

THE NEED FOR CLIMATE ACTION

All this sets the stage for an advance at the Bali climate change summit. We need a breakthrough, an agreement to launch negotiations for a comprehensive climate change deal that all nations can embrace. It will be difficult but I am optimistic. We are in a different place today than yesterday.

© TA HAI-UNEP / Still Pictures

At the September high level meeting in New York, the international community made a clear commitment to change. Governments will pursue their own solutions: from mandatory emissions controls to market mechanisms, such as carbon trading, to new fuel efficiency technologies and conservation. This is as it should be; there are many paths to Rome. The important thing is that all agree that national policies should be coordinated within the United Nations, so that our work together is complementary and mutually reinforcing.

“We need a breakthrough, an agreement to launch negotiations for a comprehensive climate change deal that all nations can embrace.”

A POLITICAL IMPERATIVE

No less important is the shared sense of urgency. Henceforth, climate change will no longer be a primarily environmental concern. It has become a matter of strategic consequence, a core political issue for every government on Earth. This is a message the United Nations has been emphasising for some time. The United Nations has been emphasising this message for some time, and it was reinforced this year by the award of the Nobel Peace Prize to climate campaigner Al Gore and the United Nations Intergovernmental Panel on Climate Change.

© NGUYEN DONG-UNEP/Still Pictures

Unusually severe El Niño weather cost Peru 4.5 per cent of its GDP in 2006.

“ Climate change will no longer be a primarily environmental concern. It has become a matter of strategic consequence, a core political issue for every government on Earth. ”

Understanding that addressing climate change is a political imperative represents a turning point with enormous implications. As a political issue, climate change becomes closely linked to economic development and the achievement of the Millennium Development Goals.

The World Bank and UNDP will begin to explore ways of financing energy efficiency and anti-pollution programmes in developing countries. Trade and technology transfer incentives will be part of the equation. Wealthy nations must help provide incentives to poorer ones to take steps that help us all. For instance, UN member states are discussing an Adaptation Fund that supplements international aid with money for climate change projects that will benefit the whole world, not merely the countries that initiate them.

And this, again, is where I detect a tipping point. It has become abundantly clear that we are all in this together. Some countries are more vulnerable to climate change, and some are more responsible for the emissions that are driving it, but no community is immune from its effects. Furthermore, no one lacks the ability to do something positive to prevent the situation getting worse.

“ Wealthy nations must help provide incentives to poorer ones to take steps that help us all. ”

CARBON NEUTRAL UN

Mindful of this, I have asked my senior executives throughout the UN system to make sure our own house is in order. In October, each Executive Head of a UN programme or agency committed, by the end of 2009, to take an inventory of their greenhouse gas emissions.

Brazil has reduced deforestation in the Amazon basin by 50 per cent.
© Leeman/iStockphoto

They will start immediately to reduce emissions to the extent they can and to assess the cost of offsetting the remainder via the purchase of Clean Development Mechanism credits.
The goal is to announce a target date for a carbon neutral UN. In this we are not alone. This publication sets out a number of actions that organisations and governments can take, and in many cases are already taking, backed up by clear examples of how reducing greenhouse gas emissions can be achieved.

> **"With political will, comes results. Our job is to translate this spirit into deeds in Bali, and beyond."**

The trend for going carbon neutral is growing. To facilitate its spread, the UN Environment Programme is establishing a carbon neutral network so businesses, governments, organisations and communities can pledge themselves to creating a carbon neutral future, and share the important lessons and tools we will all need to achieve it.
Having committed myself when I was appointed Secretary-General to making climate change a priority, I am glad that I can approach the end of my first year with a sense of optimism. As one environmental expert noted at the high level meeting in September: "curbing climate change may not be as hard as it looks." With political will, he suggested, comes results. Our job is to translate this spirit into deeds in Bali, and beyond.

Author
Ban Ki-moon, of the Republic of Korea, is the eighth Secretary-General of the United Nations, bringing 37 years of service, both in government and on the global stage.
At the time of his election as Secretary-General, Mr Ban was his country's Minister of Foreign Affairs and Trade. His long tenure with the ministry included postings in New Delhi, Washington DC and Vienna, and responsibility for a variety of portfolios, including Foreign Policy Advisor to the President, Chief National Security Advisor to the President, Deputy Minister for Policy Planning and Director-General of American Affairs. Throughout this service, his guiding vision was that of a peaceful Korean peninsula, playing an expanding role for peace and prosperity in the region and the wider world.

Organisation
In 1945, representatives of 50 countries met in San Francisco at the United Nations Conference on International Organization to draw up the United Nations Charter. The UN officially came into existence on 24 October 1945, when the Charter had been ratified by China, France, the Soviet Union, the United Kingdom, the United States and a majority of other signatories.
The purposes of the United Nations, as set forth in the Charter, are to maintain international peace and security; to develop friendly relations among nations; to cooperate in solving international economic, social, cultural and humanitarian problems and in promoting respect for human rights and fundamental freedoms; and to be a centre for harmonising the actions of nations in attaining these ends.

Enquiries
Website: www.un.org/sg/

Climate change is unequivocal

R K PACHAURI, CHAIRMAN, INTERGOVERNMENTAL PANEL ON CLIMATE CHANGE (IPCC) & DIRECTOR GENERAL, THE ENERGY AND RESOURCES INSTITUTE (TERI)

Special reports published by The Intergovernmental Panel on Climate Change (IPCC) provide a comprehensive assessment of knowledge on all aspects of climate change. These include advances in scientific understanding related to climate change and major findings from working groups. Together, these reports have, and are having, a major impact on public opinion and awareness among world leaders. The IPCC's most recent Fourth Assessment Report sets out to integrate and compact this wealth of information into a readable and concise document explicitly targeted at policymakers.

The Intergovernmental Panel on Climate Change (IPCC) has been in existence since 1988 having been established by the World Meteorological Organization (WMO) and the United Nations Environment Programme (UNEP). During this period the Panel has not only brought out four assessment reports, which present a comprehensive assessment of knowledge on all aspects of climate change, but has also produced several special reports dealing with specific issues related to climate change. As part of the Fourth Assessment Report (AR4), three volumes representing the contributions of Working Group I, II, and III have already been completed and released. The Synthesis Report bringing together the major findings of the three working groups was released in November 2007. The AR4 has incorporated several advances in scientific knowledge related to climate change. This has had a major impact on public opinion and awareness among world leaders on knowledge related to observed climate change in recent decades as well as projections for the future.

> **"Climate change is now unequivocal and the human influence on the Earth's climate is now very well established."**

OBSERVED IMPACTS

One major finding of the AR4 has been the fact that climate change is now unequivocal and the human influence on the Earth's climate is now very well established. The average temperature increase across the globe during the 20th century has been around 0.74°C and sea level rise around 17cm. These changes have been accompanied by changes in precipitation as well, including an increase in precipitation in the higher latitudes and a decline in the lower latitudes as well as the Mediterranean region. However, these changes have also been accompanied by an increase in the frequency and intensity of extreme precipitation events.

Coastal flooding is increasingly affecting cities with high density populations.

> "At the lower end of these scenarios, the best estimate of temperature increase by the end of the 21st century is placed at 1.8°C and at the upper end of the scenarios, the best estimate is approximately 4°C."

While any single event that occurs cannot be linked to human induced climate change, the pattern observed across the globe provides abundant evidence of this trend. It is also observed that floods and droughts are increasing in frequency and intensity and this trend is likely to continue.

FUTURE TRENDS

Projections of temperature increase in the 21st century have been made on the basis of established and plausible scenarios of the future. At the lower end of these scenarios, the best estimate of temperature increase by the end of the 21st century is placed at 1.8°C and at the upper end of the scenarios, the best estimate is approximately 4°C. These projections are of course based on the assumption of no specific action towards mitigation of emissions of greenhouse gases (GHGs).

UNEQUAL IMPACTS

The impacts of climate change vary across regions and communities. In general, these are most severe in those regions which display high vulnerability on account of a number of factors including poverty. For instance, during the 20th century, glaciers and ice caps have experienced widespread mass losses, which are likely to have major consequences for water supply in several areas of the world. In fact, new data clearly confirm that losses from the ice sheets have contributed to sea level rise over the period 1993-2003.

ECOSYSTEM DISRUPTION

Other observed impacts of climate change include more intense and longer droughts, widespread change in extreme temperatures as well as changes in biodiversity. In general, it was assessed that climate change will reduce biodiversity and perturb the functioning of most ecosystems and therefore compromise on the services that they currently provide. Of the species that were assessed it was found that 20 to 30 per cent of plant and animal species would be at risk of extinction if increases in global annual temperatures exceed 1.5 to 2.5°C. Some ecosystems are particularly vulnerable, such as coral reefs and marine shell organisms.

FOOD INSECURITY

Another major area of serious impacts is in the field of agricultural production. Malnutrition, for instance, would be further exacerbated by the reduced length of the growth seasons in the Sahelian region of Africa. In some countries, yields from agriculture could be reduced by up to 50 per cent by 2020. With the growing scarcity of water in different regions of the world, it was also assessed that 75 to 250 million people would be exposed to increased water stress in 2020 in Africa alone. In the case of sub- tropical and tropical regions, a temperature increase above 1.5 to 2.5° C is expected to lead to a decline in agricultural productivity of crops such as maize and wheat. In fact, there is growing evidence in South Asia, for instance, that the wheat crop is already being affected adversely by climate change.

COASTAL FLOODING

Another serious impact of climate change is seen most prominently in the Asian region where the megadeltas such as Kolkata, Dhaka, and Shanghai are at risk of coastal flooding and negative impacts. There are other cities in the world, also located in coastal areas, with high density populations at high risk of coastal flooding and other impacts with the prospect of high sea levels.

> "One scenario that would limit temperature increase at equilibrium to 2.0–2.4°C would actually cost the world less than three per cent of the global GDP in 2030."

A temperature increase above 1.5 to 2.5° C is expected to lead to a decline in agricultural productivity of crops such as maize and wheat.

COST EFFECTIVE CLIMATE MITIGATION

While an understanding of these impacts of climate change provides a rationale for mitigation of GHGs which could help to avoid or postpone some of these impacts, such an approach is strengthened by the relatively low cost and general attractiveness of mitigation measures available around the world. One scenario assessed for the future that would limit temperature increase at equilibrium to 2.0–2.4°C would actually cost the world less than three per cent of the global GDP in 2030. But if such a scenario was to be realised, the level of global emissions will have to peak by 2015 and decline thereafter.

“ In some countries, yields from agriculture could be reduced by up to 50 per cent by 2020. ”

CONCLUSION

Overall, the IPCC's AR4 has had a major impact on awareness and knowledge on various aspects of climate change across the globe. While climate change represents a growing challenge to human society, there are several other areas where the creation and dissemination of knowledge by specialists and experts, and accepted by the governments and decision makers, may have great merit in these areas of human endeavour as well. Quite apart from the specific knowledge provided by the IPCC, it is the example it sets which merits application perhaps in other important areas where we may be jeopardising the future of living systems on this planet and the welfare of future generations. There is need perhaps for clones of the IPCC in other areas of human activity and public policy.

Author

Dr. Rajendra Kumar Pachauri was elected Chairman of the Intergovernmental Panel on Climate Change (IPCC) in 2002. The IPCC, established by the World Meteorological Organization (WMO) and the United Nations Environment Programme (UNEP), is the leading body for the scientific assessment of climate change. Dr. Pachauri has also been the head of TERI, The Energy and Resources Institute, since its establishment in New Delhi 25 years ago. TERI is an Indian institute of excellence working on scientific and technological research and strategic thinking in the fields of energy, environment, forestry, climate change, biotechnology, conservation of natural resources and sustainable development. Dr. Pachauri has taught at various universities in India and the USA, including the School of Forestry and Environmental Studies, Yale University, in 2000. He was Research Fellow at The World Bank, Washington, DC in 1990. He was adviser to the Administrator of the United Nations Development Programme (UNDP) in the fields of energy and sustainable management of natural resources from 1994 till 1999.
To acknowledge his immense contributions to the field of environment, he was awarded the Padma Bhushan by the President of India in January 2001, one of India's highest civilian awards. Among other awards, he was honored as "Officier De La Légion D'Honneur" by the Government of France in 2006. Along with his colleagues at the IPCC, the organisation's work was also awarded the 2007 Nobel Peace Prize for its contributions to climate change understanding.

Organisation

The IPCC, established by the World Meteorological Organization (WMO) and the United Nations Environment Programme (UNEP), is the leading body for the scientific assessment of climate change. It is also the organisation that jointly won, with Al Gore, the 2007 Nobel Peace Price for its efforts to build up and disseminate greater knowledge about man-made climate change, and to lay the foundations for the measures that are needed to counteract such change. The organisation assesses scientific, technical and socioeconomic information relevant for the understanding of climate change, its potential impacts and options for adaptation and mitigation.

Enquiries

IPCC Secretariat
c/o World Meteorological Organization
7bis Avenue de la Paix
C.P. 2300, CH-1211 Geneva 2, Switzerland
Tel: +41 22 730 8208 | Fax: +41 22 730 8025
E-mail: ipcc-sec@wmo.int

Global capability. Personal accountability.

There's no reason why business and environmental objectives can't go hand-in-hand. With Verizon Business, you get a communications team with the resources to deliver a secure global IP network, the expertise to create advanced collaboration solutions, and the dedication to help going green make good business sense.
verizonbusiness.com

Verizon Enables Green Growth

GOING GREEN MAKES GOOD BUSINESS SENSE

For Verizon "going green" is more than a bumper sticker. We understand our enormous opportunity—and responsibility—to positively impact our planet's environment. Verizon's positive influence doesn't end with simply how we conduct our own business. By virtue of our leadership in global communications, we also help our customers and, the customers of our customers, to go green.

With 240,000 employees worldwide, Verizon operates one of the Earth's most interconnected and expansive IP networks. In the U.S., we are the first to deploy a large-scale, all-fiber network to the home, and we operate the most reliable broadband wireless network. These state-of-the-art networks allow us to offer consumers, businesses, government agencies and educational institutions advanced communications services. These network-based solutions enable our customers to live their lives and operate their businesses more productively, while also mitigating adverse environmental impacts.

ENABLING GLOBAL E-COMMERCE

Our networks, services and expertise enable global 'e-commerce' and distance learning. Audio, video and web conferencing can foster collaboration among workers and students in various locations. Along with secure e-mail and instant messaging, these solutions dramatically reduce the need for travel by allowing people to come together via our advanced technology offerings. Electronic billing saves untold numbers of trees while providing even more detailed billing information to help our customers control costs.

EDS, a leading global technology services company, is among the companies that are using Verizon's advanced communications services to go green and also work smarter. EDS is using Verizon Video Conferencing to help meet its goal of achieving a 25 percent reduction in its carbon emissions in Australia and New Zealand by 2010. By increasing the use of video conferencing to connect its 10 locations in these countries, EDS will improve collaboration across its operations, while also reducing employee air travel by nearly a third.

"Encouraging our employees to embrace video conferencing to provide face-to-face collaboration enables us to foster teamwork across our diverse office sites, and also supports our corporate social responsibility initiatives," said Chris Mitchell, EDS managing director for Australia and New Zealand. "Verizon Business gave us a complete end-to-end solution that streamlined our infrastructure while enhancing our overall communications capabilities."

In addition to reduced travel and paper bills, Verizon's broadband networks enable remote monitoring and adjusting of energy use at "smart buildings," remote reporting of traffic congestion to "smart cars," and remote monitoring of patient health to "smart doctor's offices."

PRACTICING WHAT WE PREACH

Verizon practices what we preach. In addition to using our own communications services, the company has implemented a wide range of initiatives to conserve energy and protect environmental resources. We are piloting hybrid-powered vans in our service fleet, pioneering the use of alternative technologies such as fuel cells and small natural-gas turbines to power our network equipment, and reducing energy usage in many of our buildings with new, highly reflective roofing materials.

Earlier this year, in celebration of Earth Day, Verizon Business teamed up with Computershare and American Forests to plant a tree for each large business or government customer that chooses electronic billing over paper. This program is modeled on a successful eTree program Verizon implemented for shareholders opting for electronic versions of annual reports, proxies and other shareholder communications materials. Please visit our web site at **verizon.com/responsibility** to learn more about our corporate responsibility strategy and what we are doing to reduce our environmental impact.

Environmental conservation is everybody's business. Each individual and organization contributes to the production of carbon emissions when we heat and cool our homes or offices, drive our cars or fly on airplanes. At Verizon, we're harnessing the power of our networks and the expertise of our people to deliver environmentally friendly services that help enhance productivity.

The cost of climate action

SIR NICHOLAS STERN, IG PATEL PROFESSOR OF ECONOMICS AND GOVERNMENT AT THE LONDON SCHOOL OF ECONOMICS AND FORMER CHIEF ECONOMIST OF THE WORLD BANK

A basic conclusion of the Stern Review is that the costs of strong and urgent action on climate change will be less than the costs of inaction and the impacts of climate change under business as usual (BAU). Although they are in the minority, there remain a number of people who deny the importance of strong and urgent action on climate change and they essentially offer one of, or a combination of, four arguments. These arguments are correspondingly, absurd, reckless, irresponsible and ethically indefensible.

THE CASE FOR ACTION ON CLIMATE CHANGE

A basic conclusion of the Stern Review is that the costs of strong and urgent action on climate change will be less than the costs of inaction and the impacts of climate change under business as usual (BAU). Although they are in the minority, we have come across a number of people who deny the importance of strong and urgent action on climate change and they essentially offer one of, or a combination of, the following arguments. First, there are those who deny the scientific link between human activities and global warming. Second, there are those who, while accepting the science of anthropogenic climate change, argue that the human species is very adaptable and can make itself comfortable whatever the climatic consequences. Third, there are those who accept the science, are sceptical about the idea that humans can adapt to anything but believe the costs involved are just too great and are despondent about the challenge of mitigating climate change. Finally, there are those who accept the science of climate change and the likelihood that it will inflict heavy costs, but simply do not care much for what happens in the future beyond the next few decades.

NO SCIENTIFIC BASIS

For those who fall under the first category of deniers, clearly it is for the science to persuade them otherwise, but the weight of evidence is now such that most people would see this point of view as simply absurd. The February 2007 Fourth Assessment Report of the Intergovernmental Panel on Climate Change (IPCC) sets out the evidence in a very convincing and clear way. It confirmed that there is now very high confidence that human activity is warming the climate and that human influences are likely to have been at least five times greater than those due to solar variations. There is now very little justification for believing that the scientific understanding of climate change is fundamentally flawed, or that the remaining areas of uncertainty imply current knowledge is inadequate as a basis for drawing conclusions for policy.

HUMANS CAN ADAPT TO CLIMATE CHANGE

Human beings are incredibly adaptable. Our ability to adapt to changes in climate, and the impacts this will cause, is likely to increase as countries become richer and more technologically advanced. However, there is a limit to adaptation. Under a BAU scenario, the stock of greenhouse gases could more than treble by the end of the century, giving at least a 50 per cent risk of exceeding 5°C global average temperature change during the following decades. This would take humans well into unknown territory and would involve very large movements of population away from the equator. A 5°C increase may not sound like much but it masks significantly higher temperature increases on most land masses. A temperature increase of that magnitude would dramatically change the physical and human geography of the world.

Adaptation will become increasingly difficult as temperatures rise with the costs of adaptation rising at a faster rate than the temperature. While higher GDP should increase adaptive capacity, vulnerability may increase with GDP as more expensive capital is at risk and populations are forced to inhabit increasingly hazardous areas. Some changes in climate are also not amenable to technological improvement in the same way that, say, energy generation or fighting disease is. The economics of climate change are fundamentally about the economics of risk. If you act on climate change and invest in bringing forward new technologies, and it turns out to be the biggest hoax ever perpetrated on mankind, as a US senator has described, you will still have acquired a lot of new technologies that are probably quite useful. If, on the other hand, you do nothing, you may quickly end up with a lot of irreversible and severe damage from which it will be very hard to extricate yourself. Given the scale of the outcomes that we now have to regard as possible or likely under BAU, maintaining this view must be regarded as reckless.

“Under a BAU scenario, the stock of greenhouse gases could more than treble by the end of the century, giving at least a 50 per cent risk of exceeding 5°C global average temperature change during the following decades.”

ACTION IS TOO COSTLY

The third group accept the case for action, but argue that the sacrifice required would be prohibitive, requiring a radical change in our lifestyles. The likely cost of urgent and coordinated action is around one per cent of GDP by mid century. This is not a trivial amount and represents a very significant change in the patterns of energy investment towards low carbon energy technology. But in terms of growth, it amounts to a fraction of a fraction taken off

© Lya Cattel/iStockphoto

the annual rate. Estimates of the cost of action to reduce greenhouse gas emissions published by the IEA and McKinsey since the Review have been below our estimates.

> "The likely cost of urgent and coordinated action is around one per cent of GDP by mid century. In terms of growth, it amounts to a fraction of a fraction taken off the annual rate."

The one per cent of GDP reflects the likely costs under a flexible global policy, employing a variety of economic instruments in cost effective ways to control emissions of a broad range of greenhouse gases. It would require clear long term price signals and policy frameworks that encourage technological innovation. In the absence of these factors, or were action to be delayed, the costs could be significantly higher. However, as we argued in the Review, the decision on whether to act now hinges on the question of irreversible outcomes and risks. Decisions taken today will have potentially large and irreversible consequences in terms of climate change impacts; this is not true to the same extent of mitigation costs. Moreover, policymakers can keep cost estimates under review and revise policy in the light of new information. By contrast, the impact from global warming will become increasingly costly to reverse. In part this reflects the fact that damages are caused by the stock of GHGs and not the annual flow, but it also reflects the risks associated with irreversible thresholds and discontinuities. Despondency over the cost of mitigating climate change compared to the cost of business as usual must be viewed as irresponsible.

FUTURE GENERATIONS WILL BE RICHER AND MORE ABLE TO ADAPT

The attitude of the fourth group relates to views on valuing the future and so called 'discounting'. In the Review, we do discount future damages for the likelihood that future generations will be richer than we are and we account for the risk of extinction. However, applying substantial discounting to the future simply because it is the future would be unacceptable to many, including many of the great economists, such as John Maynard Keynes, Robert Solow and Amartya Sen. This is what some economists would call pure time discounting, and we are convinced that such time discounting at a heavy rate would be viewed by most people as unethical. It involves discrimination

© T. BALABAADKAN-UNEP/Still Pictures

between individuals by date of birth and is as though a grandparent is saying to their grandchild, because you will live your life 50 years after me, I place a value of one-quarter on your well being relative to myself and my current neighbours (as would be the case with a pure time discount rate of three per cent), and therefore I am ready to take decisions with severe and irreversible implications for you. Most would regard this as unethical.

Further, it is not possible to read off these ethics from the behaviour of markets, as many recommend. We cannot see a collective expression in the markets of what, acting together, we should do for 100, 150 years' time. Current market interest rates tell us only about individuals' willingness, possibly irrational, to invest today for benefits tomorrow through their daily consumption, saving and investment decisions.

“Is it worth paying one per cent of GDP to avoid the additional risks of higher emissions?”

How much should society, acting together, spend on unborn generations is a somewhat different question. To use pure discount rates of around three per cent implies valuing impacts on lives in the middle of next century at $^{1}/_{300}$ the weight of lives today. And note that the Stern analysis does include treating a dollar to a richer generation as worth less than a dollar to a poorer generation. Indeed this approach to discounting, which takes account of a society's income levels, reminds us that discounting depends on both growth and distributional outcomes. Various combinations of these can lead to discount rates at four or five per cent, while also showing the very big damages from BAU.

In any case, pure discounting, at these rates, presupposes the conclusion: if you don't care about the future and apply that kind of discounting, you won't care about climate change. You don't need the science or the economics; further analysis becomes redundant.

“The case for strong and timely action, supported by well designed economic policies, is overwhelming.”

THE BASIC QUESTION

It is not necessary to understand the science or economics of climate change in great detail to have a view on the case for robust and urgent action. Our central estimate of mitigation costs for stabilising emissions below 550 ppm CO_2e (parts per million of carbon dioxide equivalent – all greenhouse gases expressed as a common metric relative to their warming potential) is one per cent of GDP per annum. The basic question is thus: is it worth paying one per cent of GDP to avoid the additional risks of higher emissions as described in the disaggregated story of consequences? We should recognise the balance of risks. If the science is wrong and we invest one per cent of GDP in reducing emissions for a few decades, then the main outcome is that we will have more technologies with real value for energy security, other types of risk and other types of pollution. However, if we do not invest the one per cent and the science is right, it is likely to be impossible to undo the severe damages that will follow. The case for strong and timely action, supported by well designed economic policies, is overwhelming.

Author
Sir Nicholas Stern, IG Patel Professor of Economics and Government at the London School of Economics and former Chief Economist of the World Bank, led the UK Government commissioned Stern Review on the Economics of Climate Change published in October 2006. He currently holds the IG Patel Chair at the London School of Economics and Political Science.
Enquiries
Website: www.sternreview.org.uk

© Millarovic/iStockphoto

The road from Bali:

TOWARDS A BINDING POST-2012 CLIMATE FRAMEWORK

EILEEN CLAUSSEN
PRESIDENT,
PEW CENTER ON GLOBAL CLIMATE CHANGE

After years of struggle to mobilise an effective international response to climate change, 2007 may prove to be a critical turning point. From the G8 summit in Heilegendamm to the APEC summit in Sydney, from a high level UN gathering in New York to a meeting of major economies in Washington, climate change has suddenly vaulted to the top of the global agenda. This burst of climate diplomacy culminates in December in Indonesia, where the goal will be a Bali 'road map' pointing the way to a new global agreement. The questions to ask are whether this road map will set governments on a course towards binding multilateral commitments and will some form of commitment be given by all the world's major economies?

INTRODUCTION

In 2007, the stream of summits, dialogues and ministerial huddles has produced plenty of ideas and, on occasion, points of concurrence. But a grand consensus has yet to emerge. Indeed, the newly energised debate has exposed fundamental rifts and sharply contrasting visions of the way forward. As governments head into the Bali conference, they face a fork in the road. Down one path lies a lasting solution; down the other, only false hope.

The critical issue in Bali is whether the new road map sets governments on a course towards establishing binding multilateral commitments. The Kyoto Protocol, while a step in that direction, is limited in both scope and duration. Without the United States and Australia, its targets (which expire in 2012) encompass only about a third of global emissions. The next stage of the climate effort will be effective only if it entails some form of commitment by all the world's major economies.

> "Why are international commitments essential? Because without them, countries cannot be confident that others are contributing their fair share to the global effort."

The alternative vision, championed by the Bush administration, eschews international commitments. Countries may agree on an 'aspirational' long term goal, defined perhaps as a limit on global temperature increase, a stabilisation level for greenhouse gas (GHG) concentrations, or a reduction in global emissions. But each country would decide independently what contribution it would make towards this collective goal. There would be no binding commitments. And for that reason, this path cannot produce an effort nearly sufficient to the task.

THE NEED FOR INTERNATIONAL AND BINDING COMMITMENTS

Why are international commitments essential? Because without them, countries cannot be confident that others are contributing their fair share to the global effort. And without that confidence, they will not be prepared to adopt the policies and make the investments needed to build low carbon economies. In the long term, the benefits of addressing climate change will far outweigh the costs. But in the short term, governments, business, and private citizens will be far more willing to bear those costs if they know their counterparts and competitors are bearing them as well.

PROVIDING FOR A RANGE OF COMMITMENTS

While binding commitments are needed from all the major economies, they need not take the same form. There are huge differences among the world's major economies, a group that includes industrialised countries, developing countries, and economies in transition. As a consequence, the kinds of policies that effectively address climate change in ways consistent with other national priorities will vary from country to country. If it is to achieve broad participation, the post-2012 framework must accommodate different national circumstances and strategies and allow countries to take on different types of commitments.

Economy-wide emissions

The Kyoto approach, ie economy-wide emission targets coupled with emissions trading, should remain a core element of the climate framework. Emission targets provide some environmental certainty, while emissions trading harnesses market forces to deliver those reductions at the lowest possible cost. Kyoto and the EU's Emissions Trading Scheme have spawned a multi-billion dollar carbon market, mobilising private capital for emission reduction and driving climate-friendly investment in developing countries. This market can be sustained and expanded only if industrialised countries commit to a new round of more ambitious targets.

But not all countries are prepared to take on economy-wide targets. China, India and other emerging economies fear binding targets that hold them to specific emission levels would unduly constrain their development. Economy-wide targets may also be infeasible for some countries. To accept a binding target, a country must be able to quantify its current emissions reliably and project its future emissions, which few if any developing countries currently have the capacity to do.

Policy-based commitments

Under this approach, a country would commit to a national policy or set of policies that moderate or reduce their emissions, without being bound to an economy-wide emissions limit. This would allow countries to put forward commitments tailored to their specific circumstances and consistent with their core economic or development objectives. A country such as China, for instance, could commit to implement fully its existing energy efficiency targets, renewable energy goals and auto fuel economy standards. Such commitments would need to be credible and binding, with mechanisms to ensure close monitoring and compliance. Developed countries may also need to provide incentives for developing countries to adopt and implement stronger policies. Such incentives could include policy-based emissions crediting, similar to the Kyoto Protocol's Clean Development Mechanism, granting countries tradable emission credits for meeting or exceeding their policy commitments.

Sectoral agreements

Governments could also commit collectively to a set of targets, standards, or other measures to reduce emissions from a given sector. Such agreements could include both developed and developing countries, offering a practical alternative to economy-wide commitments.

> **"If it is to achieve broad participation, the post-2012 framework must accommodate different national circumstances and strategies and allow countries to take on different types of commitments."**

In sectors most vulnerable to a loss of competitiveness due to carbon constraints (such as energy-intensive industries where goods are traded globally), sectoral agreements can help ensure a more level playing field. Sectoral approaches are actively being explored by global industry groups in both the aluminium and cement sectors, and could be helpful in sectors, such as power and transportation, where competitiveness is less an issue but large scale emission reduction efforts are most urgent.

Other commitments

If stringent enough, any of the commitments discussed would help draw low carbon technologies into the marketplace. But the post-2012 framework could include additional commitments to drive technology more directly, such as those aimed at joint research and development of 'breakthrough' technologies with long investment horizons. Such agreements could build on the Asia Pacific Partnership and similar initiatives but go much further, committing governments to the levels of funding needed to accelerate critical technologies such as carbon capture and storage. On the deployment side, the framework also could help ensure broader access to existing and future technologies by addressing finance, international property rights and other issues impeding technology flows to developing countries.

Assisting developing countries

The post-2012 framework must also do more to help vulnerable countries adapt to the impacts of climate change. Within the climate regime, there are two priorities: helping countries develop comprehensive national strategies to integrate adaptation across the full range of development activities; and providing reliable assistance to the poorest and most vulnerable countries for their most urgent implementation needs. One way to integrate adaptation into development planning is to perform routine climate risk assessments on projects up for bilateral and multilateral development assistance, and funding only those that score well on vulnerability criteria.

TOWARDS A COMMON FRAMEWORK

While the post-2012 climate effort must allow for a diversity of approaches and commitments, it is critical that all are integrated in a common framework. If the only point of agreement is an aspirational long term goal, the aggregate effort is likely to be modest at best. By linking actions and negotiating them as a package, nations are likely to undertake a higher level of effort than they would if acting on their own. Such a negotiation could take the form of sequential bargaining, with countries proposing what they are prepared to do under one or more different tracks, and then adjusting their proposals until agreement is reached on an overall package. To help ensure a balanced and stronger outcome, it may be necessary to agree that certain countries will negotiate particular types of commitments most appropriate to their circumstances at the outset. The objective must be an integrated agreement that is flexible enough to accommodate different types of commitments, and reciprocal enough to achieve a strong, sustained level of effort.

THE BALI ROAD MAP

What kind of map might lead from Bali to this goal? On its own, the Kyoto Protocol track launched in Montreal in 2005 to establish new commitments for the post-2012 period leads to a dead end. These negotiations contemplate commitments only for those countries with existing targets under Kyoto: Europe, Japan, Canada, Russia, and a handful of smaller countries. While the EU has unilaterally pledged to reduce emissions 20 per cent below 1990 levels by 2020, others are holding back. The political reality is that few if any of those with Kyoto targets will be prepared to assume binding post-2012 commitments without some form of commitment from the US and the major emerging economies.

“The political reality is that few if any of those with Kyoto targets will be prepared to assume binding post-2012 commitments without some form of commitment from the US and the major emerging economies.”

©Ugonna Emerole/UNEP/Still Pictures

What is needed in Bali is a decision to launch negotiations under the 1992 UN Framework Convention on Climate Change, Kyoto's parent agreement, which includes all the Kyoto parties plus the United States, Australia and others. These talks should run in parallel with, or encompass those under Kyoto, with the goal of a package deal setting new commitments under both the convention and the protocol. The Bali road map should then lay out a process and timeline for establishing new incentives and commitments under the convention that, in concert with new commitments under the protocol, form a comprehensive post-2012 framework.

> "There is growing political momentum in the US for mandatory GHG limits. California, New York and other states, many led by Republican governors, have enacted mandatory controls."

The prospects for such a road map emerging from Bali hinge, first and foremost, on the willingness of the US to enter talks on commitments. Only then will China, India and other emerging economies entertain the idea of commitments. The Bush administration says the aim of its major economies initiative is to forge a consensus by the end of 2008 contributing to a new global agreement under the Framework Convention in 2009. But the administration's vision of this global agreement does not entail binding international commitments and, barring a sudden change of heart, it is unlikely the administration will agree to a road map. Still, the United States is unlikely to be in a position to block a new process altogether. Perhaps the best plausible outcome is a new Convention process that, while not a negotiation *per se*, has the potential to become a *de facto* negotiation once the politics allow.

The objective from Bali must be an integrated agreement, flexible enough to accommodate commitments tailored to a country's specific circumstances, and reciprocal enough to achieve a strong, sustained level of effort.

BEYOND BALI

Looking beyond Bali, and beyond the Bush administration, prospects are considerably brighter. There is growing political momentum in the US for mandatory GHG limits. California, New York and other states, many led by Republican governors, have enacted mandatory controls. The US Climate Action Partnership (USCAP), a broad coalition of major companies and nonprofit organisations, including the Pew Center, is calling for a national cap and trade system and other mandatory policies to reduce emissions 60 to 80 per cent by 2050. In Congress, serious efforts are underway in both the House and the Senate to draw up mandatory cap and trade legislation. There is a chance such legislation could be sent to the White House with President Bush still in office, but it is more likely to happen with a new administration in place. The odds are good that mandatory controls will be enacted by 2010.

Once the US resolves what it will do at home, it will know far better what it is prepared to commit to abroad. To ensure that other major economies contribute their fair share to the global effort, the US will have a strong incentive to help build an inclusive and effective post-2012 framework. The global politics of climate change will be thoroughly transformed, and a new set of multilateral commitments may then be in reach.

Standing today at a fork in the road, some governments may not yet be prepared to head down the path of commitments. But they must resist the illusion that the alternative path of no commitments can lead to an ambitious and sustained global effort. For now, it is better to remain at the crossroads than choose the wrong path.

Author

Eileen Claussen is the President of the Pew Center on Global Climate Change and Strategies for the Global Environment. Ms Claussen is the former Assistant Secretary of State for Oceans and International Environmental and Scientific Affairs. Ms Claussen has also served as a Special Assistant to the President and Senior Director for Global Environmental Affairs at the National Security Council; as Chairman of the United Nations Multilateral Montreal Protocol Fund; and as Director of Atmospheric Programs at the US Environmental Protection Agency, where she was responsible for activities related to the depletion of the ozone layer; Title IV of the Clean Air Act; and the EPA's energy efficiency programmes.

Organisation

The Pew Center on Global Climate Change is an independent, nonprofit, nonpartisan organisation dedicated to promoting practical and effective climate change policies in the US and internationally. The Pew Center produces expert analysis of climate science, economics, solutions and policy issues; facilitates dialogue among policymakers, stakeholders, and experts; and contributes directly to the policymaking process.

Enquiries

Tel: +1 (703) 516 0630
E-mail: pewcenter@pewclimate.org
Website: www.pewclimate.org

Framework for a new global agreement on climate change

MOHAMED T EL-ASHRY
FORMER CEO AND CHAIRMAN, GLOBAL ENVIRONMENT FACILITY (GEF), SENIOR FELLOW, UN FOUNDATION

Climate change is a complex, long term problem, two centuries in the making. There are few human activities that do not contribute to it and as such, the scale of response needed for an ultimate solution is so large that widespread, collective action is essential. The Global Leadership for Climate Action, facilitated by Mohamed El-Ashry, was set up to mobilise political action and offer a framework for collective action. The framework and its recommendations for mitigation and adaptation policies, technological development, finance and global collaboration are discussed here.

Satellite images from NASA show sea ice changes in the Arctic from 1970 (above) and 2003 (below).

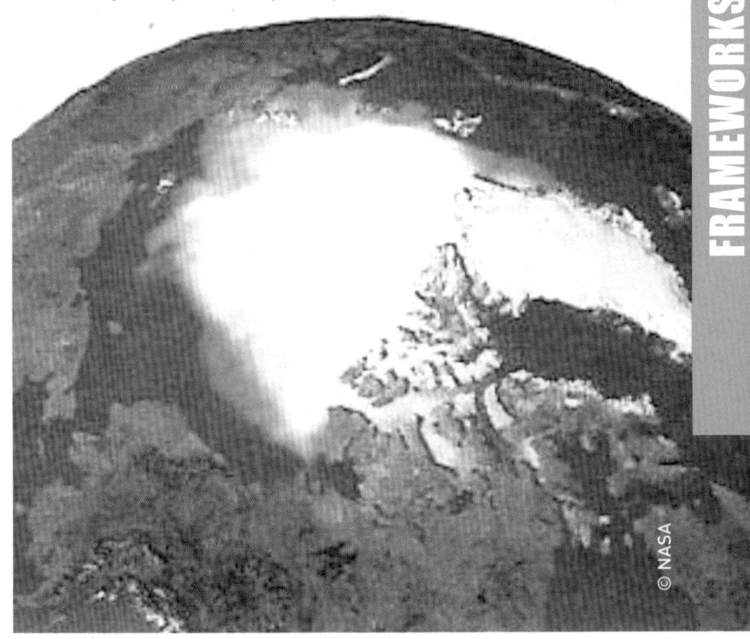

INTRODUCTION

Scientists believe that a temperature rise above 2º–2.5ºC risks serious and intolerable impacts. With rising temperatures, the Intergovernmental Panel on Climate Change (IPCC) predicts that frequency of heat waves, droughts, and heavy rainfall events will very likely increase, affecting agriculture, forests, water resources, industry, human health and settlements. Developing countries, where greater poverty and vulnerability limit the capacity to act, will be the most seriously harmed, particularly their poorer segments.

Avoiding such a future requires global greenhouse gas (GHG) emissions to peak in the next 10-15 years, followed by substantial reductions of at least 60 per cent by 2050 compared to 1990. This is a formidable task requiring unprecedented international cooperation and collective action. There is now a convergence of science, technology, economics, and finance to guide collective international action on climate change, but the window of opportunity for staying within an acceptable range of atmospheric GHGs (450-550 parts per million (ppm)) is rapidly closing, and an agreement to begin international negotiations has been elusive.

GLOBAL LEADERSHIP FOR CLIMATE ACTION

The Global Leadership for Climate Action (GLCA) was established in January 2006 to address two objectives: to mobilise political will and invigorate negotiations towards a post-2012 agreement, and to develop a framework for such an agreement.

The GLCA is a collaborative effort of the Club of Madrid and the UN Foundation consisting of 25 members from 20 different countries: 13 former heads of state and government and 12 leaders from business, intergovernmental organisations, and civil society. In September 2007, the GLCA agreed on a framework for a post-2012 agreement on climate change. The following highlights this framework and its recommendations.

COMPREHENSIVE AGREEMENT

Any new global climate change agreement must be comprehensive and negotiated under the auspices of UNFCCC, including all countries, sectors, sources and sinks, and mitigation as well as adaptation. A comprehensive emissions-based agreement sends a clear signal to the market and offers countries the flexibility to implement emissions reduction strategies most appropriate to their national circumstances. Smaller, targeted agreements, eg on industrial sectors, energy efficiency, technology cooperation, offer the potential of early action by countries that are not ready to accept emissions limits and should be encouraged. GLCA proposes four interconnected pathways for future negotiations:

- Mitigation targets, timetables, and market-based mechanisms.
- Adaptation.
- Technology development and cooperation.
- Finance.

MITIGATION – TARGETS AND TIMETABLES

A new agreement will be successful only if all participating countries perceive it to be equitable. Requiring all countries to achieve the same percentage reduction in emissions in the next commitment period would be unfair.

Developed countries should take the lead in global emissions reduction, given their historic responsibility and capability to act.

GLCA recommends:

All countries should commit to reduce global emissions collectively by at least 60 per cent below the 1990 level by 2050.

Even an 80 per cent reduction of GHG emissions in all developed countries by 2050 would not avoid the most adverse impacts without emissions reductions by developing countries. But not all developing countries are alike. Some are rapidly industrialising and some are least developed. Their engagement should be differentiated by their responsibilities and capabilities.

GLCA recommends:

As a first step, developed countries should reduce their collective emissions by 30 per cent by 2020; rapidly industrialising countries should commit to reduce their energy intensity by 30 per cent and agree to emissions reduction targets after 2020. Other developing countries should commit to energy intensity targets differentiated by their responsibilities and capabilities.

Targeted agreements on energy efficiency and renewable energy

Energy and climate security are intertwined and should be addressed simultaneously. Renewable energy and energy efficiency can contribute to such a strategy. Both are win-win propositions for all countries. Renewable energy generates employment, contributes to reducing air pollution and climate change, and enhances energy security through reliance on domestic energy sources. The technical and economic potentials of improving energy efficiency are also enormous although this has not been pursued by countries as aggressively as new supply in spite of experience showing the opportunities for gains.

GLCA recommends:

Long term policies, measurable and verifiable targets, should be adopted by all countries to increase significantly the use of renewable energy and promote greater efficiency in energy production and use. In addition, global standards for end-use efficiency should be developed and adopted.

Carbon sinks

Land-use changes, mainly deforestation, account for more than one-fifth of global emissions. With increasing emphasis on biofuels for transport, pressure to convert remaining forests will increase. Strategies to reduce deforestation have other benefits beside reducing atmospheric carbon dioxide. These include conserving biodiversity, providing ecosystem goods and services, especially water resources, and improving livelihoods for neighboring communities.

GLCA recommends:

In order to capture the many co-benefits of reducing deforestation, a full range of interventions to create and maintain biological sinks of carbon is recommended.

Market-based mechanisms

As the Stern Review said, "Establishing a carbon price, through tax, trading or regulation, is an essential foundation for climate change policy." GLCA agrees with most economists that the preferred mechanism is a system of harmonised universal carbon taxes, which reduce emissions and generate financial resources. Carbon taxes are easier to implement than cap and trade schemes and are economically efficient.

Cap and trade schemes are generally welcomed by industry, as they tend to reduce the cost of complying with targets. The cap is generally set at a level below the national allowance because small sources and those difficult to monitor are excluded. But without binding targets and a clear policy framework, a formal system cannot function. Tradable allowance systems can target either upstream sources, ie fossil producers and importers, if they are based on carbon content, or downstream if they focus on end uses and emissions. If tradable allowances are issued at no cost, the problem is one of distributing initial allowances among recipients. If the allowances are sold or auctioned, these schemes can raise revenue that can be used for other purposes.

GLCA recommends:

A price on carbon should be set through a system of harmonised, universal carbon taxes, although many in industry prefer a cap and trade system. For a well functioning cap and trade system, carbon markets need to be financially linked. In general, emissions allowances should be auctioned, raising resources that can be allocated by national governments for other purposes.

ADAPTATION

Substantially reducing global emissions of GHGs will not avoid the serious impacts of climate change to which the world is already committed, with the poor in developing countries being the least able to adapt. Adaptation is not simply a matter of designing projects or putting together lists of measures to reduce climate change impacts. A national policy response would increase resilience to climate vulnerability and should be anchored in a country's framework for sustainable development and integrated in its poverty reduction strategies. Responses to climate change need to encompass several levels including access to clean energy for vulnerable populations, crop and farm level adaptations, national level agricultural and supporting policies and investments.

Future agricultural systems will have to be more resilient to a variety of stresses to cope with the consequences of climate change. Technologies for adaptation, eg salt- and drought-resistant crop cultivars, should be developed and disseminated widely.

Effective adaptation will require broader planning capacity in all relevant departments in developing countries. Local scientists should be supported for monitoring and research on climate impacts on various sectors in their own countries.

GLCA recommends:

Financial support should be provided for vulnerability assessments, enhancing resilience to climate impacts, access to information and best practices, building human and institutional capacity, and making public and private investments in developing countries less susceptible to climate change. Centres for adaptation in agriculture should be established, particularly by the Consultative Group on International Agriculture Research (CGIAR) in Africa.

ACCELERATE TECHNOLOGY RESEARCH AND DEPLOYMENT

If the world continues on its current energy path, energy-related CO_2 emissions in 2050 will be two and a half times their current levels. According to the International Energy Agency, these emissions can be returned to their current levels by 2050 through a combination of all-country actions including strong energy efficiency gains, increased renewable energy deployment, natural gas, nuclear, and coal with CO_2 capture and storage (CCS), and increased biofuels use.

Reducing global emissions by at least 60 per cent from 1990 levels at acceptable costs will require a technology revolution, at least as large as those in space and telecommunications, to make clean energy technologies more efficient and affordable. Technologies such as solar, wind, biofuels, hydrogen, energy efficiency, and CCS need breakthroughs that will only be made possible by public funds. Unfortunately, investments in energy R&D programmes have been declining for the past two decades. These declines need to be halted and reversed.

GLCA recommends:

Clean energy technologies should be made available and utilised by all countries. Barriers that hamper dissemination of such technologies in developing countries, such as intellectual property rights and competitive rules, should be overcome. Recent declines in investment for R&D should be reversed and the aggregate amount of public funds doubled to US$20 billion per year.

Consultative group on clean energy research

Forming a consultative group on clean energy research (CGCER) could facilitate international collaboration in developing low cost, low carbon technologies and the exchange of information about clean energy technologies. Initially, the CGCER could be established as a virtual institution, linking centres of excellence in developed and developing countries. It could support research, act as a catalyst for South-South cooperation, and pay for patents or licensing fees to enable cleaner technologies to be deployed in the South.

GLCA recommends:

A consultative group on clean energy research should be formed to encourage a clean technology revolution. Innovation targets to bring new technologies to market, as well as incentives for meeting them, should be considered.

FINANCE

Both public and private finance are essential for adaptation, to avoid deforestation, increase technology transfer and implement a long term strategy to combat climate change. Climate friendly investments need to be multiplied through national and international frameworks, and the current international carbon market needs to be enhanced in order to scale up private flows. However, external funding must be additional to national resources obtained through domestic savings and taxation. Governments have an obligation to establish a supportive framework for investment. Local capital markets should facilitate long term investments in adaptation measures. Carbon taxes or the auctioning of allowances can also raise resources for other purposes.

Reform Clean Development Mechanism

The Clean Development Mechanism (CDM) was created under the Kyoto Protocol to support low carbon investments in developing countries. For the developed countries, its purpose is to lower emissions reduction costs and provide flexibility in carrying out national obligations. For developing countries, CDM's purpose is to promote sustainable development and contribute to GHG stabilisation. The CDM has encountered administrative and technical hurdles with initial projects limited to a few countries and a few gases and plagued by bureaucratic procedures.

To promote policy reform, underwrite technology development, and stimulate investment flows at a transformational scale, an additional market mechanism must take a sectoral approach. The distinction between a sectoral approach and project-based or programmatic approach (the current emphasis of CDM) is that a developing country could set sector-wide baselines for carbon-intensive sectors at levels that coincide with its economic interest while meeting commitments to reduce the energy intensity of its growth.

Enhance public finance

Public finance can demonstrate new approaches for building human and institutional capacity and for mitigation and adaptation in developing countries. However, the existing funding sources for these purposes, eg the GEF and multilateral development banks, are too small for the scale of assistance required. However, they should be strengthened and their funds enhanced.

Sustainable development is not possible without making energy systems more sustainable. Rapidly industrialising countries need to grow in a climate-friendly manner. However, the current costs of cleaner and more efficient technologies are higher, as much as US$100 million for an average 1 GW coal-fired power plant. It is estimated that between US$30-35 billion a year is required to 'green' energy sectors in developing countries.

A new US$50 billion per year climate fund

The average net public financial flows from all developed countries, including loans, amounted to about US$58 billion per year between 1996 and 2005, or about 0.23 per cent of GDP, of which about US$7 billion was for energy. GLCA estimates that about US$50 billion per year will be needed for activities in developing countries to support a comprehensive climate change agreement. Since commitments and actions to meet a 60 per cent reduction by 2050 will have to be undertaken in phases, the first phase could be about US$10 billion per year. Funding sources could come from a combination of public finance and the carbon market, especially the auctioning of emissions allowances.

GLCA recommends:

A climate fund of additional resources, starting at US$10 billion and growing to US$50 billion per year, should be established to support climate change activities in developing countries and should include both public and private resources. It should have an innovative structure and transparent and inclusive governance.

CONCLUSION

As we embark on a more comprehensive and inclusive agreement, we need to build on the experience gained from Kyoto and from cities, states, communities, businesses, and individuals who have voluntarily undertaken important steps to address climate change. Above all, we need to build trust between North and South and establish an equitable basis and new modalities for genuine international cooperation to address the linked challenges of energy and climate security.

Author

Mohamed T El-Ashry is currently Senior Fellow at the UN Foundation in Washington, DC. He was CEO & Chairman of the Global Environment Facility (GEF) from 1991-2003 and previously Chief Environmental Advisor to the President and Director of the Environment Department at the World Bank. Prior to that, he served as Senior Vice President of the World Resources Institute and Director of Environmental Quality at the Tennessee Valley Authority (TVA). He has also served as Special Advisor to the Secretary General of the 1992 Rio Earth Summit.

Organisation

The Global Leadership for Climate Action (GLCA) is a task force of world leaders committed to addressing climate change through international negotiations. A joint initiative of the United Nations Foundation and the Club of Madrid, the GLCA consists of former heads of state and government as well as leaders from business, government and civil society from more than 20 countries.

Enquiries

UN Foundation
1800 Massachusetts Ave, NW
Washington DC 20036 USA
Tel: +1 (202) 887-9040
Fax: +1 (202) 887-9021
E-mail: mel-ashry@unfoundation.org
GLCA website: www.globalclimateaction.org
UN Foundation website: www.unfoundation.org
Club of Madrid website: www.clubmadrid.org

A framework for climate change and global equity

PROFESSOR JEFFREY D SACHS
DIRECTOR,
THE EARTH INSTITUTE AT COLUMBIA UNIVERSITY

With the world initiating new climate change negotiations in December 2007 to succeed the Kyoto Protocol expiring in 2012, there is still no shared framework for agreement between developed and developing countries. There has been a sea change in global attention to this issue in the past three years, reflected in this year's 2007 Nobel Peace Prize which is shared by former Vice-President Al Gore and the Intergovernmental Panel on Climate Change. Yet there is still no consensus on how to proceed. This brief note sketches out a framework for action, building on the underlying core facts of the climate change challenge itself.

The 1992 UN Framework Convention on Climate Change (UNFCCC), under whose auspices the December negotiations will proceed, is a remarkably sound framework for global action. Fifteen years ago, the world's nations agreed on the proper core objective, "to stabilise greenhouse gas concentrations at a level to avoid dangerous anthropogenic interference in the climate system." They agreed that all countries must adopt plans of action consistent with this objective, but that the developed countries (called the Annex I countries in the Kyoto Protocol) must take the lead. The Treaty's doctrine on cross-country responsibilities, familiar from other agreements as well, is one of "common but differentiated responsibilities."

CARBON DIOXIDE IS INCREASING RAPIDLY

The basic arithmetic is roughly as follows. The current atmospheric concentration of carbon dioxide (CO_2) is 380 parts per million (ppm). It is now rising by roughly two ppm per year, based on global emissions of roughly 36 gigatonnes ($GtCO_2$) of manmade CO_2 emissions. These 36 $GtCO_2$ are divided between energy based emissions of 29 $GtCO_2$, and seven $GtCO_2$ caused by deforestation. While there are several other greenhouse gases that are important to control, including methane, nitrous oxide, and various industrial chemicals, carbon dioxide is at the centre of the story.

There is widespread agreement that the world will enter a very severe danger zone if the CO_2 concentration reaches 560 ppm; indeed the threshold of safety could be 450 ppm or even lower. With CO_2 concentration rising by two ppm per year, the current emission flow would take us past 560 by end of the century. Yet the threat is even more dire and immediate than that. The world economy is growing rapidly, at around five per cent per year, with energy growth of four per cent per year or higher. If this rate of growth is sustained for a half century, first as Asia booms, and then as other parts of the developing world, including

Africa and Latin America, achieve rapid economic growth, total emissions will skyrocket. The world could easily reach 560 ppm by 2050 or even earlier.

DECOUPLING GROWTH WITH EMISSIONS

The core challenge is to combine the stabilisation, then reduction, of annual global carbon dioxide emissions without sacrificing the developing countries' right to achieve significant economic growth, and presumably without affecting economic progress in today's high income countries either. Decoupling economic growth with greenhouse gas emissions is the basic challenge. And the core solution will be technologies which reduce energy use per unit of GNP and carbon emissions per unit of energy. Such technologies exist, but need a global framework for their development and global diffusion.

> "There is widespread agreement that the world will enter a very severe danger zone if the CO_2 concentration reaches 560 ppm; indeed the threshold of safety could be 450 ppm or even lower."

A GLOBAL FRAMEWORK FOR TECHNOLOGY DEVELOPMENT

Here's how this can work. Four sectors dominate as sources of CO_2 emissions: electricity production at independent power plants, motor vehicles, heavy industry (steel, cement, refineries, and petrochemicals), and deforestation. These sectors account for approximately 80 per cent of total carbon dioxide emissions. Emissions directly from residential and commercial buildings (mainly for heating) account for another 10 per cent or so, and many lesser emitting industries make up the remainder. Our focus should be on the big four categories.

ENERGY EFFICIENCY AND ENERGY DE-CARBONISATION

Part of the answer will be energy efficiency. We can economise on energy at the end user, such as through better insulation to reduce home heating, low watt illumination, and greater kilometres per litre of gasoline, or through more efficient power plants, which use less primary energy per kWh of delivered electricity. These steps will be crucial, but they will not be enough. Indeed, if the world economy grows at five per cent per year, energy efficiency might reduce energy growth to two-three per cent per year, but would still mean a global increase of energy use of perhaps two to four times by 2050.

We must therefore combine energy efficiency with a de-carbonisation of the energy system. Part of the answer can be a shift to nuclear and renewable energy sources to produce electricity, perhaps rising from roughly one-quarter of the total energy for electricity production (with fossil fuels now constituting three-quarters), up to one-half by 2050. Even so, with overall electricity demand likely to rise five fold by 2050, the fossil fuel-based electricity production could well double or even triple despite a massive increase of renewable and nuclear power production.

CARBON CAPTURE AND SEQUESTRATION

A central challenge, then, will be to use fossil fuels at a higher rate than now but in a manner that does not emit carbon dioxide. The most promising route for this is carbon capture and sequestration (CCS). The idea is to capture the CO_2 emitted from fossil fuel combustion and then geologically dispose of it in stable underground deposits, eg in basalt rock, or saline aquifers, or under the ocean bottom. Most engineering is now working on CCS at the power plant site, capturing the carbon dioxide from exhaust gases. Another possibility is to capture the carbon dioxide directly from the air (at the low concentration of 380 ppm), decoupled from the production of the CO_2.

The technological options will have to be used in combination. For example, today's coal-fired power plants, at one-third energy efficiency, ie they use three units of heat energy as coal to produce one unit of delivered electricity, can be converted to power plants with much higher efficiency and which capture their CO_2. The combination of higher efficiency plus CCS could reduce the emissions per kWh of electricity by 90 per cent or more. Probably the most promising technology is an Integrated Gasification Combined Cycle (IGCC) power plant with CCS capability.

Clean energy produced at highly efficient IGCC-CCS power plants would open up vast possibilities. For example, today's hybrid vehicles, which already get 50 to 100 per cent more miles per gallon than conventional internal combustion engines, could be converted into plug-in hybrids, in which the automobile is recharged by connection to the power grid, which itself is fueled by low emitting electricity (combining renewable, nuclear, and IGCC-CCS power plants). As a result, by 2050 there could be an automobile that gets roughly five times the mileage per gallon of gasoline achieved today, and whose overall emissions are reduced by four to five times compared to today's vehicles, taking into account the low emissions at the electricity plant.

Home heating could similarly be converted from home furnaces to electricity, using heat pump technology, and resulting in much lower overall emissions as long as the power grid is fed by low emission electricity. Industrial users would remain the big emitters in this scenario, and the largest industrial plants would be called upon to reduce their own emissions through CCS as well. In the end, every major energy using sector would contribute to achieve:

- Broad efficiency gains in end use.
- More efficient power plants.
- A partial shift to nuclear and renewable sources.
- CCS at the remaining coal and gas fired plants (or through air capture).

- CCS at big industrial facilities.
- Plug-in hybrids
- Electricity based home heating.

Together, these steps could cut total emissions by a quarter or more compared with today, despite a four to six fold increase in the size of the world economy by 2050.

COSTS

The added costs of this low emission package are likely to be modest. Renewable and nuclear power are already cost competitive with coal-fired power plants up to a point, and their competitiveness will make possible a further partial move in their direction. CCS is unlikely to cost more than two or three cents per kWh of electricity, but this needs to be proved. Plug-in hybrids, or alternative high mileage and low emission vehicles, are within a few years of production, depending now on technical solutions mainly to battery performance and reliability. Direct and relatively low cost air capture of carbon dioxide may soon prove to be feasible. Back of the envelope calculations suggest that the overall package needed to keep total emissions below current levels as the world economy grows sharply, is likely to cost under one per cent of global income each year and, quite possibly, much less.
Given that the risks and known costs of unconstrained climate change will be vastly larger, cost-benefit logic should push the world strongly towards an aggressive demonstration and diffusion of low cost, low emission technologies.

TOWARDS A WORLDWIDE ADOPTION OF LOW EMISSION TECHNOLOGIES

How then can we best get to where we need to be: a worldwide adoption of low emission technologies? The following steps are vital. First, we recognise that all countries, both high income and developing, must participate in the transition, since the basic arithmetic shows that all regions of the world will have to play their part in a rapidly growing world economy. Second, the rich countries should pledge to the poor countries that no actions on the part of the developing countries will be made to jeopardise their legitimate aspirations for continued rapid economic growth. The global agreement must be against emissions, not against economic development.

Third, the world should agree on the 2050 targets, for example of emissions no larger than 20 $GtCO_2$ through fossil fuel use (two-thirds of today's target), and a concentration at or below 450 ppm of CO_2, and then adopt a global framework of action to achieve those targets. Fourth, in addition to stabilising and reducing emissions from fossil fuels, deforestation and its resulting emissions should be ended, not only for the sake of climate change, but for the sake of the vast biodiversity now being threatened with extinction through the rampant destruction of tropical forests. Rich countries should help poor countries avoid deforestation through a system of incentive payments for taking forest land out of deforestation.
Fifth, rich countries should immediately finance a dozen or so large scale demonstration power plants with CCS technology in China, India, Indonesia, Brazil, the US, Europe, Australia and other major coal producing and coal using countries. We need to know that CCS is viable, both technically and economically. The technology looks great on paper, but we'll never know its real possibility and performance until it is tried in many parts of the world. Furthermore, countries such as China and India will not agree to sharply binding targets on emissions until they are confident in low cost solutions, such as CCS that will enable them to preserve their rapid economic growth. Sixth, rich countries should immediately undertake a major scale up of public funding and public private partnerships for other key technologies, such as plug-in hybrids, cellulosic ethanol (which does not directly compete with arable land for food production), concentrated solar power and others.
Seventh, each country should be prepared to put economic incentives to work around the adoption of low emission technologies. A global tradable permits system is unlikely to be the way to go. It would be highly cumbersome and very difficult to regulate, and it would miss the point that the vast bulk of solutions can come through a focus on adoption of a few key technologies. Success is more likely if countries each adopt a subsidy on carbon sequestration and a tax of some sort on carbon emissions, combined with global agreements on basic minimum industrial standards in emissions at power plants, automobile mileage, and performance at steel mills, cement factories, refineries, and other high CO_2 emitters. A realistic starting point would be a US$30 per tonne subsidy on carbon sequestration, and a small tax, perhaps US$10 per tonne, on carbon emissions

(amounting to roughly one cent per kWh for delivered electricity), which would rise gradually over time.
Eighth, at the same time that the world is adopting a strategy to curtail emissions, it must also adopt a strategy of adaptation to the climate change that is already underway, and will intensify in the future. Specific and crucial measures will be needed in the face of growing threats of droughts and floods, extreme weather events, temperature stress on crop productivity, the expanded range of malaria and other tropical diseases, and the loss of biodiversity. As with mitigation, the rich countries will have to bear the bulk of the financial responsibilities for funding adaptation measures.

LOOKING FORWARD

The world is entering the new round of negotiations looking for a 'grand bargain,' a single agreement that will chart the new course on emissions reduction and adaptation. This is probably the wrong mind set, as it could lead to years of frustrating negotiations without a global agreement. Just as with the Doha Trade Round, the world could miss its target by several years and fall into worsening finger pointing and acrimony. An alternative approach would be to view the climate change challenge as a matter of global problem solving, in which a sequence of steps is adopted year by year, based on the emerging evidence and technological possibilities.
In this approach, the world would agree on certain less controversial matters now, and reserve some of the other detailed and crucial matters for later years. As an example, the world's governments could agree now to the following: a fund to avoid deforestation; the launch of several CCS demonstration plants; an increased global R&D effort around a set of core technologies; a mid century objective of keeping CO_2 below 450 ppm; the launch of a new adaptation fund to help the poorest countries. We could agree to set country-by-country emissions trajectories until 2030 within the coming three to five years, based on the evidence on best performance low emissions technologies, especially the early results on CCS and plug-in hybrid technologies.
Surely the world would rally with enthusiasm to early and concrete measures now, even in the absence of a complete and overarching framework. The good news for the world is a framework that combines emissions reduction at modest economic costs, and allows the world to combine its twin objectives of sustainability and development is feasible. Now is the time for constructive agreements and a fair sharing of responsibilities.

Author

Jeffrey D Sachs is the Director of The Earth Institute, Quetelet Professor of Sustainable Development, and Professor of Health Policy and Management at Columbia University. He is also Special Advisor to United Nations Secretary-General Ban Ki-moon. From 2002 to 2006, he was Director of the UN Millennium Project and Special Advisor to United Nations Secretary-General Kofi Annan on the Millennium Development Goals, the internationally agreed goals to reduce extreme poverty, disease, and hunger by the year 2015. Sachs is also President and Co-Founder of Millennium Promise Alliance, a nonprofit organisation aimed at ending extreme global poverty.
For more than 20 years Professor Sachs has been in the forefront of the challenges of economic development, poverty alleviation, and enlightened globalisation, promoting policies to help all parts of the world to benefit from expanding economic opportunities and wellbeing. He is also one of the leading voices for combining economic development with environmental sustainability, and as Director of the Earth Institute leads large-scale efforts to promote the mitigation of human-induced climate change.

Organisation

The Earth Institute's overarching goal is to help achieve sustainable development primarily by expanding the world's understanding of Earth as one integrated system. We work toward this goal through scientific research, education and the practical application of research for solving real world challenges. With 850 scientists, postdoctoral fellows and students working in and across more than 20 Columbia University research centers, The Earth Institute is helping to advance nine interconnected global issues: climate and society, water, energy, poverty, ecosystems, public health, food and nutrition, hazards and urbanization. With Columbia University as its foundation, The Earth Institute draws upon the scientific rigour, technological innovation and academic leadership for which the University is known.

Enquiries

The Earth Institute at Columbia University
405 Low Library, MC 4335
535 West 116th Street
New York, NY 10027
E-mail: director@ei.columbia.edu
Website: www.earth.columbia.edu

Managing risks of climate change for human well being

THE CASE FOR ADAPTION

© saulgranda/Fotolia

IBRAHIM THIAW
CO-AUTHORED WITH **RENATE FLEINER** AND **ANANTHA DURAIAPPAH**, DIVISION OF ENVIRONMENTAL POLICY IMPLEMENTATION, UNEP

Impacts of climate change will inevitably influence ecosystem services and human well-being, and increase the pressure on the most vulnerable regions and communities that depend more on natural resources and ecosystems and have lower adaptive capacity. Taking appropriate adaptation measures is urgent to manage the risks and alleviate the impacts of climate change. Various mechanisms and instruments can be used, while concerted action from public and private sector and the international community is required in order to take equity and fairness into consideration.

CLIMATE CHANGE IMPACTS ON ECOSYSTEM SERVICES AND HUMAN WELL BEING

Climate change was identified by the Millennium Ecosystem Assessment as one of the five main direct drivers causing ecosystem services decline across the globe. Climate change can have significant impacts on ecosystems and their services and consequently on human well being, see Table 1.

Ecosystem services influenced by climate change encompass water, food production, provisioning

Table 1: Climate change impacts on ecosystem services and human well being.

Ecosystem services	Impact areas	Human well being
Water	Water availability, quality, regulation (resulting from precipitation pattern changes, increased groundwater salinity due to increased sea level and overexploitation, decreased river flows due to melting glaciers)	Health, material wealth and livelihoods
Food production	Drought, flood, salinisation, desertification of agricultural land	Adequate nourishment, health and security
Provisioning services, such as timber and fibre	Scarcity (due to migration caused by floods and famines)	Material wealth and livelihoods
Buffering natural disaster risks	Exposure, magnitude, recurrence, type	Security, health, material wealth and livelihoods

services and natural disaster risk management. The availability of these ecosystem services is mainly affected by precipitation pattern changes and temperature increase. Effects of changes in these ecosystem services on human well being can have impacts on security, basic material for life, and health. Effects can be various and complex, depending on vulnerability but also on the magnitude of impact and the specific context.

Climate change impacts will vary depending on regions and sectors but will be more severe where vulnerability to climate variability and change is higher, stress factors are multiple and adaptive capacity is low. In systems already exposed to increasing resource demands, unsustainable management and pollution, exposure to climate change constitutes an important additional pressure. Poor communities tend to be more vulnerable especially when located in high risk areas, as they have lower adaptive capacity and depend more on local ecosystem services, such as water and food. Climate change has the potential for exacerbating poverty.

Historically, Africa, Asia and Latin America have been the regions most at risk and vulnerable to climate change impacts. Major climate change impacts observed in these regions include:

- Reduced length of growing season in the Sahel region in Africa with detrimental effects on crops.
- Increased frequency of intense rainfall events in many parts of Asia causing severe floods, landslides, and debris and mud flows.
- Increased occurrence of climate related disasters in Latin America increasing mortality and spread of diseases in the affected areas.

Vulnerability can be further increased by existing stress factors, such as endemic poverty, limited access to capital, ecosystem degradation, disasters and conflicts and lack of effective response from the side of the government.

INSTRUMENTS AND MECHANISMS TO COPE WITH CLIMATE CHANGE

Adaptation depends on the adaptive capacity or adaptability of a system, region or community to cope with risks or impacts of climate change. If appropriate, adaptation can reduce negative impacts and even create benefits from new opportunities provided by changing climate conditions. However, if climate changes are incorrectly perceived, adaptation may even result in increased vulnerability (maladaptation).

Adaptation is an integral part of managing the risks caused by climate change, while risk reduction activities should be embedded in an integrated approach to ensure sustainability. Among international agencies there is growing consensus that climate change adaptation and therefore also climate change risk management should be integrated into all development planning and implementation activities to address sustainable development and poverty reduction in a more appropriate and effective way.

A variety of mechanisms and instruments involving different actors can be used to manage climate change risks that affect ecosystem services and human well being, see Table 2.

Adaptation can be part of economic development that makes use of financial risk sharing and market based mechanisms to address climate change through

Table 2: Mechanisms and instruments for managing climate change risks.

Mechanism	Sector	Instruments (examples)
Financial risk sharing	Insurance and reinsurance	▸ Alternative insurance schemes ▸ Risk transfer mechanisms ▸ Assistance for risk management (expertise, procedures, databases, tools) ▸ Practical risk handling measures
Market	Finance and investment, private sector, public-private partnerships, public sector and policy making institutions	**Regulation** ▸ User fees and taxes ▸ Investor friendly policies to promote climate friendly, resilient facilities ▸ Increasing regulation of access to ecosystem services for improved operating efficiencies of companies using land, energy and water resources **Business** ▸ Compensation payment for ecosystem services, eg through governmental incentive programmes that compensate landowners for protecting ecosystem services ▸ Carbon emissions trading markets with cap and trade systems for pollutant reduction, eg voluntary non-regulated schemes (Chicago Climate Exchange in US) that impose caps on emissions and allow companies to trade their pollutant rights to meet defined targets, 'compliance' market (EU's emissions trading scheme that is based on the Kyoto protocol)

Continued overleaf

Continued from previous page

Mechanism	Sector	Instruments (examples)
		▸ Ecolabeling/environmental certification ▸ Flexible business strategies and practices to address risks and uncertainties by taking environmental aspects into account to improve cost of capital and investor risk, business tools and methods to anticipate impacts of own activities on ecosystems ▸ Technologies that are environmentally friendly and improve availability of ecosystem services ▸ Precautionary environmentally sound approaches ▸ Tradable development rights (marketable rights given to landowners in areas reserved for conservation) ▸ Credits for diverse commodities (e.g. aquifer recharge, renewable energy, wasteload allocation, wetlands mitigation, water exchanges) ▸ Low input systems, eg organic farming to improve sustainability of production systems and agricultural biodiversity ▸ Industry clusters (closed loop) where waste of one industry becomes resource of another ▸ Eco-/Agro-/Cultural tourism
Technology transfer	Science and research, experts, private sector	**Agriculture/food production** ▸ Drought resistant high yielding (crop) species ▸ Improved cropping calendar, crop rotation ▸ Sustainable technological applications **Ecosystems/natural resources management** ▸ Integrated ecosystem management ▸ Habitat fragmentation reduction through migration corridors and buffer zones ▸ Mixed use strategies ▸ Conservation of forests and natural habitats in climatic transition zones with genetic biodiversity for ecosystem restoration ▸ Soil and water conservation ▸ Wetlands/coastal and marine resources protection ▸ Land use planning **Natural disaster risks management** ▸ Early warning systems ▸ Risk assessment and monitoring tools ▸ Flood and drought control management systems ▸ Contingency plan for response to emergencies **Human health** ▸ Health care systems, eg public infrastructure ▸ Financial and public health resources, eg public health training programmes, research for effective surveillance and emergency response, monitoring and information dissemination, sustainable prevention and control programmes ▸ Technological/engineering solutions to prevent vector borne diseases and epidemics ▸ Waste disposal infrastructure ▸ Water supply and sanitation facilities
Capacity building	Various sectors including science and research, experts, development agencies, meteorological institutes, etc	▸ Improving adaptive and coping capacity ▸ Knowledge sharing/training ▸ Interlinking with research centres and experts, eg through monitoring and climate risk information platforms
Humanitarian and development aid	International, national, non-governmental and governmental agencies	▸ Integrated climate change risk management in development programmes, covering land use planning, settlement development, infrastructure planning, building codes and related policies, or diversification of income generation in high-risk areas to reduce vulnerability towards disasters ▸ Response and relief

various approaches in the private and public sector. Businesses traditionally dealing with risk management, such as insurance and reinsurance, are leveraging their capacities in order to integrate climate change issues. Business sectors, such as finance and investment are beginning to develop innovative approaches to tackle these new challenges and opportunities arising from the increasing awareness of the importance of environmental, social and corporate governance for businesses and markets.

The public sector and policy making institutions provide the necessary pillars for market regulation, while public-private partnerships can facilitate governmental institutions the access to capital and resources. Interlinkages and dependencies of businesses on ecosystem services and environment factors can be important triggers in the pursuit of innovative market based and financial risk sharing approaches; see for example the consideration of the high economic value of biodiversity for potential new business opportunities.

CHALLENGES AND OPPORTUNITIES

The mechanisms described offer a variety of opportunities to cope with climate change risks. Since uncertainty is a major characteristic feature of climate change and clarifications are needed on how to put values on ecosystem services with public good characteristics that cannot be allocated to clear property rights, many of these opportunities still require research and support to become viable.

Further, the realisation of these opportunities requires large amounts of new and additional investment and financial flows. National policies could be supportive by promoting optimised resources use, providing incentives for private investors, and integrating climate change adaptation in key ministries and institutions as well as policies. Also, the insurance sector could attract private sector investment through alternative insurance schemes and risk transfer mechanisms, while an international fund could be established to backstop reinsurance schemes, support public-private partnerships or backstop national disaster funds. Despite of all these possibilities, further funding will be needed which could come through international cooperation from the developed countries, but most appropriately from polluting agencies based on the principle that the party responsible is liable for damages.

Another handicap is the urgency of taking appropriate actions in the fields of adaptation and mitigation in time to avoid increasing costs in the future. Economic losses due to climate change could reach US$1 trillion in a single year by 2040. Effective action requires a multistakeholder approach with the active involvement of governmental and non-governmental institutions at different levels. This must be based on the integration of climate change adaptation as a priority topic into all levels of decision making and operations, and on joint and coordinated efforts, together with other involved agencies and partners, at different levels to facilitate the use of synergy effects and improve cost effectiveness.

Governments should provide policy frameworks to facilitate adaptation and provide stability for business operations including subsidies and regulations to encourage appropriate actions, while providing accurate information on climate change. Inputs from the private sector could be in form of knowledge and resources as well as innovative products and services for new markets. The international community could undertake a concerted effort and provide a coordinated approach towards integrating climate change risk management into development and addressing underlying vulnerabilities, based on the integration of development, climate change and disaster risk management expertise.

While markets may be effective in determining efficient allocation of scarce resources, they do not take into account issues relating to equity and fairness. In developing countries, the poor depend heavily on ecosystem services for their well being, and the transfer and use of these resources is usually done through non-market channels. Therefore, bringing these ecosystem services into the formal market may cause some groups of individuals to be pushed into destitution. Moreover, by placing a price on a service which, previously, had been free and which people believe should always be free, such as clean water for personal consumption, clean air and flood regulation, may raise issues of ethics and rights.

The views expressed in this article do not necessarily reflect those of UNEP or the editors, nor are they an official record. A full version of the article with the references is available on: www.climateactionprogramme.org

INVESTING IN BIODIVERSITY FOR THE FUTURE

Climate change is projected to have the second largest global impact on biodiversity by the year 2100. Biodiversity loss can have far reaching implications causing increased water loss and altered ecosystems and a loss to ecosystem services.

Biodiversity is of high economic value as a reservoir of potential active pharmaceutical ingredients and of genes for growing crop species, for biotechnological processes or for bionic developments.

Examples of market based mechanisms that apply to biodiversity as an ecosystem service include tradable quota systems and bioprospecting. The former addresses the over exploitation by allocating quotas to extractive companies so that the sum of all quotas does not exceed the carrying capacity of the extracted resources, while quotas can be traded between producers, for example in fisheries management. Bioprospecting is the process of scientific research for the useful application of genetic resources in various commercial markets, while also providing incentives for biodiversity conservation.

Financial institutions that lack information about companies depending on biodiversity issues run the risk of getting exposed to significant reputational and other risks if loans, investments or other products are provided to companies with negative impact on ecosystems or that depend on ecosystems for their services. Goldman Sachs for example adopted a biodiversity benchmark that was developed by Fauna and Flora International and Insight Investment. This allows for a performance assessment of extractive companies in terms of biodiversity. Such instruments will become increasingly important as stricter biodiversity laws are passed. Still, awareness of the financial sector towards the economic relevance of biodiversity conservation needs to be further increased. In this regard a review on costs of biodiversity loss and benefits of biodiversity protection, and linking biodiversity business concerns to tangible financial metrics such as risks, shareholder value, or market capitalisation, could be useful.

Sources: Conservation Biology 2006, Nature 2000, Business.2010, 2007, UNEP 2005.

Author

Ibrahim Thiaw is the Director of the Division of Environmental Policy Implementation (DEPI) in UNEP. Mr Thiaw has more than 22 years expertise natural resource management and environmental policy – both at national, Pan African and global levels.

Organisation

DEPI is responsible for the implementation of environmental policy in order to foster sustainable development at global, regional and national levels. It leads UNEP's activities on ecosystem management and services.

Enquiries

Website: www.unep.org/DEPI

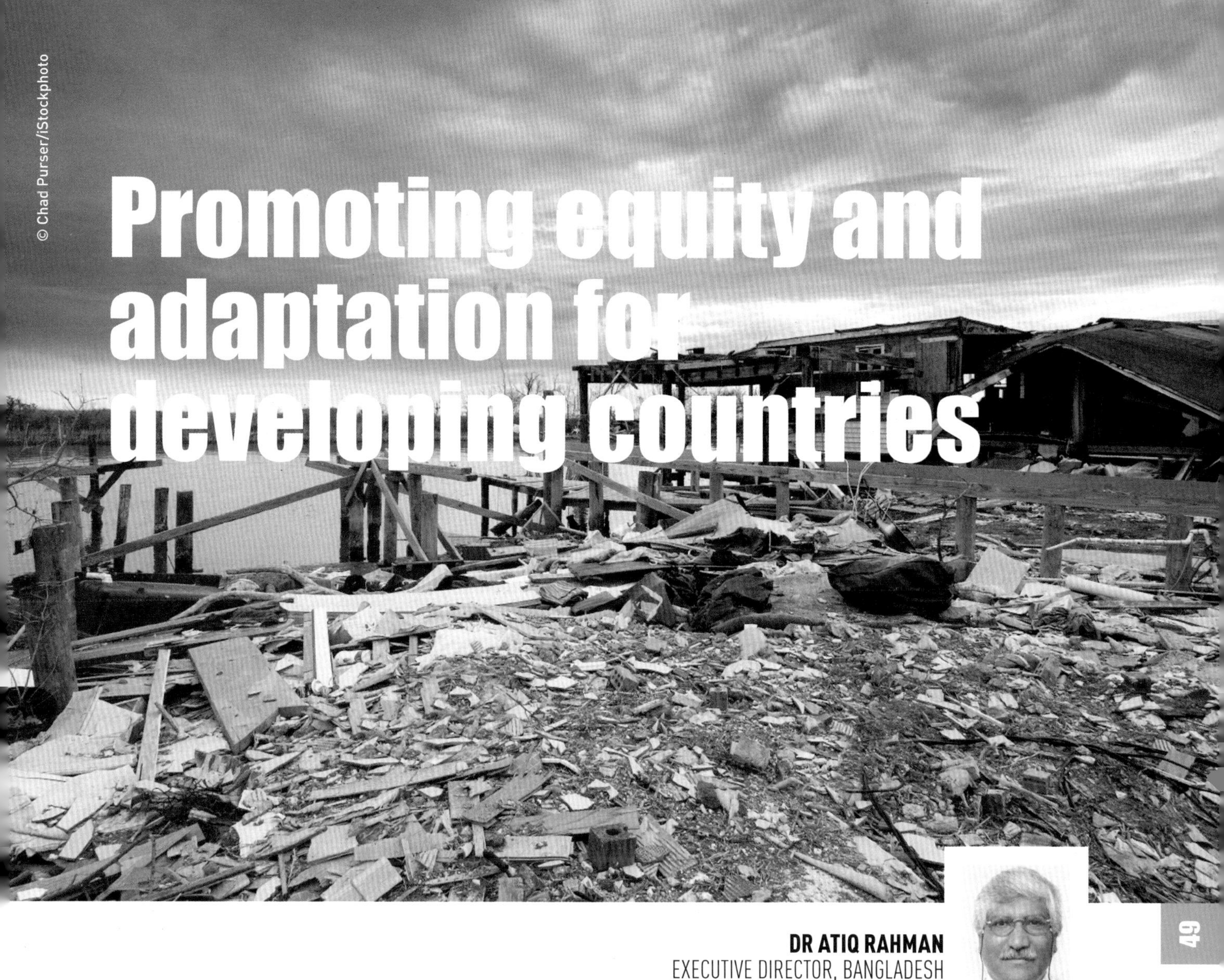

© Chad Purser/iStockphoto

Promoting equity and adaptation for developing countries

DR ATIQ RAHMAN
EXECUTIVE DIRECTOR, BANGLADESH
CENTRE FOR ADVANCED STUDIES (BCAS)

The IPCC Fourth Assessment report demonstrates evidence of human induced global climate change. The poor are the worst victims of a global problem that has been created mainly by the rich and industrialised countries. Climate change is a harsh reality now for many of us, and this is demonstrated through various recent extreme climatic events around the world. These include: prolonged and devastating floods in Bangladesh, India, China and in the UK in 2007; severe droughts in Africa and Asia, extreme heatwaves in central Europe, devastating cyclones and tidal surges across the coasts. We live in an unequal world economically, politically and socially, and climate change impacts will increase poverty, food insecurity, hungers, inequity and social conflict in the coming decades if we don't take urgent action now before it escapes our control.

Climate change is the result of global inequity. It is the result of unsustainable production, wasteful consumption and burning of fossil fuel in both developed and developing countries. The rich have created the problem while the poor are taking most of the burden. This burden is in the form of food insecurity and hunger, water stress and health risks, increasing poverty and greater disaster risks.

Rising sea levels and natural disasters may displace millions of people from their homes and livelihoods. In Bangladesh, 35 million people from the coastal districts may be displaced if there is 45cm sea level rise by 2050, which may result in rural to urban migration and put tremendous pressure on scarce resources in the country. The poor in Bangladesh and India, in Asia and Africa are not responsible for this. Climate change will also increase both intragenerational and intergenerational inequity. We are borrowing the environment, atmosphere and resources from our children and next generations. They will have to pay higher costs for environmental regeneration and climate change adaptation in the future.

COST OF DELAYED ADAPTATION

The UN Framework Convention on Climate Change, the Kyoto Protocol and the recent Nairobi Work Programme (NWP) on adaptation have enshrined the simultaneous actions of mitigation and adaptation. But scientists and development practitioners strongly suggest that mitigation is the best form of adaptation. But adaptation is not a substitute for mitigation. The primary commitment of the industrialised (Annex-1) countries is to mitigate, ie reduce greenhouse gas emission. Any delay in emission reduction will only increase the need and cost of adaptation.

Any delay in mitigation increases the risk of runaway global climate change with severe regional impacts. The primary responsibility of UNFCCC, the Kyoto Protocol and all its member states and signatories and non-signatories is to reduce emission today and now. Industrialised countries must take the lead; the developing countries must do their utmost to reduce their emissions and seek a path of development with a lower GHG emission.

Isolated US

In this rapidly globalising world it is neither desirable nor helpful to leave the greatest economy out of the Kyoto process as the US is not only the greatest GHG emitter, it also can play a key role in solving the problem of climate change with its scientific and financial capabilities.
If the US Administration is serious about freedom across the world, then freedom from the increasing threat of climate change, particularly for the poor and vulnerable countries and communities, must be part of the equation. Just as the US would not like to see a free rider in the global trade regime, it must not be allowed to be a free rider in a climate regime.

PROGRESS AND BARRIERS TO ADAPTATION

People always take action to adapt to the variability in weather and climate and try to reduce their risks and vulnerability based on their resources, own knowledge and experiences accumulated through generations. Sometimes vulnerable communities are also forced to react to and recover from weather and climatic extremes, such as floods, drought, salinity and cyclones. Existing practices and patterns of adapting to such events would not be effective in the context of devastating and long term climate change impacts. Planned and informed coping and anticipatory adaptation is needed, with greater resistance and diversity in human, social and natural systems to reduce risks and vulnerability as well as to protect lives, livelihoods, wealth and well being from climatic disasters.
The UN Framework Convention on Climate Change (UNFCCC) provides the moral, ethical and scientific basis as well as the structure for advancing adaptation. Parts of the UNFCCC clearly indicate the need and importance of adaptation to climate change to ensure food and water security, disaster risk reduction and promotion of sustainable development. Article 4.4 of the Convention calls for actions and resources from developed countries to assist the developing country parties that are vulnerable to the adverse effects of climate change in meeting the costs of adaptation.
The Nairobi Work Programme (NWP) on impacts, vulnerability and adaptation to climate change aims to assist particular developing countries, including those least developed countries and small island states, to improve their understanding and actions on climate adaptations. The main objective is to promote informed decisions on practical adaptation actions and measures to respond to climate change on a sound scientific, technical and socioeconomic basis taking into account current and future climate change and variability.
The recent IPCC Fourth Assessment Report suggests that adaptation to climate change is taking place across the world in limited scale and adaptive capacity is uneven among various communities and societies. The report also identifies a number of barriers to adaptation, including physical and ecological barriers, financial, institutional and technological barriers, and information and cognitive barriers. The least developed countries and poor in developing countries need greater support for adaptation.

> **"Any delay in mitigation, increases the risk of runaway global climate change with severe regional impacts."**

Though mitigation has been stumbling, there has been some progress in adaptation efforts, particularly in developing countries who are already suffering from various impacts of climate change. Such adaptation has a direct and reinforcing relationship with poverty, food security and health, livelihood promotion of the poor, disaster risk reduction and sustainable development. The Least Developed Countries group and many of its member countries have formulated NAPA (National Adaptation Plan of Action) but need resources and technology support from the developed countries and multilateral

development agencies to implement the adaptation activities identified. These NAPAs should adopt a holistic and livelihood approach instead of a sectoral approach. The process, in both planning and implementation, must ensure integration of indigenous knowledge and practices in advancing adaptation that might support sustainable development locally and regionally.

THE POOR NEED GREATER ADAPTATION

Since the poor have the least capacity, they need greater adaptation support. Some southern institutions are very keen to promote adaptation. The Bangladesh Centre for Advanced Studies (BCAS), an independent research and policy institute in Dhaka, has a great interest in understanding adaptation needs as well as promoting climate change adaptation with various stakeholders and actors including governments, development agencies and community people. BCAS, in association with development partners, has organised a community-based adaptation workshop in Dhaka to enhance understanding among scientists and practitioners. BCAS is also advancing community adaptation with partners under the SSN (South-South-North) initiative through increasing resilience and adaptive capacity of the communities in drought and salinity affected areas of Bangladesh. The project has completed vulnerability and needs assessment and is now developing actions for community adaptation with vulnerable groups, partners and key stakeholders. Project activities may include both adaptation and mitigation measures, such as irrigation with solar energy, afforestation, water security and livelihood promotion. However, there is further need for research and collective action to explore the possibility of blending adaptation and mitigation measures.

BEYOND 2012: COMMITMENT TO EQUITY AND FAIRNESS

During the first commitment period of the Kyoto Protocol, the world should have seen an absolute 5.2 per cent reduction in greenhouse gas emissions in the atmosphere relative to 1990 level by 2008-12. Unfortunately, what we have seen is a greater increase of GHG emissions. Let us remember the Kyoto target was the first small and inadequate step. It is an absolute priority that the UNFCCC parties achieve deeper GHG cuts to meet the prime objective to save the planet from the threat of runaway climate change. COP-13 in Bali must strive to achieve this goal. The 2007 Conference of the Parties should be a fertile ground for discussing the second commitment period with a focus to follow up the Kyoto Protocol's first commitment period with deeper cuts in the North and decarbonised economic development in the South.

CONCLUSION

Freedom from the threat of climate change is one of the key freedoms for which each and every country will have to work and take their responsibility in the 21st century. Freedom from the threats of severe impacts of climate change are as required as the freedom from hunger, injustice, terrorism and dictatorship. If the freedom from climate instability is not ensured, other freedoms such a poverty, food insecurity and hunger, could be equally threatened.

This year the Norwegian Nobel Committee awarded the Peace Prize to Al Gore and the IPCC for their great efforts and contribution to build awareness and generate scientific knowledge about man-made climate change as well as to lay the foundations for the urgent and long term measures to halt dangerous climate change and address the devastating impacts of global climate change. The Nobel Committee highlighted the potential risk between accelerating climate change and the risk of violent conflicts for various scarce resources including freshwater across the regions, countries and communities. This award encourages us and gives greater responsibilities to the policy and decision makers, development agencies and the actors at global, regional, national and local levels to take bold steps and actions with strong political will and commitment to reduce the threat of dangerous climate change for humankind. Actions are required now from various actors and stakeholders before climate change goes beyond our control.

Author

Dr Atiq Rahman is the Executive Director of the Bangladesh Centre for Advanced Studies (BCAS) and a Lead Author of the Intergovernmental Panel on Climate Change (IPCC). Dr Rahman is a Visiting Professor of International Diplomacy and Sustainable Development at Fletcher School of Law and Diplomacy, Tufts University, Boston, USA. He is the Chairman of Climate Action Network South Asia (CANSA) and of the Coalition of Environmental NGOs in Bangladesh and continues to represent this vibrant community as its key decision maker.

Organisation

The Bangladesh Centre for Advanced Studies (BCAS) is an independent research, policy and implementation institute in Dhaka, Bangladesh established in 1986. BCAS works at the local, national, regional and global levels on: Environment Development Integration; Poverty Alleviation and Sustainable Livelihoods; Good Governance and People's Participation and Economic Growth, Public Private Partnerships and Sustainable Markets.

Enquiries

Tel: (880-2) 8851237, 8852217, 8851986
Fax: (880-2) 8851417
E-mail: atiq.rahman@bcas.net
Website: www.bcas.net

EU Corporate Leaders Group representatives meet with EU President Barroso.

© Aled Jones

DR ALED JONES
DEVELOPMENT DIRECTOR AT THE
CAMBRIDGE PROGRAMME FOR INDUSTRY

Partnerships and climate change

With an increased understanding about the scale of change required to tackle global warming, business, government and civil society need to work in partnership to address such change. Building these partnerships allows governments to implement policies that will drive the necessary change, and business to respond rapidly to these drivers in a positive way.

INTRODUCTION

There is no one 'silver bullet' that can halt climate change in its tracks. Action across all industry sectors and society is key, with barriers to progress being similar across each sector. For example, concern about the impact of action on competitiveness and growth is a major factor in organisations' and countries' unwillingness to commit to more ambitious goals and strategies. In many countries, businesses are lobbying hard against unilateral action to tackle climate change preferring to wait for international frameworks to set the agenda for long term strategies.

As presented in *The Economics of Climate Change: The Stern Review*, the costs of adapting to unconstrained climate change far outweigh the costs of mitigation. Action is required now to prevent the most dangerous aspects of climate change from becoming an unstoppable reality. Therefore, while an ambitious international

Business and governmental partnerships, such as the CLG, aim to strengthen domestic and international progress on reducing greenhouse gas emissions as the effects of climate change become more evident in the world around us.

© Antonio Magras Martinez/UNEP/Still Pictures

framework is key to undertaking serious global action, lack of progress in this area should not prevent early action by leading organisations and countries. Such a leadership position in a developing market could result in huge rewards and benefits for those willing to take early risks.

There are several signs of this early leadership taking place, and opportunities in global markets are already leading to significant benefits. Finding ways to reduce the risk of taking a leadership position has been key and new and innovative partnerships between businesses and governments are taking shape around the world.

THE TREND TOWARDS GREATER TRANSPARENCY

Over the past few years there has been a noticeable shift towards greater transparency within the corporate sector. This is due to a number of different factors and influences including increased investor and media scrutiny over how issues, such as climate change, affect the long term value of an organisation as well as that organisation's responsibility to customers and society as a whole.

The Carbon Disclosure Project, for example, now represents over US$40 trillion worth of investment and is encouraging multinationals to take a serious and open look at their exposure to carbon risk (both regulatory and physical). In 2006, 72 per cent of the world's 500 largest companies responded to the Project's request for information regarding greenhouse gas emissions data and the risks and opportunities presented by climate change. In addition to this, in 2005 the *Influencing Power* report by SustainAbility and WWF surveyed 100 of the world's top companies and found a significant increase in transparency around lobbying activity over the previous five years.

> **" A growing awareness of climate change among consumers has resulted in a significant increase in corporations reporting how they are helping to tackle climate change. "**

A growing awareness of climate change among consumers has also resulted in a significant increase in corporations reporting how they are helping to tackle climate change. A UK and USA consumer survey by Accountability and Consumers International found that 40 per cent of consumers distrust what they hear about global warming from businesses while a further 50 per cent do not know whether to believe corporate claims or not. However, 66 per cent of consumers felt that everyone needs to take responsibility for their personal contribution to global warming and there is a big opportunity for business to be more open and work with customers to enable them to be more 'green'.

A STRATEGIC PRIORITY FOR BUSINESS

Being more transparent about actions concerning climate change and looking strategically at opportunities to take an early lead in this area should be a priority for business. Over the past three years all of the major consultancy firms have produced reports on the importance of businesses looking at climate change as a strategic issue.

There is clearly a potential for missed opportunities with most environmental and clean technology sectors seeing double digit growth rates over the past couple of years, or savings to the bottom line through energy efficiency measures. The importance that staff, customers and shareholders are increasingly placing on climate change issues also makes it a strategic priority for business.

Forum for the Future and the Universities and Colleges Admissions Service in the UK surveyed 54,240 young people applying to higher education in 2006. Of those surveyed, 80 per cent believed they will be personally affected by climate change in a negative way and only 20 per cent believed little or no change is necessary for

for human survival. With this in mind, attracting and keeping the best staff around the world is becoming increasingly difficult without a clear and proactive stance on climate change.

Evidence of companies moving towards such a proactive stance is borne out in a report by the audit, tax and advisory firm, KPMG, *Climate Change Business Leaders Survey,'* April 2007. It showed that 51 per cent of the FTSE350 senior executives surveyed are now starting to develop strategies to deal with concerns around global warming.

> **"There is clearly a potential for missed opportunities with most environmental and clean technology sectors seeing double digit growth rates over the past couple of years, or savings to the bottom line through energy efficiency measures."**

BUILDING PARTNERSHIPS

To help businesses reduce the perceived risk in taking a leadership position on climate change, an increasing number of partnerships are being set up. Whether these call for national policies to underpin an organisation's efforts (such as the US Climate Action Partnership), or enable a sector to share its knowledge and to work on combined strategies (such as the Institutional Investors Group on Climate Change), partnerships offer opportunities for real leadership to develop.

Two examples are summarised here. Both are part of the Prince of Wales's Business & the Environment Programme (BEP) which was set up over 10 years ago to engage with senior leaders from across the business community and give them a global forum for exploring and debating the business case for sustainable development.

Creating a common platform

In late 2006, BEP and the Prince of Wales met with a group of Chief Executives from the insurance sector. The intention was to provide a trusted forum to enable them to consider the role they might play in encouraging more climate-friendly behaviour both within the industry and among its customers. While it was acknowledged that the insurance industry has already been playing a leadership role in understanding the risks associated with climate change as well as developing responses to that risk, it was clear there was an opportunity for an even stronger leadership position to emerge.

On September 13th, 2007 the ClimateWise principles were launched in London with 37 companies signing up. These principles aim to enable development of innovative products and solutions across the whole spectra of insurance products and to share insurance-sector best practice. This will enable the sector to work with its customers (both business and individual) at reducing the effects associated with climate change over the long term. Working together as an industry to develop these types of products and to establish an open position on climate change reduces the overall exposure to business risk while clearly communicating to others the reasons behind the need for a change in the industry.

Creating political space

The Prince of Wales's Corporate Leaders Group on Climate Change (CLG) brings together business leaders from major UK and international companies who believe there is an urgent need to develop new and longer term policies for tackling climate change. The CLG thinks that taking serious action is not only important to prevent the worst aspects of climate change but also makes good business sense. The first output from the group was a letter to the UK Prime Minister in the run up to the G8 Summit in Gleneagles in 2005. The letter argued that investing in a low carbon future should be, "a strategic business objective for UK plc as a whole" and pointed out that at present, "the private sector and governments are in a 'Catch 22' situation with regard to tackling climate change, in which governments feel limited in their ability to introduce new climate change policy because they fear business resistance, while companies are unable to scale up investment in low carbon solutions because of the absence of long-term policies".

© Aled Jones

The current members of the UK CLG are:

ABN AMRO	BSkyB	Johnson Matthey	Sun Microsystems
Anglian Water Group	Centrica	LloydsTSB	Tesco
AXA Insurance	EOn	Reckitt Benckiser	Thames Water
B&Q	F&C Asset Management	Shell	Unilever
BAA	John Lewis Partnership	Standard Chartered Bank	Vodafone

The CLG aims to break the 'Catch-22' deadlock. In partnership with the UK Government, it works to strengthen domestic and international progress on reducing greenhouse gas emissions – seeking the backing for this effort from other British businesses, the UK public and international governments and businesses.

By developing a high level political strategy with well-positioned messages, the CLG has emboldened senior politicians to make decisions on climate policy which go further than they would have otherwise done. Indeed, UK Government insiders report that the CLG has had a direct impact on policy-making decisions relating to the UK's EU Emissions Trading Scheme National Allocation Plan targets and the 2006 UK Energy Review.

ENABLING A STEP CHANGE IN RESPONSE

The CLG was set up to create political space for Government action on climate change and the ClimateWise principles were developed to reduce the risk of an entire industry sector in changing working practices to help tackle this issue. Both initiatives allow governments to build on the case for change and develop policies that push forward this change.

The CLG has succeeded in creating political space and has started to dismantle the 'Catch 22'. While the CLG has not been, and should never be, solely responsible for getting policy decisions made, the group, in conjunction with other organisations, has started to make a significant positive impact on policy making in many of the areas addressed by its June 2006 letter to the UK Prime Minister. The test of the ClimateWise principles will be in their implementation over the next couple of years.

> "Working together as an industry reduces the overall exposure to business risk while clearly communicating to others the reasons behind the need for a change in the industry."

Bold leadership on climate change has the potential to deliver significant economic benefits – through improvements in economic performance, increased energy efficiency, improved growth as a result of technological innovation, greater energy security and access to significant global export markets for low carbon technologies. There are a range of technologies and services that could give any organisation a leading position in these new markets. While there is no one 'silver bullet' that can halt climate change in its track, there is more than one opportunity to be taken.

Author

Aled Jones is Development Director at the Cambridge Programme for Industry where he is in charge of Climate Change and Energy. He is part of the Secretariat for the Prince of Wales's Corporate Leaders Group on Climate Change, sat on the development group for the ClimateWise Principles and is Director of the Climate Leadership Programme.

Organisation

The University of Cambridge Programme for Industry (CPI) works with senior leaders internationally to help them understand and respond effectively to the most significant social and environmental challenges that face their organisations now and in the future. Our work supports the mission of the University of Cambridge, which is to contribute to society through the pursuit of education, learning and research at the highest international levels of excellence. Our alumni network of leaders extends to nearly 2,000 individuals from over 400 organisations in more than 40 countries. CPI has managed the Prince of Wales's Business & the Environment Programme for the past 14 years.

Enquiries

Dr Aled Wynne Jones MA FIMA
Development Director, Climate Change
Cambridge Programme for Industry
University of Cambridge
1 Trumpington Street, Cambridge, CB2 1QA
Tel: +44 (0)1223 342100
E-mail: climate@cpi.cam.ac.uk
Website: www.cpi.cam.ac.uk/climate

Developing partnerships across developed and developing countries –

AN AGENDA FOR INCLUSION OF THE POOR

JYOTI PARIKH
EXECUTIVE DIRECTOR, INTEGRATED RESEARCH AND ACTION FOR DEVELOPMENT (IRADE)

© Gerald Bernard/Fotolia

Increasingly, businesses in developed countries are recognising their responsibilities regarding greenhouse gas (GHG) emissions and climate change. But as well as clearing up their own houses, they have the potential to help mitigate the global impacts of climate change, by working with NGOs, governments and developing country businesses to ensure the mistakes of developed countries are not repeated in new growth. In addition, businesses, through an agenda of corporate social responsibility, can help alleviate the hardship of the most vulnerable groups who stand to bear the brunt of much of the impacts of climate change.

INTRODUCTION

Businesses wishing to adopt a corporate social responsibility agenda can approach this from a number of different fronts:

- Reduce GHG emissions in their own countries by introducing green procurement, energy efficient processes and systems and rigorous GHG accounting.
- Provide assistance to developing country businesses through:
  - Offering best technologies and best practices.
  - Purchasing carbon credits to ensure developing country businesses are active in reducing emissions and making correct technology choices during their growth paths. The Clean Development Mechanism (CDM) is one such instrument.
  - Joint Technology Development suited for large populations that does not repeat the mistakes made by the developed countries but charts new growth courses.
- Create an agenda for Corporate Social Responsibility (CSR) and:
  - Help vulnerable groups in the developing countries who are subject to much hardship through no fault of their own. Those who earn US$1-2/day emit as little as 100 to 200 Kg of CO_2 per year as compared to 4,000 Kg per person as global average and 8,090 kg by the EU.
  - Provide risk protection and insurance to vulnerable groups.

A SECTORAL APPROACH WITH FINANCIAL INCENTIVES

Mainstreaming climate issues in development opens a vast number of opportunities for partnerships and collaborations. A sectoral approach with financial incentives and technology inputs is one favoured by many and especially suited for large countries, such as India, China and Brazil. Sectors can be classified into three categories: production, energy and consumption. Production sectors are those with large energy consumption, such as steel, aluminium, fertiliser and cement. Energy sectors include coal, power, renewable and nuclear. Consumption sectors include transport and building, including houses.

Production sector

Emphasis is on upgrading technology, Best Available Technology (BAT) or best practices (BP). Incentives may be required for such international technology cooperation initiatives for carbon savings.

In the steel sector, China is the world's largest producer, accounting for nearly 35 per cent of world production. Indian consumption of steel has risen 14 per cent in the past 15 years, while global consumption grew at six per cent. Steel is a large GHG emitter; one tonne of steel

Water shortages mostly affect the rural poor, such as these villages in Thailand.

releases around 2.7 tonnes GHG. The corporate sector can help develop acceptable baselines and there is no reason why global benchmarks cannot be established for steel, cement and aluminium.

Cement contributes five per cent of total global CO_2 with Asia Pacific countries accounting for 61 per cent of the world's cement production. Opportunities to reduce CO_2 from cement are found through the introduction and/or replacement of technology, primarily the wet kiln process, in favour of dry processing technologies, energy efficient technologies, process improvements, power generation from waste heat recovery and enhanced coprocessing of low grade primary fuels and industry wastes.

> "Improving the efficiency of the mining and processing of coal and improving the monitoring and control of coal mine methane gas can make a significant contribution to emissions reductions and workplace safety."

Energy sector

World total coal consumption is 6,100 million tonnes; the USA consumed 18 per cent of the world's total coal production in 2004 with Europe consuming 17 per cent. Asia Pacific Partners collectively generate approximately 65 per cent of world primary coal production. Improving the efficiency of the mining and processing of coal and improving the monitoring and control of coal mine methane gas can make a significant contribution to emissions reductions and workplace safety. Renewable energy and energy efficiency promotion can be a major area of cooperation.

Consumption sector

Partnerships are needed to build nations. Good planning and design will be needed to fortify bridges, flyovers and major roads to ensure they withstand extreme events, high precipitation and dry periods. Construction companies and building material manufacturers must address this challenge in partnership with planners in developing countries.

Energy efficiency must become the prime criteria for manufacturing processes, products and purchase decisions for household appliances, industrial machinery and large scale infrastructure. Green procurements will have to be introduced to government procurement policy which normally favours the lowest bidder.

FRAMEWORK FOR A SOCIOECONOMIC AGENDA FOR POOR

Corporate social responsibilities need to expand to address climate change impacts. On one hand, the poor have to be lifted above poverty levels and on the other, GHG emissions must be significantly reduced. The conscientious corporate sector needs to recognise the debt it owes to the poor. But just who are the most vulnerable groups? Rural female populations in developing countries are still largely responsible for securing food, water and energy for cooking and heating. Drought, deforestation and erratic rainfall force this group to work harder to secure these resources. These are the most vulnerable groups to climate change.

Energy efficiency

Rural poor are already playing a role in using biomass, biogas, plantation and solar devices.

However, the poor should not be denied the use of fossil fuels, such as LPG or kerosene in the name of climate change. If the climate change problem requires the reduction of use of fossil fuels, then the wealthier have a key role to play. Disadvantaged people, without access to modern energy, are faced with problems relating to indoor air pollution and bear huge health burdens. Projects promoting renewable options should not leave poor people on the margins of decision making. Reducing vulnerability to climate change would mean finding appropriate technologies that take into account the specific socioeconomic realities of different rural areas, reduce workload, free up time and enable them to become micro or macro entrepreneurs.

Agriculture and food security

Climate change is predicted to reduce crop yields and food production in some regions, particularly the tropics. Women are responsible for 70–80 per cent of household food production. Traditional food sources may become more unpredictable and scarce as the climate changes thus invariably affecting women. Stronger public distribution systems will be needed that target such vulnerable groups.

Water and other resource shortages

Climate change may exacerbate existing shortages of water. Women, largely responsible for water collection in their communities, are more sensitive to the changes in seasons and climatic conditions that affect water quantity and accessibility, which makes its collection even more time consuming. The large river basins of the Niger, Senegal and Lake Chad as well as those of South Asia have experienced a 40-60 per cent reduction in the water level, according to UNEP figures. Water harvesting and storage systems will be needed to deal with shortages.

RISKS AND INSURANCE

Currently, the insurance sector is the world's largest industry, generating about US$4 trillion in 2006. Hundreds of new insurance initiatives, including 'green' building credits, drought protection in developing countries and incentives for investing in renewable energy and carbon emissions trading are being offered to tackle climate change and weather related losses in the US and globally, according to a major new report announced at the annual conference of the International Association of Insurance Supervisors.

The report, commissioned by the nonprofit group Ceres, outlines more than 400 climate related activities in the US and abroad – double the number of products and services identified in a similar report just 14 months ago. The report observes that, despite the impressive increase in recent activity, most insurance companies are still not focused on the climate change issue and fewer still are offering climate related products.

Among the recent offerings that show promise for customers and insurers alike are:

- Renewable energy related insurance products that allow more companies and investors to participate in renewable energy projects.
- Pay as you drive insurance products are now being offered by 19 insurers worldwide, who recognse that reduced driving means reduced accident risk, as well as reduced energy use. Tests have shown that PAYD products can reduce overall miles driven by 10-15 percent or more.
- Munich Re and Swiss Re are offering microinsurance to parts of the developing world where insurance did not previously exist. Swiss Re created, the Climate Change Adaptation Program in 2007 that uses climate models and satellite data to determine when weather related claims are to be paid in response to severe drought conditions causing food shortages in villages in Kenya, Mali and Ethiopia. Swiss Re has also sold weather risk products to 320,000 small farmers in India.

These need to be extended to the developing world where such measures enable them to take risks to reduce GHG emissions. The dearth of innovative products that would reduce climate risks and preserve insurability for developing courtiers is a particular concern. A large number of people live on coastal zones and high temperature zones vulnerable to disasters.

PROGRAMMES, INSTITUTIONS AND POLICIES

There is a need for an integrated approach to climate change monitoring and adaptation based on livelihoods of vulnerable communities. An integrated approach would:

- Make and demonstrate a compelling case for alternative approaches to climate change adaptation based on reducing the danger from climate change to vulnerable groups.
- Promote natural resource-based approaches for the reduction of vulnerabilities and mitigation of climate change effects. Such approaches would provide multiplebenefitssuchasgeneratingimmediateeconomic returns to poor people, sustaining and diversifying their livelihoods and conserving ecosystems.
- Offer convincing demonstrations of how on the ground livelihood activities can be linked with policy processes to reduce existing and future climate related vulnerabilities of poor.
- Identify multistakeholder, participatory processes for selection, implementation and appraisal of adaptation strategies.
- Critique and analyse the prevalent policy approach for addressing adaptation and the assumption that adaptation needs to focus on global rather than local processes.

The ultimate goal is to implement a socioeconomic model for sustainability as well as poverty reduction and conservation of biological diversity. Adaptation and vulnerability need to be mainstreamed into partnerships while maintaining the tempo for mitigation.

Adversity can bring people together and forge new partnerships among business, NGOs, governments and communities. Let us hope that conflicts give way to cooperation.

Author

Professor Jyoti K Parikh is Executive Director of Integrated Research and Action for Development (IRADe), New Delhi. She has worked for IPCC as a convening lead author and reviewing editor, served on Scientific and technical panel (STAP) for GEF and is on the Prime Minister of India's Climate Change Council. She has served as energy, environment and climate change consultant to the World Bank, the US Department of Energy, EU and UN agencies such as UNDP, FAO, UNU and UNESCO.

Organisation

Integrated Research and Action for Development (IRADe) serve as a hub of network among various stakeholders. It carries out multidisciplinary, multistakeholder analysis in local areas such as Energy System, Climate Change, Natural Resource and Environment Management, Infrastructure, and Rural and Urban Management. IRADe works with government ministries and departments, bi-lateral and multi-lateral agencies like UNEP, UNDP, World Bank, national and international non-government organisations like SEWA, ENERGIA and Stanford University.

Enquiries

IRADe, C-50, Chhota Singh Block
Asian Games Village, Khelgaon
New Delhi - 110049, India
Tel: +91 11 26495522, 26490126
Fax: +91 11 26495523
E-mail: jparikh@irade.org
Website: www.irade.org

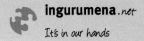

MINDY S LUBBER
PRESIDENT OF CERES AND DIRECTOR OF INCR, AND
RUSSELL READ, CHIEF INVESTMENT OFFICER,
CALIFORNIA PUBLIC EMPLOYEES' RETIREMENT SYSTEM (CALPERS)

Investors:

A NEW WORLD OF RISK AND OPPORTUNITY

Climate change presents enormous risks and opportunities for corporations. Increasingly, as investors make decisions about the value of their investments, they are factoring in how well a company is prepared to manage those risks and act on those opportunities. To do so, they are increasing pressure on companies to make climate change a priority governance issue and using their leverage to encourage companies to analyse and disclose their plans for managing climate risk and their strategies for seizing climate-related opportunities.

CLIMATE CHANGE: A STRATEGIC IMPERATIVE

The debate over whether human behaviour is causing climate change is over. The question is no longer 'if', but 'how' climate change is changing the world and the scale and urgency needed to reduce greenhouse emissions globally in the near future.

For investors, the question is how well companies will prepare for and adapt to this changing world. Which companies have the management savvy and strategic vision to manage the legal, physical and competitive risks associated with climate change? Which are best positioning themselves to seize what legendary venture capitalist John Doerr has called "the biggest economic opportunity of the 21st century" – renewable energy and clean technologies? Which companies are best prepared to function in a changing regulatory environment? In short, there will be winners and losers as some companies adapt to the new world of risk and opportunity and others do not.

© Mario Hornik/iStockphoto

> “When industry is willing to be regulated, it is a good bet that policy changes will follow.”

REGULATORY RISKS AND OPPORTUNITIES

There is no doubt that the imperatives of climate change are already bringing about regulations that seek to slow the pace of climate change, especially by limiting greenhouse gas emissions (GHGs). The Kyoto Protocol is now in force, the European Union operates under a carbon cap and trade system and many US states, including California, are imposing limits on GHGs. Such regulatory changes will surely continue, especially as leading corporations and investment firms appeal for clear national policies that will establish the rules under which business must operate in the future. For example, Duke Energy, a leading US utility, has called for "a national, economy-wide greenhouse gas mandatory policy as soon as possible," and JP Morgan Chase has called for "mandatory carbon constraints." When industry is willing to be regulated, it is a good bet that policy changes will follow.

How companies, especially those in the energy, auto and energy intensive manufacturing sectors, integrate regulatory changes into capital investment decisions and strategic planning will have a significant impact on their long term financial health. For example, utilities that continue to build coal-fired power plants without taking into account carbon reducing regulations, which will make CO_2 emissions more costly, are likely to find their shareholder value under serious pressure. "It is naive to believe that CO_2 legislation will fail to pinch profit margins," wrote Citigroup in a July 2007 research report downgrading coal company stocks. But it's not only carbon intensive industries that will be affected. No sector of the economy will be immune.

Regulatory changes are creating significant economic opportunities, as well. For example, automakers such as Toyota and Honda that have invested significantly in hybrids and fuel efficient vehicles have already reaped the benefits of their foresight. They are also better positioned to exploit China's burgeoning demand for cars than US automakers under China's stringent new auto-emission standards. Similarly, agricultural firms that are producing crops for exploding biofuel markets are well positioned to thrive in a carbon constrained economy.

Indeed, according to a June 2007 report by the United Nations Environment Programme (UNEP) climate change concerns, high oil prices and increasing government support are fueling a dramatic increase in investments in the renewable energy and energy efficiency industries. According to the report, investment capital flowing into renewable energy rose from US$80 billion in 2005 to US$100 billion in 2006, a trend that "shows no sign of abating." The McKinsey consulting firm is equally bullish about energy efficiency, saying aggressive investments in this area could reduce the global growth rate of energy consumption by more than 50 per cent over the next 15 years.

LITIGATION RISKS

Much as tobacco companies have been held to account for the enormous health care costs their products have imposed on state governments in the United States, legal accountability for climate change is an emerging risk for companies. It is too soon to tell whether major contributors to GHGs will be held liable for damages resulting from climate change, but the risk of litigation alone is a significant bottom line concern for companies.

A number of high profile lawsuits now pending in US courts seek to hold corporations liable for damages resulting from their contribution to global warming. For example, eight states and the City of New York have sued American Electric Power, Southern and several other power providers regarding their carbon emissions. A class action suit on behalf of victims of Hurricane Katrina seeks billions in damages against several oil and gas companies, claiming their GHGs contributed to the storm's severity.

According to a subcommittee of the American Bar Association, the leading US professional association of attorneys, "The prospect of liability is a serious matter for people who understand climate change and take it seriously." For investors, the prospect of such liability should also be taken seriously.

© Marc-André Besel

A GROWING CONSENSUS

In March 2007, US and European institutional investors managing more than US$4 trillion in assets called on the US Congress to enact strong legislation to curb the pollution responsible for global climate change. These investors, part of the Investor Network on Climate Risk (INCR), include some of the largest pension funds and investment firms in the world, the US$230 billion California Public Employees Retirement System (CalPERS) and UK-based F&C Asset Management, among them. Joining INCR in this 'Call to Action' were dozens of the world's leading investment houses and corporations, including Merrill Lynch, Alcoa, BP America, Allianz SE, DuPont and Pacific Gas & Electric.

This unprecedented statement was a strong message from the corporate and investment communities that climate policy uncertainty and the lack of federal regulations in the US may be undermining their long term competitiveness because they inhibit the level of investment in clean energy and climate-friendly technologies needed to address the climate change challenge. Speaking for many institutional investors in connection with the Call to Action, Fred R. Buenrostro, chief executive officer at CalPERS, said, "To tap American ingenuity and drive business to a leadership position in the low-carbon future, we need regulations to enable the markets to deploy capital and spur innovation."

The INCR-inspired Call to Action urges:

- Leadership by the US government to achieve sizable, sensible long-term reductions of greenhouse gas (GHG) emissions in accordance with the 60-90% reductions below 1990 levels by 2050 that scientists and climate models suggest is urgently needed to avoid worst case scenarios. Wherever possible, the national policy should include mandatory market-based solutions, such as a cap and trade system, that establish an economy-wide carbon price, allow for flexibility and encourage innovation.
- A realignment of national energy and transportation policies to stimulate research, development and deployment of new and existing clean technologies at the scale necessary to achieve GHG reduction goals.
- The Securities and Exchange Commission (SEC) to clarify what companies should disclose to investors on climate change in their regular financial reporting.

PHYSICAL RISKS AND OPPORTUNITIES

Perhaps the easiest of climate risks to understand are severe weather events, drought, floods, forest fires, rising seas, thawing permafrost and temperature extremes, all of which will have profound effects on many companies in the decades ahead. Property insurers are already heading for the hills, so to speak, with many no longer willing to insure coastal and other high-risk property. Single storms such as Hurricane Katrina resulted in US$40 billion in insured losses in 2005 and Lloyd's of London predicts a US$100 billion hurricane will hit the United States in the coming years. (Whether Katrina's strength was directly related to climate change is unknowable, but many climate scientists predict an increase in severe hurricanes as the earth warms.)

Oil and gas companies are among those that operate in areas at high risk for climate-related damage – hurricane-prone waters threaten off shore oil rigs and thawing permafrost damages pipelines, for example. Infrastructure damaged by such events can hinder their ability to get products to market. Following hurricanes Katrina and Rita, which caused widespread refinery outages, retail gas prices rose by 50 cents a gallon. Forest product companies face increased risk from drought and warmer temperatures that are making their primary asset more fire and disease prone.

Again, the question for investors is how well prepared are companies in these sectors, and others vulnerable to the physical effects of climate change, to manage their risk? Which among them are creating innovative products and services to meet the needs of customers who may be most impacted by climate change?

COMPETITIVE RISKS AND OPPORTUNITIES

To thoroughly assess a company's ability to compete in a changing world, investors and analysts must weigh a company's climate risk preparedness. Which automakers are making the capital investments to be competitive in a carbon constrained economy? Which energy producers are diversifying sufficiently to operate profitably in a world that turns increasingly to alternative fuels? Which major retailers and real estate developers are saving millions on their energy bills by becoming energy savvy in their operations?

Investors stand to benefit from identifying those companies that are anticipating the new economic playing field. Those that are adapting and those that are seizing the abundant competitive opportunities of climate change will undoubtedly fare better than those that are not. It is worth noting that many leading companies are already identifying ways to profit from this transition. For example, General Electric has made a major commitment to development of clean technologies with sales of energy efficient and environmentally advanced technologies. The company's 'ecomagination' programme already has a backlog of green technology orders worth US$50 billion, more than twice the company's projected revenue of US$20 billion by 2010. DuPont has reduced its GHGs by 70 per cent since 1990, saving US$2 billion in the process, while developing climate-friendly products such as Tyvek insulation that reduces home energy use by 13 per cent. Pacific Gas & Electric (PG&E) has invested billions in energy efficiency programmes that have made California 50 per cent more energy efficient than the rest of the country. The programme also reduced California's CO_2 emissions by 61 million tons and helped the state avoid having to build two-dozen new power plants.

> "Which companies have the strategic vision to manage the legal, physical and competitive risks associated with climate change?"

In addition to the bottom line impacts of seizing the competitive edge on climate change, companies that are responding to climate change also reap unquantifiable good will and reputational benefits from leading the movement towards a low carbon future. Such benefits increase competitiveness in many ways: attracting new customers, attracting and motivating a more dynamic work force and cementing a company's identity as a market leader and innovator in its field. "We've found that there are more high quality prospects that are interested in joining the company because of the very positive approach we've taken on the environment," said PG&E CEO Peter Darbee.

CLIMATE RISK DISCLOSURE: TRANSPARENCY IS KEY

A fundamental operating principle for any market system is transparency. To make sound investment decisions, investors must have access to a wide range of information about a company – its operations, management and finances. Securities laws emphasise disclosure of all material facts that are reasonably likely to affect a company's financial performance.

© D JUNTAWONSUP/UNEP/Still Pictures

To enable investors to assess a company's climate risk, generally accepted standards for disclosure, similar to that which already exists for other types of financial and governance information, are needed. Because climate change is a relatively new issue, however, industry norms for such disclosure have not yet evolved and corporate disclosure in this area has been spotty and inadequate. Indeed, many shareholders, including members of INCR, are introducing shareholder resolutions to try and force such disclosure, sometimes using such resolutions as a tactic to engage management in discussions about climate change preparedness. More than 40 such resolutions were filed with US companies in 2007, many of which received record high voting support.

To address the lack of disclosure standards, INCR, the Carbon Disclosure Project, the United Nations Environment Programme and other investor groups developed the Global Framework for Climate Risk Disclosure. While adherence to this reporting framework is voluntary, there is a basis in existing US securities law to require disclosure of material information related to climate risk. Accordingly, INCR members have urged the US Securities and Exchange Commission to require disclosure of the information specified in the framework.

The framework calls for corporate disclosure in four key areas that would allow analysts and investors to evaluate companies' climate change preparedness: (1) emissions disclosure, (2) strategic analysis of climate risk, governance, and emissions management, (3) physical risks and (4) regulatory risks.

Emissions disclosure. The framework encourages use of the Greenhouse Gas Protocol's Corporate Accounting and Reporting Standard, the most widely agreed upon emissions accounting standard, to enable analysts and investors to examine a company's total GHG output. This data is helpful in approximating a company's risks from current and future GHG regulations.

Strategic analysis. The framework encourages companies to undertake a thorough analysis of their climate change risks and opportunities to include (a) a statement of the company's position on climate change, its responsibility to address it, and its engagements with governments and advocacy groups to affect climate

INVESTOR NETWORK ON CLIMATE RISK MEMBERS (AS OF JULY 2007)

1. AFSCME PENSION FUND
2. AIG Global Investment Group
3. Boston Common Asset Management
4. British Columbia Investment Management Corporation
5. Bullitt Foundation
6. California Public Employees' Retirement System (CalPERS)
7. California State Controller
8. California State Teachers' Retirement System (CalSTRS)
9. California State Treasurer
10. Calvert Group
11. Christian Brothers Investment Services
12. Communications Workers of America/ITU Negotiated Pension Plan
13. Connecticut State Treasurer
14. Connecticut Retirement Plans and Trust Funds
15. Domini Social Investments
16. Doris Duke Charitable Foundation
17. Ethical Funds
18. Evangelical Lutheran Church of America
19. F&C Asset Management
20. Florida State Board of Investment
21. Florida State Treasury
22. Fred Alger Management
23. Green Century Funds
24. Illinois State Board of Investment
25. International Brotherhood of Teamsters, Affiliates Pension Plan
26. Kentucky State Treasurer
27. Maine State Treasurer
28. Maryland State Treasurer
29. Municipal Employees Retirement System of Michigan
30. Nathan Cummings Foundation
31. New Jersey State Treasurer
32. New Jersey State Investment Council
33. New York City Employees Retirement System
34. New York City Comptroller
35. New York State Common Retirement Fund
36. New York State Comptroller
37. New York State Teachers' Retirement System
38. North Carolina State Treasurer
39. Oregon State Treasurer
40. Parnassus Investments
41. Pax World Funds
42. Pennsylvania State Treasurer
43. Presbyterian Church, USA
44. Rockefeller Brothers Fund
45. Rockefeller & Co. Socially Responsible Investing Group
46. Service Employees International Union
47. Sheet Metal Workers Pension Fund
48. Stark Investments
49. State Retirement Agency of Maryland
50. State Street Global Advisors
51. TIAA-CREF
52. Trillium Asset Management
53. Tri-State Coalition for Responsible Investment
54. UNITE HERE
55. United Methodist Church General Board of Pensions and Health Benefits
56. Vermont Community Foundation
57. Vermont State Treasurer
58. Walden Asset Management
59. Washington State Investment Board
60. Winslow Management

change policy, (b) disclosure of steps being taken to reduce the company's GHGs, including benchmarks and timetables, and (c) how the company is addressing climate change as a matter of corporate governance. For example, have members of the management team been tasked with identifying the company's climate change risks and opportunities? Has compensation of key officers or managers been linked to meeting climate change goals?

Physical risks. Companies should examine and report on how the physical risks associated with climate change could affect the company's performance. What physical assets are vulnerable to extreme weather events such as hurricanes or melting of permafrost? Is the company's supply chain vulnerable to disruptions caused by severe weather? What steps have been taken or are planned to mitigate these risks?

Regulatory Risk. To determine whether companies are prepared for various regulatory futures, analysts and investors need to know whether management has assessed how possible regulatory changes could impact financial condition or operating performance. For example, has the company anticipated (a) possible increased energy or transportation costs, (b) how various national and local GHG emission regulations will impact the bottom line and (c) the future cost of carbon fuels? In short, how does management view the regulatory future and is the company prepared to operate effectively in it?

CONCLUSION

How a corporation manages the risks and takes advantage of new business opportunities presented by climate change will have important consequences on its competitiveness, economic health, and, in some cases, its survival. Companies must improve disclosure of their strategies so their owners can evaluate them. To successfully manage their portfolios for the future, analysts and investors must factor into their decision-making a thorough understanding of how effectively the companies they invest in, or choose not to invest in, are managing climate risk and seizing the opportunities.

Authors

Mindy S Lubber is the President of Ceres, a leading US coalition of investors and environmental leaders working to improve corporate environmental, social and governance practices. She also directs the Investor Network on Climate Risk (INCR), an alliance that coordinates US investor responses to the financial risks and opportunities posed by climate change.

Russell Read is the Chief Investment Officer for the California Public Employees' Retirement System (CalPERS). He oversees all asset classes in which CalPERS invests and is responsible for the strategic plan for CalPERS' Investment Office.

Mr Read joined CalPERS in 2006 after previously serving as Deputy Chief Investment Officer for Deutsche Asset Management and Scudder Investments. He was responsible for more than US$250 billion of retail and institutional investments in equity, fixed income and commodity-based products. He previously served as Global Head of Quantitative Investing and Research at Zurich Scudder and has held senior investment positions with the OppenhiemerFunds, CNA Insurance, Prudential Insurance and First Chicago.

Organisation

The Investor Network on Climate Risk is a network of institutional investors and financial institutions that promotes better understanding of the financial risks and investment opportunities posed by climate change. Much of INCR's focus is aimed at improving corporate disclosure and governance practices on climate change. The four-year-old network, coordinated by Ceres, includes more than 60 investor members with collective assets totalling more than US$4 trillion.

Enquiries

Investor Network on Climate Risk, c/o Ceres, Inc.,
99 Chauncy Street, 6th Floor Boston, MA 02111 USA
Tel: +1 617 247 0700 | Fax: +1 617 267 5400
E-mail: lubber@ceres.org | Website: www.incr.com

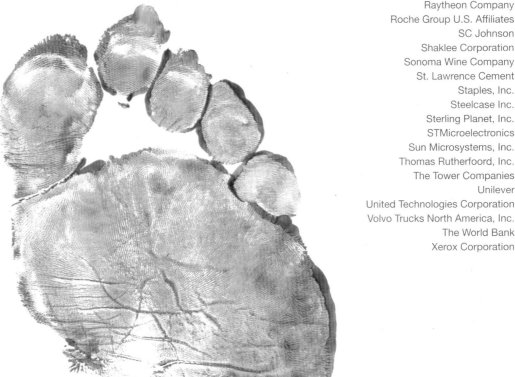

TakeAction! – unlimited access to renewable energy

At Novo Nordisk, we recognise that there are limits to growth. It is only when we respect the principles of sustainable development that we can grow a healthy business. A sustainable business.

We base our business on the Triple Bottom Line: striving to conduct our activities in a financially, environmentally and socially responsible way. This principle is anchored in our actions – from board decisions to daily practices among all our employees.

TakeAction! is the employee engagement program that enables everyone to make the Triple Bottom Line their business. It unleashes creativity, harnesses innovation and inspires action. A key element is the principle that activities must support business objectives. Examples are fundraisers for underprivileged people in need of medical care, educational programs that inspire healthier living among children and environmental initiatives in the local community.

Building a sustainable business is about overcoming the constraints of current practices. It's about nurturing new skills, supporting disruptive innovation, practicing deep listening and learning with stakeholders. It's about valuing people's motivation to drive change and to keep pushing forward.

There may be a limit to growth, but there is no limit to ideas. That is our most precious source of renewable energy. That is the foundation for our business. That is what makes people at Novo Nordisk TakeAction! To defeat diabetes. To combat climate change. To chart a road towards a healthier future for the next generations.

Read more about TakeAction! and other initiatives to put our values into action at novonordisk.com

Kristina Kempf, Daniela Rimpf and Lisbeth Juhl Christensen, Novo Nordisk Germany.

© iStockphoto/Nicholas Monu

Investor expectation on company climate change risk disclosure

IAN WOODS
SENIOR RESEARCH ANALYST,
AMP CAPITAL INVESTORS

Investors expect three key aspects of a company's climate change risk approach to be disclosed. They expect a clear articulation of the nature of material climate change risks and the impact on both short and long term strategy and competitiveness. Secondly, they expect a quantification of the risk and level of uncertainty that surrounds any risk assessment. Finally, investors expect a description of the risk management, mitigation and adaptation actions identified and the level of any residual/acceptable risk. Investors recognise the area of climate change risk is a rapidly evolving one; one that continues to demand a greater level of sophistication in disclosing those risks to investments.

CLIMATE CHANGE: A SYSTEMIC, LONG TERM RISK FOR INVESTORS

Understanding and managing an effective response to climate change is one of the greatest challenges facing modern economies. The scale of the structural changes within and between economies as a result of the drive to significantly decrease greenhouse gas (GHG) emissions and adapt to new climatic conditions will leave almost no sector in the economy untouched. Some will be impacted by government policy or regulation to reduce emissions, such as emissions trading schemes; others will be affected by the increase in climate variability, while others will be affected by long term climate changes. Some sectors will be affected by all three.

As with other risks facing companies, investors need to understand the nature of climate change risk as well as how companies, on behalf of their shareholders, are managing this risk.

Although institutional investors, such as pension and superannuation funds, invest in all sectors of the domestic and international economy and across all assets classes, the systemic nature of the climate change risk can neither be diversified away, nor managed effectively, through active investment management. Being so limited, investors are reliant on individual companies to manage the collective risk climate change represents to investors.

The importance of this risk to institutional investors is reinforced by its long term nature and how this matches potential long term liabilities or responsibilities to provide superannuation returns.

INVESTORS RESPONSE TO THE CLIMATE CHANGE RISK

Recognising the importance of climate change risk, investors have set up a number of collective investor initiatives, such as:

- Investor Group on Climate Change
- Institutional Investor group on Climate Change
- Investor Network on Climate Risk
- UNEP FI Climate Risk working group.

These groups work to develop a better understanding of climate change risk as it applies to investors, contribute to the public policy response and encourage the timely and appropriate disclosure of climate change risk by companies.

Increasingly, sell-side analysts are researching the potential impact of climate change on company value and individual investors are assessing exposure of portfolios to climate change risk, eg VicSuper in Australia and Henderson Investors in the UK. (A sell-side analyst works for a brokerage firm and evaluates companies for future earnings growth and other investment criteria.)

DISCLOSURE NEEDS OF INVESTORS

The investor need to make informed investment decisions and manage investment risk drives the demand on a company to disclose how it is managing its climate change risk.

The disclosure expectations of investors are different from other stakeholders in both their scope and detail. This has meant that emissions disclosure and associated discussion through company sustainability reports is often insufficient for investors. There are a number of reasons why this should be so:

- The disclosure of emissions from managed operations is not aligned to the exposure of investors, who are exposed to the emissions based on their equity interest in operations.
- The greatest exposure to investors may be through suppliers or customers and the management of the value chain.

Physical climate risks, such as floods, forest fires and rising seas will all have profound effects on companies in the decades ahead.

© iStockphoto/Shaun Lowe

- The mechanism for managing the risk, through financial risk products may be inappropriate for the normal audience of a sustainability report.
- Financial risk to investors may be, in the short term, limited to emissions covered by an emissions trading scheme, which may not cover all company emissions.

INVESTOR EXPECTATIONS

There are a number of frameworks developed by investors to facilitate the appropriate level of disclosure on climate change risk by companies. These include the Carbon Disclosure Project and the Global Framework for Climate Risk Disclosure. All frameworks focus on three key areas of disclosure:

- Clear articulation of the nature of material climate change risk or opportunity facing the company and, in particular, the impact on both short and long term strategy and competitiveness
- The quantification of the risk and the level of uncertainty that surrounds any assessment of the risk
- The description of the risk management, mitigation and adaptation actions identified to address the risks and the level of any residual/acceptable risk.

This framework will be familiar to many companies as it mirrors the structure of risk disclosure required by investors of other company risks.

Clear articulation of material climate change risk

The key issue for investors is material climate change risk and this is the starting point for any disclosure by companies. To demonstrate to investors that a systematic approach to climate change risk has been undertaken, companies should disclose which risks they have considered and why they believe a particular climate change risk is not material, noting that current uncertainty in either understanding of climate change impacts or regulation are not reasons *per se* for discounting the risks.

Similarly, companies may often see opportunities that might arise as a result of climate change. As noted, the three likely areas of risks faced by a company are:

- Regulatory risk
- Physical risk
- Value chain risk.

Investors recognise that each risk tends to be multidimensional and look for companies to demonstrate a clear and systematic analysis of all dimensions of the risk both as they may exist at present and in the future.

Regulatory risks may be as a result of the new performance standards, eg for energy efficiency standards for properties, appliances or vehicles, or result in a cost on emissions or other input costs, eg through an emissions trading scheme or carbon tax. In some cases, climate change may increase an existing regulatory risk, eg planning approval for an energy intensive or greenhouse intensive operation, such as a coal mine.

Physical risk may be as a result of direct weather related impacts of extreme weather events and an increase in climate variability. It may also be indirect, such as effects on water supplies or human health, eg heat related illness or an increase in pests and diseases.

Value chain risks are probably the least well understood or assessed by companies. Investors are looking for an assessment of both the supply and customer side of the value chain, considering both physical and regulatory risks and the impact of potential climate change adaptation or risk management options that the value chain players may adopt.

> "The disclosure expectations of investors are different from other stakeholders in both their scope and detail. This has meant that emissions disclosure and associated discussion through company sustainability reports is often insufficient for investors."

In disclosing the analysis of the risks, investors are looking for an analysis of the strategic and competitive implications of a particular risk and how this may affect the growth or risk profile of the company.

Many companies, particularly service sector companies, focus on the risks associated with their own direct or indirect greenhouse gas (GHG) emissions. While this may be the most recognisable, quantifiable and perhaps most easily managed risk, the emissions are not the principal climate change risk facing the company.

Investors recognise that addressing climate change may be part of a broader strategic positioning for a company, eg brand enhancement or employee values alignment. If this is so, then this should be articulated. The danger for companies is that focusing on potentially less important climate change risks, without putting them within a strategic/competitive advantage context may mean that more significant risks are not addressed or actions are ineffective.

Quantification of climate change risk

The challenge for companies is to demonstrate to investors that they have adequately quantified the climate change risk. Investors recognise this is an evolving area as a result of uncertainty in regulation and uncertainty in the physical impacts of climate change. However, companies need to demonstrate that they understand the risks, notwithstanding these uncertainties.

Clear disclosure of assumptions used in scenarios, eg price of carbon, time for introduction of a trading scheme or increased frequency/intensity of weather extremes or interruptions, is necessary to enable investors to assess the veracity of the scenarios.

For those companies significantly, directly, or indirectly impacted by a price on carbon, the most important disclosure is the emissions profile. Investors are looking for disclosure on an equity basis:

- Total actual historic direct and indirect absolute emissions by GHG
- Total current direct and indirect emissions by GHG
- Expected changes in emission profile in future years
- Direct and indirect GHG emissions by division for diversified or vertically integrated companies
- Direct and indirect GHG emissions covered by emissions trading scheme or other carbon price
- The basis for and level of accuracy of reported emissions
- Operations or gases not covered by data
- Current and future permits/credits given through any emissions trading allocation process or through CDM/ Joint Implementation projects.

Investors are looking for companies to use the most widely agreed upon international accounting standards. This is the Corporate Accounting and Reporting (revised edition) of the Greenhouse Gas Protocol, developed by the World Business Council of Sustainable Development and World Resources Institute.

A key step is quantifying the consequence on the growth and risk profile of the company and the company's assessment of the likelihood and timeframe of the particular climate change risk scenario. The basis for the company's assessment of the materiality or significance of the risk of the scenario should also be disclosed.

Risk management, mitigation and adaptation

There are three additional aspects that companies should disclose to investors:

- The relationship of climate change risk with other risk management processes within the company, in particular responsibility and reporting
- The specific current and future risk mitigation and adaptation actions or options
- The potential residual climate change risk, including the potential for stranded assets.

In discussing risk mitigation and adaptation actions, investors are interested in both the strategic risk mitigation and adaptation actions with respect to new investments and existing operations.

CONCLUSION

The systemic nature of climate change risk to investors and, in particular, long term institutional investors means usual diversification or active investment tools are inadequate for these investors to manage their climate change risk. As a result, these investors rely heavily on the climate change risk management of their investments to manage the risk.

This reliance leads to increased investor expectations on the level and detail of climate change risk disclosure by companies. Investors have worked together to outline their expectations through initiatives such as the Carbon Disclosure Project and the development of the Global Framework for Climate Change Risk Disclosure. Investors expect three key aspects of a company's climate change risk approach to be disclosed as already discussed.

The area of climate change risk is a rapidly evolving area as a result of international and national regulatory changes, societal expectations and increased confidence in the accuracy of predicted physical impacts of climate change. As a result, investors will continue to expect a greater level of sophistication in the disclosure of the climate change risks of their investments.

Author

Dr Ian Woods is the Senior Research Analyst for the AMP Capital Investors and Deputy Chair of the Investor Group on Climate Change Australia/New Zealand. Dr Woods undertakes assessment of GHG risk issues for the wider AMP Capital Investment teams. He is co-author of the report *Climate Change and Company Value – A Guide for Company Analysts.*

Organisation

AMP Capital Investors is an Australian based specialist investment manager with over $111 billion (AUD) in funds under management. AMP Capital Investors is the investment arm of AMP Ltd, a leading Australian based wealth management company.

The Investor Group on Climate Change (IGCC) is a collaboration of Australian and New Zealand investors focussing on the impact that climate change has on the financial value of investments. The IGCC aims to ensure that the risks and opportunities associated with climate change are incorporated into investment decisions for the ultimate benefit of individual investors.

Enquiries

Dr Ian Woods, Senior Research Analyst,
AMP Capital Investors, AMP Centre, 50 Bridge Street
Sydney, NSW, 2000, Australia
Tel: (612) 9257 1343
Fax: (612) 9257 1399
E-mail: ian.woods@ampcapital.com
Website: www.ampcapital.com

Using financial indices for responsible investment

WILL OULTON, HEAD OF RESPONSIBLE INVESTMENT FTSE GROUP AND
JAYN HARDING, PRINCIPAL ADVISOR, RESPONSIBLE INVESTMENT FTSE GROUP

The Stern Review suggests that climate change threatens to cause the greatest and widest ranging market failure ever seen. Stern states, "...our actions over the coming few decades could create risks of major disruption to economic and social activity, later in this century and in the next, on a scale similar to those associated with the great wars and the economic depression of the first half of the 20th century." As a result, investors, governments and wider society expect companies to take rapid action in response to the varied and considerable risk that climate change poses. Financial indices, such as the Dow Jones Sustainability Index, and the FTSE4Good Index can help investors find companies doing just that. Here, the FTSE4Good Index is explored in more detail.

TWIN-TRACK STRATEGY

Climate change is the environmental challenge that has the clearest and most tangible economic implications. Costs are increasingly being allocated to carbon through regulatory and voluntary economic instruments, such as the European Union (EU) Emissions Trading Scheme, which are also leading to significant market opportunities arising for innovative companies. Investors expect companies to develop appropriate climate strategies, of which there are two key components:

- **Manage down risks and impacts** – protect the value of existing company assets by reducing climate change risks and impacts through appropriate management systems and controls.
- **Maximise business and environmental benefits** – identify and develop related business opportunities through value creation.

RISK MANAGEMENT – AND VALUE PROTECTION

Stern's report highlighted that tougher and more comprehensive cap and trade international frameworks are essential and will lead to far more radical and far-reaching impacts on company valuations over the years ahead.

> "Detailed climate change requirements have already been added to FTSE4Good inclusion criteria to ensure the index series is aligned to one of the major concerns facing business and investors today."

A robust management system is vital in implementing aspects of a climate change strategy, see Table 1.

There are a range of well known standards for managing environmental and social risks, eg the Environmental Management System ISO 14001; SA8000. The FTSE4Good Index Series provides an extra-financial overlay to the FTSE All World Developed markets index using a range of environmental and social inclusion criteria. Detailed climate change requirements have already been added to FTSE4Good inclusion criteria to ensure the index series is aligned to one of the major concerns facing business and investors today.

Beginning with widespread market consultation with final criteria ratified by an independent committee of experts, the climate change criteria for the FTSE4Good index series are designed to encourage companies to mitigate climate change based on their operational and product impacts, see Table 2.

As with all FTSE4Good criteria, the climate change criteria are set to be challenging but achievable, rather than leading edge practice for a small number of companies. The index aims to encourage companies to address key environmental and social issues and work to improve performance continuously.

Table 1: Overview of elements of a good management system.

- **Policy and strategy**: publishing a clearly defined public policy and strategy with Board responsibility, which establishes broad goals and sets objectives and targets for improved performance.
- **Processes and structures**: establishing procedures which define internal processes and structures to ensure policies are implemented and risks managed effectively.
- **Monitoring and performance**: setting up appropriate performance monitoring mechanisms and key performance indicators to measure continually, quantify and improve corporate responsibility performance.
- **Reporting** regularly to stakeholders on impacts, policies, management systems and performance.
- **Consulting** with key stakeholders on these issues.

Table 2: Climate change criteria for the FTSE4Good index series.

	High operational impact	Medium operational impact	Additional high product impact
Policy and Governance	▸ **Board level or senior executive responsibility** for climate change related issues (individual or committee). ▸ **Public statement/policy** identifying climate change as relevant to business activities and the need to address climate change as a key concern.	▸ **Board level or senior executive responsibility** for climate change related issues (individual or committee). ▸ **Public statement/policy** identifying climate change or energy consumption as relevant to business activities and the need to address climate change as a key concern.	▸ **Responsibility**: no additional requirement. ▸ **Public statement/policy** should also include a commitment to reduce product-related emissions or climate change impact.
Management and Strategy	**At least one of the following must be met (unless the company meets the performance requirements):** ▸ **Long term strategic goal** of significant quantified reductions of operational GHG emissions or carbon intensity improvement over more than five years, which should be publicly available. ▸ **Short/medium term management targets** for quantified GHG operational emissions reduction over less than five years.	*No requirements yet: criteria for medium impact companies are initially focused on disclosure.*	*No requirements yet: for companies with product-related emissions reduction management targets are currently regarded as impractical.*
Disclosure	**Public disclosure of both the following:** ▸ Total operational CO_2 or GHG emissions as tonnes of CO_2 equivalent. ▸ Sector metric where established as an industry norm, eg. for cement companies, kg CO_2 per tonne of cement; or efficiency ratio.	**Public disclosure of one of the following:** ▸ Total operational CO_2 or GHG emissions as tonnes of CO_2 equivalent or operational energy consumption. ▸ Sector metric where established as an industry norm. eg kg CO_2/t cement; or efficiency ratio.	**Public disclosure of** product-related emissions/efficiency. This will vary for different sectors: ▸ Oil & gas: end user emissions. ▸ Coal mining: end user emissions. ▸ Automobiles: fuel efficiency. ▸ Aerospace: fuel efficiency.
Performance	**At least one of the following must be met:** ▸ **At least a five per cent reduction** in carbon intensity over the last two years. ▸ The company is able to demonstrate that for the previous two years it is in the **top quartile** of companies in its sub-sector when assessed on accepted carbon efficiency metrics. ▸ A **Transformational initiative** or a combination, providing they are quantified and significant.	*No requirements yet: as understanding of how to measure performance increases, performance requirements will be introduced for medium impact companies.*	Automobile and aerospace companies must meet one of the following: ▸ **Emissions reductions**: fuel efficiency improvements above average for sub-sector. ▸ **Eco-efficiency**: above average fuel efficiency relative to sub-sector peers. ▸ **A Transformational initiative** to reduce product emissions. NB: Oil & gas and mining: no further "product" requirement at this time, but still need to meet performance criteria for their operational impact.

There are a variety of standards and initiatives that investors and companies can refer to in addressing climate change elements. Examples include the Greenhouse Gas (GHG) Protocol by the World Business Council for Sustainable Development, and the recently established Climate Disclosure Standards Board which aims to further standardise disclosure of GHG emissions. However, no single comprehensive standard has yet been established for the management of GHG reduction. The FTSE4Good climate change criteria form a useful standard for companies to aspire to and provide a reference point for investors.

There are a number of ways companies can achieve performance in the area of climate change reduction. Many are beginning to reduce their emissions by using the systems, support and infrastructure that already exists within the company. However, measuring emissions reduction is difficult as there is a lack of industry eco-efficiency metrics. The need for metrics presents a significant but worthwhile challenge for a number of sectors and there are still alternative means for companies to demonstrate performance and commitment to emissions reductions.

TRANSFORMATIONAL INITIATIVES AND VALUE CREATION

Some companies are undertaking major projects and investments to develop methods to reduce GHG emissions. These could deliver more significant global GHG reductions and business opportunities than incremental company efficiencies. The FTSE4Good climate change criteria recognise transformational initiatives as a measure of performance as they play an important role alongside energy efficiency and renewables in addressing climate change. Quantification may include: percentage emissions reduction actual or predicted; expenditure as a percentage of research and development budget; or percentage of the market.

SUPPORTING TECHNOLOGY SOLUTIONS COMPANIES

Companies providing climate change solutions are those with core business in the development and operation of solutions-driven clean technologies. Clean technologies include: alternative energy or energy efficiency which uses cleaner and more efficient methods of energy production; water technologies and pollution control which employ clean technologies to reduce the contamination of air, water, and soil; recycling and sustainable waste management which help to address environmental resource problems.

Investor perceptions have changed and these environmental technology companies are now seen as providing exciting opportunities for growth. There is also significant risk in this sector and diversified tracker funds following recognised indices from respected international index providers are attractive. FTSE recently launched the global environmental technology index, the FTSE ET 50 Index in partnership with Impax, the respected boutique clean tech specialist.

“Investor perceptions have changed and these environmental technology companies are now seen as providing exciting opportunities for growth.”

Table 3: Definition and examples of transformational initiatives.

DEFINITION: *A transformational initiative is a strategic initiative that makes a significant contribution to the reduction of GHG emissions.*

Example of initiative	Description
1 Buying low carbon electricity and fuel switching	Where there is a substantial reduction in the carbon intensity of electricity that is purchased/ used by the company, or an alternative, less carbon-intensive, fuel is selected.
2 Demand side management	The company has a programme to reduce the energy use of its clients. This would include development of new products that helped clients reduce energy use.
3 Research and development in technologies	Total R&D investment in low carbon technologies, eg carbon sequestration, renewables, or technologies low carbon which significantly reduce the energy needs of current technology.
4 Production of low carbon technologies	Energy and technology companies bringing new low carbon technologies to market.
5 Generation of renewable energy	Utility/electricity companies – build significant renewable generation capacity. Other companies may generate own renewable energy.
6 Product/service innovation	Quantified reductions in energy requirements of new products and services (GRI EN7).
7 Carbon capture and storage (sequestration)	Total CO_2 sequestered.
8 Supply chain/upstream	Reducing the upstream carbon intensity, eg reducing the 'embedded' CO_2 emissions (reducing energy input into the processing of components); switching raw material input to less carbon intensive one, eg from steel to aluminium; or reducing energy used by suppliers.
9 New business model	Owning and addressing the whole lifecycle and GHG footprint of a product or service, eg switching from products to services with associated GHG savings. Provision of mobility services, or owning and running a vehicle that is then used by consumers instead of selling vehicles and leaving the fuel consumption/emissions responsibility to the consumer.
10 Breakthrough project	Any single initiative that dramatically reduces GHG emissions. Redesign of production process, building redesign or redesign of transport fleet.

Investors are increasingly expecting companies to develop appropriate climate change mitigation strategies.

THE ROLE OF BENCHMARKING

Benchmarking is a useful tool for both companies and investors. The FTSE4Good indices are used as benchmarks for a range of investment funds, and as a tool for identifying best in class companies. Many companies also use FTSE4Good indices in developing strategy and policy, designing programmes and initiatives, and setting goals and targets.

Through FTSE's expanding responsible investment portfolio of index products, investors have a variety of tools to assess company performance on corporate responsibility. Conversely, these instruments provide companies with a means of demonstrating risk management and creating stakeholder confidence.

CONCLUSION

Setting international targets for global CO_2 concentrations is a task for governments, not for the financial services sector. However, among the investment community there is expectation of increasing costs associated with carbon and that those companies with the greatest carbon efficiencies will be better positioned to deliver greater returns. The FTSE4Good climate change criteria, as well as the FTSE4Good ratings service will help investors identify those companies. More importantly, they will encourage improvements in management systems across hundreds of companies and in this way support international efforts to curb emissions.

Authors

Will Oulton heads up FTSE's Responsible Investment Unit. In this role, he provides a consultative role to FTSE's Chief Executive in defining FTSE's role in the responsible investment arena, including managing and developing socially responsible investing (SRI). He is widely recognised as a leading thinker in the SRI world and is a regular commentator in the media.

Jayn Harding leads the direction and development of criteria and standards in corporate responsibility for FTSE's responsible investment products. Prior to joining FTSE Group, Jayn was Environmental Manager at Sainsbury's, the UK based food retailer and was responsible for managing all environmental issues, as well as socially-responsible sourcing and CSR.

Organisation

FTSE4Good is an innovative series of real time indices designed by FTSE Group to reflect the performance of socially responsible equities. It covers five markets: UK, Europe, Japan, US and Global with four tradable and five benchmark indices. A committee of independent practitioners in SRI and CSR review the indices to ensure they are an accurate reflection of current CR best practice. FTSE Group contributes income including licence fees for FTSE4Good to UNICEF, the global charity.

Enquiries

FTSE Group
12th Floor
10 Upper Bank Street
Canary Wharf
London E14 5NP
Tel: +44 (0)207 866 1800
E-mail: info@ftse.com
Website: www.ftse.com/Indices/index.jsp

Responsible investment for responsible investors

© bluestocking/iStockphoto

PAUL DICKINSON
CEO,
CARBON DISCLOSURE PROJECT

A spectacular global political failure in dealing with climate change has led consumers to vote with their money, buying hybrid cars and green electricity and in turn creating huge new markets. Investors too are voting with their money and organisations such as the Carbon Disclosure Project offer a unified way for them to find out about the greenhouse gas emissions and risk management of many of the world's largest corporations.

CLIMATE CHANGE AS A POLITICAL ISSUE

It is a simple mistake to make, to imagine the world is quite safe and the changes greenhouse gases are making will not lead to catastrophe. There is a growing school of thought that suggests catastrophe is already here, now.

The World Health Organisation talks of 150,000 people dying each year from climate change. In 2007 some 66 million people were made homeless by exceptional floods in Asia. Recently a million people were evacuated from fires in California. It is difficult to imagine what kind of catastrophe we are trying to avoid that is not already happening. And it will of course get worse and worse, every year, for all our lifetimes.

Some people say such talk is environmental alarmism. But whatever the talk is, climate change has risen as a political issue in almost every country. Citizens are demanding action. When something is true in reality they figure, it must also be true in theory.

CLIMATE CHANGE AS A CORPORATE ISSUE

Corporate responsibility is now evaluated through the acute lens of what a company does in core operations. The 21st century consumer is now less inclined to fund the products and services that invest in catastrophe for their children. This is unsurprising.

Human society is failing to solve climate change through our cherished democratic tradition of putting an x on a piece of paper every few years. Rather, the major response is being enacted by consumers voting with their money every day to create a different possible future. They are buying hybrid cars and green electricity thereby creating huge new markets. And it is not just consumers using the power of markets to avert catastrophe. As Madeleine Albright said at the launch of my organisaton: "Our business is to help investors vote with their money."

Today there are many remarkable new associations seeking to protect citizens from climate change and those working towards a unified business response to climate change. The Carbon Disclosure Project is one such association with a mission to facilitate a dialogue, supported by quality data, from which a rational response to climate change will emerge. It is particularly worth emphasising the importance of a unified approach.

THE CARBON DISCLOSURE PROJECT

Representing 315 investors with assets of US$41 trillion (three years of US GDP), the CDP has found this unprecedented investor authority has led over 1,300 of the largest corporations in the world to report their greenhouse gas emissions to its website. If you want to know what Disney or Coca Cola and McDonalds emit, or Exxon and Shell for that matter, the data is available to download free of charge.

The CDP is funded by the five governments of the US, UK, France, Sweden and Australia and is pioneering some very exciting new developments. For example, Wal-Mart is sending our questions on greenhouse gas (GHG) emissions out to their supply chain. Tesco, Nestle, Cadbury, P&G, Unilever, L'Oreal and many others are following suit. This could see the CDP's work grow

Investors are increasingly expecting companies to disclose details of their greenhouse gas emissions and corporate strategy on climate change.

exponentially. And as CDP Advisor Lord Adair Turner of Standard Chartered Bank said at our launch events in New York and London in 2006: "What gets measured gets managed."

A UNIFIED APPROACH TO CLIMATE CHANGE

Businesses and investors can have considerable impact. Yet for this to happen they need to act together. It takes the union to make the force. But that force can move mountains. At a time when we have the technology of gods, but the politics of children, businesses and investors all over the world, supported by enlightened consumers, are starting to take the urgent action required. The global business community is looking with horror at the pathetic squabbling between national governments on measures to reduce greenhouse gas emissions. The political process is failing. On present trends we face certain disaster. Against this grim backdrop of government ineptitude, a softly repeating mantra of hope is coming from the business world. It goes like this. 'You do your worst. And we'll do our best'.

And the business world is changing, in more ways than one. Through powerful new initiatives such as the UN Principles for Responsible Investment, shareowners are beginning to identify how to exercise their responsibilities with regard to defending their long term financial interests.

THE PRUDENT INVESTOR

Major pension funds own so much of the economy, they have been described as having a legitimate interest not just in individual companies, but also the overall system. This very wide and long term perspective has led some people to call major pension funds 'Universal Investors'.

So what are the responsibilities of so called 'Universal Investors' with regards to the key issue of climate change? The world's many large pension funds control enormous assets and operate under a strict fiduciary duty to invest in the best interests of the beneficial owners.

Until now that duty has been expressed purely in terms of calculating financial return. However the UK Government Stern report has seen this distinguished former World Bank Chief Economist comment:

"If we do not act, the cost of climate change will be at least five per cent of global GDP each year, now and forever. If a wider range of risks and impacts is taken into account, the estimates of damage could rise to 20 per cent of GDP or more."

In the well known formulation of the prudent man rule, Justice Putnam of the Supreme Judicial Court of Massachusetts in Harvard College v. Amory (1830) held that a trustee is required only to "observe how men of prudence, discretion and intelligence manage their own affairs." The core definition of fiduciary duty identifies an obligation to invest in the 'best interest' of beneficial owners, giving reasonable consideration to the facts a 'prudent man' would pay attention to, given reasonable knowledge of the situation prevailing at the time.

This creates a complex situation for fiduciaries. Financial returns today are relatively unencumbered by taxation or regulation of GHG emissions. It is a stark but unavoidable fact that national and international government efforts to materially limit GHG emissions have simply failed.

In response to this fact, prudent investors cannot and probably should not simply divest from companies associated with heavy emissions of GHGs for two reasons. Firstly, many of them are very profitable. Secondly, divestment extinguishes the possibility to alter the behaviour of the company through acting as a responsible owner.

As corporate governance guru Bob Monks has noted, votes are an asset (voting shares always have a market premium over non-voting ones). Accordingly they should be used to further beneficiaries' interests on all occasions. In effect, the voting of all institutionally held shares should be virtually compulsory.

But ownership is about more than voting at AGMs. It is about active participation in the life of a company, where required.

In a democracy, the citizen is a shareowner in the society, but his or her duty to that society is not limited to casting a vote at elections every few years. If something is going seriously wrong in society, a good citizen gets directly involved with the problem.

So what is going wrong with regard to climate change? A sequence of events is occurring that presents us with insurmountable problems. It runs as follows:

1 Scientists warn governments of a need for urgent action on climate change.

2 Governments prepare plans to cut emissions. They have difficulties negotiating among themselves but these problems are greatly amplified by;

3 Industry groups who anticipate risk from a disruption to business and then lobby governments requiring them to desist from taking action. The application of corporate resources to such campaigns overpowers government.

4 Inaction results from this sequence of events. This inaction threatens to result in dangerous climate change, with generally accepted predictions of very significant loss of life and catastrophic negative economic impact.

Practical steps to take for a concerned shareowner

First, communicate to the corporations you own that you take climate change seriously. Participation in UN Principles for Responsible Investment and CDP serve this function.

Secondly, consider specifically how the corporations you own are influencing government. This can involve desk research, interviewing companies or inviting expert organisations to provide submissions.

Thirdly, prepare communications to be issued individually or collectively (from one or many shareowners to one

or many corporations), explaining the concerns of the fiduciary investor with regard to the best interests of beneficial owners.

Such a dialogue may seem slow and uncomfortable at first. New activities often do. But it may be useful to focus on three expert comments made at CDP launch events.

Sir Derek Higgs, author of a major report on corporate governance for the UK Government said, February 2003:

"Our hands, all our hands, are on the controls and if we don't exercise those controls appropriately we'll have nobody to blame but ourselves. That's companies, that's governments, but it's also institutional investors. All of us are the real principals in this; the companies are in a sense the agents.

".... there is too much of a circularity, a deafness of dialogue if you like, between suppliers and users of capital in terms of information, in terms of the knowledge deficit. It goes something like this. Too often investors and analysts say, well we don't have information from companies so we don't know what to ask and the response comes back from companies, well we don't actually know what they want so we don't know what to give them. That's silly, that's unhelpful, it's not necessary."

Alan Hevesi speaking at the CDP launch at JPMorgan Chase Head Office, September 2005 observed.

"I'm here as the sole trustee of the New York State Common Retirement Fund. It is the second largest in America after California.

"We have 982,000 members. We are long term investors and I'm a fiduciary. As a fiduciary I take the following position about our role as investors. I don't accept the limited role that many pension fund managers and other investors take, that it is your job to get up in the morning, find a growth stock or other financial instrument, make some money for the fund and go home. There is a larger social purpose to our responsibility as fiduciaries, not for its own sake to pursue social issues, but because as fiduciaries we have to be prepared for the long term consequences of the behaviour of our companies.

"That's the beauty of the Carbon Disclosure Project, to systematise and to organise this. The people who have sponsored this and participated, it's terrific. A quiet, mature, expert way to raise the consciousness of CEOs who are under pressure to provide immediate profits for the next quarter's reporting and by the way pressure from shareholders. This is a way to give them the sense that looking at the long term is really the smart and sensible way to go.

"Divesting means disengaging. It's the last thing you do. The bottom line is that there is a potential crisis if we don't do our job and we don't get our governments and our corporations to get serious about these issues. We must be prepared, we must take the long term view, we must see that risk is also an opportunity and we must play our role in establishing that both thoughtfulness and faith in the sustainability of our planet is our mission and we do it on behalf of all the people that we represent."

Chancellor Merkel of Germany, Chair of the G8 has put the point well:

"Global climate protection policy will only be successful, however, when it is supported by business and industry. Here, the capital market is of great importance, and it is extremely important for investors to take account of climate change in their decision making. This contributes to enhanced public perception of both the risks and the chances of climate protection."

Can shareowners collaborate to break the logjam of political inaction? The UN PRI and CDP are committed to developing systems that help shareowners take the necessary action required. We hope all reading this will join the work programme and act together to defend assets from the risks presented by unmitigated climate change.

Author

Paul Dickinson is one of the founding members of the Carbon Disclosure Project and previously was responsible for the founding and expansion of Rufus Leonard Corporate Communications. More recently he has led the development of the EyeNetwork, the largest video conferencing service in Europe. He is a member of the Environmental Research Group of the UK Faculty and Institute of Actuaries. He is also the author of several publications, including *Beautiful Corporations*.

Organisation

The Carbon Disclosure Project (CDP) is an independent, nonprofit organisation aiming to facilitate a dialogue between investors and corporations through the dissemination of high quality information regarding corporate strategy in response to climate change. Each year a questionnaire is sent to the Chairmen of the world's largest corporations by market capitalisation, asking for disclosure related to climate change risks, opportunities, greenhouse gas emissions accounting and management. It is the gold standard for carbon disclosure and methodology processes.

Enquiries

Paul Dickinson, CEO
Carbon Disclosure Project
Tel: +44 7958 772864
E-mail: paul.dickinson@cdproject.net
Website: www.cdproject.net

©Chua Kian Khoon/Fotolia

Carbon emissions reporting in Asia

MELISSA BROWN
CO-AUTHORED WITH **SCOTT LINDER**,
YK PARK AND **ALEXANDER READ-BROWN**,
ASrIA

Asia ex-Japan results from the 2007 Carbon Disclosure Project reveal that: 1) a range of Asian companies are increasingly describing strategic business initiatives to curtail carbon emissions, 2) companies working on climate change strategies often rely on industry associations and global NGOs for guidance, 3) the degree of top management engagement is a critical determinant in the disclosure of carbon emissions and the development of formal initiative, 4) pictures of regional response characteristics are becoming apparent and 5) the quality and quantity of responses from Asia's tech and telecoms companies is notable, while Asia's banks and utilities lag their global counterparts.

In 2007, the Carbon Disclosure Project's fifth report (CDP5) included responses from 44 companies from countries located in Asia, not including Japan (Asia ex-Japan). An analysis of these responses by ASrIA offers compelling insights into Asian companies' emerging policy responses to climate change. This is meaningful because most of these companies do not have the benefit of government policy guidance or carbon trading markets which would create transparent price incentives for action. Instead, they are responding to a range of pressures, from customers, competitors, investors and global regulators, which promise to shape their long term competitiveness.

A MORE STRATEGIC RESPONSE

The most positive trend in 2007's Asian CDP responses is the shift toward disclosure of more strategic corporate policies linked to climate change. The best respondents provided clear examples of a move from awareness to action, with comprehensive carbon policies driving investment, new products and technology development. Such strategies include carbon offsets, supplier accountability, green building programmes and steel production process innovation. Although climate change remains a relatively new policy area in Asia, leading companies are now taking the initiative and positioning themselves to capitalise on a shift in policy and growing consumer awareness.

NEW INDUSTRY RESOURCES

Companies in Asia typically lack consistent drivers for responding to climate change and resources to clarify emerging standards. By contrast, in the EU companies benefit from high levels of disclosure, stronger regulatory incentives, heavy monitoring requirements and growing shareholder feedback. Under the Kyoto Protocol, Asian countries, as developing markets, are recognised as vulnerable to adverse economic impacts of climate change and are therefore not required to meet carbon emissions caps. Only Japan faces binding emissions reduction targets and is included in the rules which support international emissions trading.

Most Asian government policies, directives and regulatory systems related to carbon emissions and climate change are embryonic, featuring headline policy statements but limited signs of implementation. Initial policy frameworks typically have a flexible time frame and are often overlooked if the policy focus shifts to industrial development, which may conflict with efforts to restrict carbon emissions growth or improve energy efficiency.

> "Although climate change remains a relatively new policy area in Asia, leading companies are now taking the initiative and positioning themselves to capitalise on a shift in policy and growing consumer awareness."

As a result, Asian companies, especially those in globally competitive export sectors, such as technology and building materials, are increasingly relying on industry associations to shape their emerging carbon emissions policy responses, carbon reporting and management strategies. These industry associations appear to be filling a vacuum created by limited government initiatives with a growing body of global research on best practice responses at the industry level with regard to climate change. Some of the more frequently named associations are local, but there is strong representation from international groups, many of which can provide a tested roadmap for sector-specific carbon emissions reporting.

GOVERNANCE MATTERS

The CDP Asia sample presents high variability in the extent to which top management is addressing issues of reporting and strategy. This is an important subject because corporate governance is a critical variable in assessing long term corporate environmental performance. One leading corporation, for example, indicated that top management views climate change as an issue which must be evaluated in terms of the global footprint of their operations. Rather than framing its strategies in terms of its home country standards, it is developing policies which are consistent with the greenhouse gas reduction policy framework in the various countries in which it operates. This will likely mean that trends in the EU and North American markets will be particularly influential.

One of the key differences in responses was evidence of active engagement versus statements of intention. Many companies reported recently established planning committees, initiatives under consideration and possible future reduction targets. While some of these responses may represent a management that has recently become genuinely engaged in addressing carbon emissions, it could also be that management commitments are still tentative.

SECTOR FOCUS: TWO LEADERS, TWO LAGGARDS

Asian bankers waiting for guidance

With several exceptions, the 2007 CDP report demonstrates that Asia's banks are making slow progress in addressing the long term economic impacts of climate change. It is tempting to see the limited capacity of Asian banks in assessing climate change issues as a by product of the bureaucratic restrictions faced by Asian bank regulators who often have a narrow policy remit. While global groups, such as UNEP FI, are broadening their contact base in Asia, many Asian banks lack even the environmental management capacity to address basic operational issues.

All of the financial sector respondents display at least a basic perception that climate change is a global issue, and two-thirds acknowledge potential business impacts. Only one-third, however, recognise that there might be large scale economic repercussions. For example, one Indian financial sector respondent offered a clear view that the potential long term impacts on the agricultural economy could be harsh and would have fundamental effects on its business model. Despite the still limited understanding of climate change impacts, two useful strands of reporting were evident from the responses concerning the role of energy efficiency and opportunities for new products.

> "Asian companies are increasingly relying on industry associations to shape their emerging carbon emissions policy responses, carbon reporting and management strategies."

Energy efficiency

Asian banks are gradually identifying energy efficiency as the best means to reduce carbon emissions in their industry. Four of the bank respondents refer to or have implemented energy efficiency programmes of varying depth and quality. Most seek to increase efficiency by decreasing air conditioning usage or replacing and minimising lighting elements. Only two of the financial sector respondents had clear energy reduction targets.

© Rene Drouyer/Fotolia.com

Harnessing consumer sentiment

Bank respondents primarily viewed climate change, heightened public awareness and shifting consumer sentiment as an opportunity to offer new investment products. Ironically this trend was most evident in less thorough responses from banks which did not see any direct risk to their business from climate change. In general, those respondents with a more developed grasp of the issues placed greater emphasis on carbon emissions data, energy efficiency and other initiatives.

> "Asia's leading tech companies appear to be making systematic efforts to match the breadth and depth of reporting common to their global competitors."

Asian utilities hiding under a cloud

With the exception of a few well covered respondents, Asia's fastest growing coal-fired power companies are still absent from reporting carbon emissions. Several key companies are established participants in CDP and have a developed investor dialogue covering emissions trends, strategic choices and renewables and CDM initiatives. China's most influential power companies, however, have not yet provided any material data on emissions or formal policies to CDP or the investment community at large.

The lack of disclosure by this politically sensitive sector is a by product of the ongoing, but in many instances, still opaque policy debate on climate change by Asian governments. Indeed, where governments have yet to establish a clear policy framework, it appears premature to expect government-invested or government-controlled power companies to take the initiative in disclosing emissions or articulating critical policy objectives.

Asia's tech players demonstrating new carbon sophistication

The Asian tech sector is emerging as a strong reporter on environmentally linked issues including carbon. Asia's leading tech companies appear to be making systematic efforts to match the breadth and depth of reporting common to their global competitors. For supply chain companies in the consumer electronics and components sector, engagement with international customers has provided a baseline of knowledge about the environmental management standards of global markets.

Virtually all of the tech sector respondents framed their responses in terms of the need to stay competitive in a low cost and loosely regulated sector. In this context, the development of innovative energy efficient products is a frequently cited goal for Asian tech respondents. Efficiency improvements to laptops, computers, semiconductors, LCDs, and DRAMs, as

© Mark Huls/Fotolia.com

well as linked investment and awards these products have received were all highlighted. Although many of the disclosures had a promotional flavour, several demonstrated an acute awareness of a nuanced but increasingly demanding green market place.

> "Telecom respondents understand they are one of the few sectors which can benefit strategically from carbon linked trends."

A new trend which is beginning to affect Asia's tech companies is the need to play a role in customers' carbon accounting as brands begin to evaluate their supply chain carbon footprints. These efforts also link to an emerging awareness of carbon neutrality. Of the CDP5 respondents seeking to become carbon neutral, 50 per cent are in the tech sector. Although these companies generally stated that climate change is an incidental risk to their operations, a number of statements suggest they are motivated to embrace carbon neutrality as a tool for becoming more energy efficient and raising brand equity.

Asian telecoms benefitting from carbon linked trends

Like the tech sector, telecom respondents set a generally high standard on carbon reporting. With the exception of the lone participating Chinese telecom, all of the companies disclosed carbon emissions. It is also apparent that respondents, especially the wireless players, understand they are one of the few sectors which can benefit strategically from carbon linked trends. Whether talking about wireless banking or navigation devices, Asia's wireless players are keen to be viewed as solution providers in a carbon constrained world. As one telecom response succinctly expressed it, "communications technology is a great enabler..."

A second common theme in the telecoms responses is the recognition that physical assets are at risk due to adverse weather conditions and higher ambient temperatures. This is a particular worry for operators in typhoon- and flood-prone regions in Asia. Risks to facilities and related service failures obviously pose a serious financial and reputation threat to companies, which are often heavily regulated.

REGIONAL CHARACTERISTICS

Thanks to the increase in the sample size, the 2007 Asian CDP offers a more realistic picture of trends at country level. It is becoming easier, for example, to form a picture of a Korean response versus a Singaporean response. This should not be surprising as reporting norms are usually set at the country level. Nonetheless, material reporting continues to lag in nations with few drivers, such as India, while improvement is evident in countries where leading companies have a more international footprint, eg Hong Kong and Taiwan. Korean companies are the Asian leaders in more comprehensive carbon reporting while Indian companies were the largest group of new respondents.

SETTING A BENCHMARK

Evaluating corporate carbon emissions disclosure provides valuable insights into future Asian climate change reporting and strategy development. Thanks to the increase in the Asian sample in 2007, there are more tools to assess peer group competition on carbon fundamentals. Additionally, the 2007 responses offer initial data points on the growing impact of supply chain carbon reporting initiatives and global sector initiatives affecting sectors such as airlines and technology. Taken together, these developments suggest that the Carbon Disclosure Project continues to act as a valuable catalyst and identifies important leading indicators.

Organisation

The Association for Sustainable & Responsible Investment in Asia (ASrIA) is a not for profit, membership association dedicated to promoting sustainable investment practice and corporate responsibility in the Asia Pacific region since its founding in 2001. ASrIA's members include investment institutions managing over US$4 trillion in assets. Our goal is to build market capacity for SRI by providing insightful, up to date and accessible information on the development of SRI in Asia and globally.

Enquiries

Website: www.asria.org

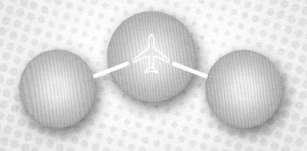

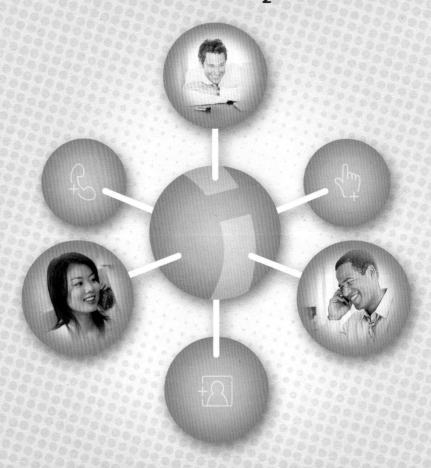

Companies are beginning to invest in emerging renewable technologies, such as wave power.

Companies:

A NEW WORLD OF RISK AND OPPORTUNITY

MINDY S LUBBER
PRESIDENT OF CERES AND
DIRECTOR OF INCR

Climate change will have far-reaching consequences on companies across all sectors of the economy. Already, billions are being invested in energy efficiency, renewable energy and development of low carbon technologies. Company disclosure on climate-related risks is increasing, carbon costs are being factored into company capital planning and corporate leaders are calling for mandatory caps on greenhouse gases (GHGs). Nevertheless, while many leading companies are taking important steps to manage climate change, bolder actions to maximise energy efficiency, develop climate-friendly products and curb greenhouse gas emissions throughout company supply chains will be needed.

INTRODUCTION

The potential consequences of climate change are generally understood, but how fast and dramatic those changes will be and how society will respond is still uncertain. Given projected physical impacts and the rapid spread of carbon-reducing regulations in the US and abroad, the risks, opportunities and challenges of climate change for business are profound (see *Investors and climate change: a new world of risk and opportunity* on p60 for more information on risks and opportunities). How companies respond will have important consequences for their competitiveness, profitability and even their very survival. Some companies and industries, by virtue of their own energy use or the goods or services they provide, face more pronounced challenges than others.

ELECTRIC POWER

Anticipated regulation of GHGs poses significant financial risks for the electricity sector, which accounts for 40 per cent of US CO_2 emissions. The recent proliferation of state-level renewable portfolio standards (RPSs), which require a certain percentage of electricity needs to be met by renewable energy sources, is another challenge for US utilities. Fuel mix is an important variable affecting electric power companies' climate risk profiles. Companies that burn large amounts of coal, the most carbon-intensive fuel, have higher emissions and greater climate risk exposure than companies that don't. Southern Co, a heavily coal-dependent utility that ranks among the nation's highest CO_2 emitters, recently estimated that a moderate scenario for controlling GHG emissions could cost the company nearly US$800 million annually.

Pacific Gas & Electric (PG&E), a California-based utility that generates more than half of its power from carbon-free energy sources, has avoided building dozens of new fossil-fuel power plants by aggressively pursuing energy efficiency to an extent not seen elsewhere in the industry. (California, unlike nearly every other state, rewards utilities financially for energy efficiency expenditures.) By 2008, PG&E will invest an additional US$1 billion to install 10 million 'smart meters' to allow its customers to monitor and conserve their energy use, especially during peak demand hours.

With renewable portfolio standards in place in 25 US states, electric power companies are moving to diversify their generation portfolio to include larger amounts of renewable energy. Florida Power and Light (FPL), the fifth largest US electric power producer, is now the country's leading wind power generator and developer of wind projects. FPL is also planning to spend US$2.4

billion on major solar energy projects, including a new 300 megawatt solar thermal generating operation in Florida.

Even heavy coal burning utilities are addressing GHG emissions. American Electric Power, the largest CO_2 emitter in the Western Hemisphere, is making major investments to retrofit existing coal-fired plants and to build new plants with cutting edge carbon capture and storage technology. Proposals to build new coal-fired power plants, especially those with no CO_2 controls, are facing tougher scrutiny. As noted in a July 25, 2007 article in the *Wall Street Journal*, coal plant proposals in Texas, Florida, North Carolina, Oregon and Minnesota have all been cancelled after "states [concluded] that conventional coal plants are too dirty to build".

OIL AND GAS

Because oil and gas production and consumption accounts for more than half of carbon dioxide emissions in the US and because the industry makes large infrastructure investments that last for decades, oil and gas companies face substantial financial risks from regulatory developments, including carbon limits. These limits will drive the US and global economy towards low carbon energy sources such as wind, solar and other renewables.

The physical changes from climate change carry particular risks for the oil and gas industry – for example, damage to critical infrastructure from the 2005 hurricanes helped trigger nationwide petroleum shortages and a surge in gas prices. Climate change also poses risks for billions of dollars of pipelines and other infrastructure built on melting permafrost in Alaska, Canada and elsewhere.

A growing number of companies are boosting their investments in alternative energy sources such as wind, solar and hydrogen. Both BP and Shell have built a large stake in wind power, with Shell being one of the top five wind power generators in the US. Chevron, a major US oil producer, recently announced plans to invest US$400 million annually into renewable energy, including emerging technologies such as wave power and geothermal energy.

Some oil companies are also aggressively pursuing energy efficiency. BP's energy efficiency initiative has been the company's most profitable project in the last decade; with a US$100 million investment, the company realised US$400 million in savings. Statoil of Norway emits less than a third of the industry average of GHGs per barrel produced thanks to an intense focus on energy conservation throughout its production chain.

Many companies are also factoring carbon costs into long term capital spending plans, setting targets to reduce GHG emissions and calling for federal policies with mandatory carbon caps. Among those urging strong federal action are ConocoPhillips, BP America and Shell. Exxon Mobil is unique in its dismissal of both renewable energy and mandatory carbon limits.

BANKING/FINANCE

No one yet knows the exact combination of measures society will need to avoid the hazardous effects of climate change, or the new technologies that will produce energy with less climate impact. Regardless, the banking and financial sector, which decides who borrows what money at what cost, will play a huge role in helping to deliver the innovation to meet the climate challenge.

Citigroup's 10 year, US$50 billion climate change commitment exemplifies strategies employed by financial firms. In addition to investments reducing its own energy use and carbon footprint, about US$30 billion will finance the development and commercialisation of its clients' low carbon technologies. Citigroup has also developed expertise to advise clients on environmental risk management, including climate risk management. At the retail customer level, Citigroup is financing home equity loans specifically for the installation of solar electric systems and encouraging the use of efficient 'paperless' statements, planting a tree for each customer who selects this option.

Bank of America has made a similar 10 year, US$20 billion commitment to achieve internal and external GHG reductions. In addition to making its own facilities more energy efficient, US$18 billion will support activities such as the development and financing of 'green' buildings, advising clients participating in carbon markets, lending to support development of low carbon and low emissions technologies and favourable lending rates for mortgage customers whose homes meet certain energy efficiency standards.

"With renewable portfolio standards in place in 25 US states, electric power companies are moving to diversify their generation portfolio to include larger amounts of renewable energy."

Reducing the GHG footprint of their lending portfolios is a key step financial institutions can take to address global warming. In 2004, Bank of America adopted an environmental policy that calls for a seven per cent reduction in GHG emissions in its energy and utility portfolio by 2008, along with a nine per cent reduction in its own operational GHG emissions by 2009. Citigroup has also set targets to reduce its operational GHG emissions, but has not addressed the climate change impacts from its lending portfolio. A growing number of investors and environmental groups are pushing Citigroup, Wells Fargo and other major banks to lower the carbon footprint in their lending portfolios.

The water supply in the Greater Mekong in Vietnam is highly vulnerable to climate change. WWF and Coca-Cola have joined forces to conserve this and other key freshwater river basins around the world.

TECHNOLOGY SECTOR

Technology companies are a major contributor to GHG emissions, mostly because they require massive amounts of energy to power their operations and servers, as well as their products. As major information hubs, technology firms also have a unique opportunity to shape and facilitate positive societal responses to global warming. For example, Sun Microsystems recently introduced a first-of-its-kind online community, OpenEco.org, which helps companies and other organisations track and compare GHG emissions and share strategies for reducing them.

Many technology companies are responding to these risks and opportunities by launching progressive climate change strategies that focus on minimising direct and indirect emissions impacts. Dell Computer is committed to reducing its operational GHG emissions 15 per cent by 2012 and is asking major suppliers to identify and report their own GHG emissions.

IBM is spending US$1 billion a year on a new energy efficiency initiative, Project Big Green, that it hopes will enable the average 25,000ft^2 data centre to cut its energy bills by nearly half. IBM has also set a goal of doubling the computing capacity of its worldwide data centres by 2010 while keeping power consumption levels steady.

Sun Microsystems is partnering with California-based utility PG&E on a programme that provides PG&E customers with US$700 to US$1,000 cash rebates for each energy efficient computer server they purchase from Sun. Similar partnerships between vendors and utilities are being explored in other parts of the country.

AUTOMOTIVE

With the transportation sector accounting for roughly 28 per cent of US CO_2 emissions, car makers have good reason to be concerned about climate change risks and opportunities. The long term value of car companies will depend in part on their ability to deliver appealing products in a competitive marketplace that is increasingly shifting towards fuel efficient, low emission vehicles and technologies.

Canada, the European Union, China, Australia, Japan and many US states are implementing legislation to reduce GHG emissions and increase vehicle fuel efficiency, ensuring that virtually all of the world's major car markets will be covered by increased fuel economy and air emissions standards. In China, which is projecting double-digit annual growth in vehicle sales, the government has set emissions standards for 2008 that less than a quarter of US manufactured passenger cars and light-duty trucks will be able to meet.

Honda and Toyota are far ahead of their US counterparts in the development of climate-friendly, fuel saving technologies. Honda builds the most fuel efficient cars in the world, while Toyota dominates the gasoline/electric hybrid market. Indeed, more than a million of Toyota's eco-friendly cars, the Prius, have already been sold. Both Toyota and Honda plan to introduce fuel efficient hybrids to China's burgeoning car market. US car makers are trying to catch their overseas counterparts, with Ford offering two hybrid models and General Motors developing cars that run on ethanol, but both companies continue to lag behind.

“ BP's energy efficiency initiative has been the company's most profitable project in the last decade; with a US$100 million investment, the company realised US$400 million in savings. ”

CONSUMER PRODUCTS

In addition to consuming prodigious amounts of energy to power their stores, retail companies impact climate change through their massive supply chains, transportation needs and the products they sell. Large chain stores such as Home Depot, Wal-Mart and Bed, Bath and Beyond operate tens of millions of square feet of space that must be heated in winter and cooled in summer. They also operate thousands of trucks that are driven millions of miles each year. Because such a large portion of retailers' expenses is related to energy consumption, they have a significant incentive

to become more energy efficient. Among those seizing this opportunity is retail giant Wal-Mart, which is re-tooling its truck fleets to cut fuel use and building energy efficient Supercenters that will use 20 per cent less energy. The programmes are saving the company tens of millions of dollars a year and most are paying for themselves within two years.

Climate change can impact consumer products companies in other ways. Manufacturers of soft drinks and computer chips, for example, are major consumers of water, a natural resource that is and will continue to be profoundly affected by climate change. Coca-Cola is partnering with the World Wildlife Fund to protect the world's seven main watersheds and become neutral in terms of water use, while Pepsico has become the nation's largest purchaser of renewable energy credits. Renewable energy credits, or RECs, allow consumers to purchase a specified amount of their energy from a renewable energy supplier, such as a wind farm. In 2006, Whole Foods, another major retailer, purchased enough RECs to offset all of the electricity consumed by all of its stores.

Consumer product companies are also beginning to use carbon labels, which disclose the quantity of GHG emissions involved in making a specific product. Timberland has included carbon labels on its shoes, making it the first major US company to do so. Carbon labels are also taking hold in the UK where they are used for products such as potato crisps and shampoo.

CONCLUSION

Business is responding to the risks and opportunities from climate change. Companies have begun integrating climate change into their business strategies by reducing the GHG emissions from their operations, products, supply chains and employees. But more companies will need to accelerate and deepen their actions if global greenhouse gas emissions are to remain at levels scientists consider safe.

Author

Mindy S Lubber is the President of Ceres, a leading US coalition of investors and environmental leaders working to improve corporate environmental, social and governance practices. She also directs the Investor Network on Climate Risk (INCR), an alliance that coordinates US investor responses to the financial risks and opportunities posed by climate change.

Ms Lubber has held leadership positions in government as the Regional Administrator of the US Environmental Protection Agency; in the financial services sector as Founder, President and CEO of Green Century Capital Management; in the private sector as the President of an environmental law and policy consulting group; and in the not-for-profit sector leading environmental and public interest law organisations, including the National Environmental Law Center, which she founded.

Organisation

Ceres is a US-based network of investors, environmental organisations and other public interest groups working with companies and investors to address environmental and social challenges such as global climate change. Its mission is to integrate sustainability into company practices and the capital markets to protect the health of the planet and its people. Ceres directs the Investor Network on Climate Risk, a network of more than 60 institutional investors managing more than $4 trillion in assets focused on the business impacts from climate change.

Enquiries

Ceres, Inc.
99 Chauncy Street, 6th Floor, Boston, MA 02111
USA
Tel: +1 (617) 247 0700 | Fax: +1 (617) 267 5400
E-mail: lubber@ceres.org | Website: www.ceres.org

Life in the low carbon economy.
Is your business ready?
© KPMG
KPMG

In order to maximise commercial value from the issues arising as a result of climate change, businesses need to act and they need to act now. Designed to help businesses respond to the challenges of the new operating environment in a way that enhances brand reputation, operating efficiency and financial performance, KPMG's Carbon Advisory Group provides a deep understanding of the issues and challenges facing businesses today.

Climate change is forcing companies of all sizes to rethink the way they do business. The tide of legislation and regulation across the globe is moving in just one direction. Whether new initiatives come from international bodies, such as the UN or the EU, or from national governments, the aim is the same: to find ways to sustain growth in a low, rather than high, carbon economy.

BECOMING A LOW CARBON BUSINESS

Increasingly, companies are under pressure from their investors, employees and customers to become low carbon businesses by reducing their emissions and accounting for their actions in a transparent and trusted way. That pressure will only increase.

But making the transition to low carbon operations is far from straightforward. Evidence suggests that many companies do not always know where to start the process of creating such low carbon strategies. Businesses need better guidance, both from government and their business advisors, to plot their approach. They need help in devising carbon strategies and support to think through how best to deal with a range of diverse issues including reporting, measuring, offsetting, trading, investing in new technologies and dealing with the complex financial, tax and insurance implications of their actions.

> **"Climate change is forcing companies of all sizes to rethink the way they do business."**

Some organisations clearly see this as a threat. Others, further along the road of planning for life in a low carbon economy and willing to see carbon management as a fundamental part of their business planning processes, take a more positive view.

They are already beginning to see that taking the right action at the right time can bring real benefits, not just in terms of brand and company reputation, but also in terms of operating efficiency and cost control. They are trying to plan a route that will allow them to comply with any future changes in legislation and regulation while making sure they increase, rather than decrease, business performance.

“KPMG's Carbon Advisory Group is providing insight and strategies to help companies understand and manage the many business implications of climate change.”

HOW TO PROSPER IN A LOW CARBON ECONOMY

KPMG's Carbon Advisory Group is providing insight and strategies to help companies understand and manage the many business implications of climate change.
It is helping businesses to make sense of the confusing and, at times, inconsistent approaches being taking to carbon measurement and accounting, strategy development and the important tax and insurance implications of their approach to dealing with climate change.

WHAT ARE THE KEY SUCCESS FACTORS?

There are four key areas where an organisation can focus in order to create an efficient low carbon business, see below.
In order to plan, measure, manage and report efficiently we have identified seven key areas where KPMG can support a business to meet its carbon reduction objectives while at the same time maintaining business performance and operating efficiency, see box overleaf.

Four key areas in creating an efficient low carbon business.

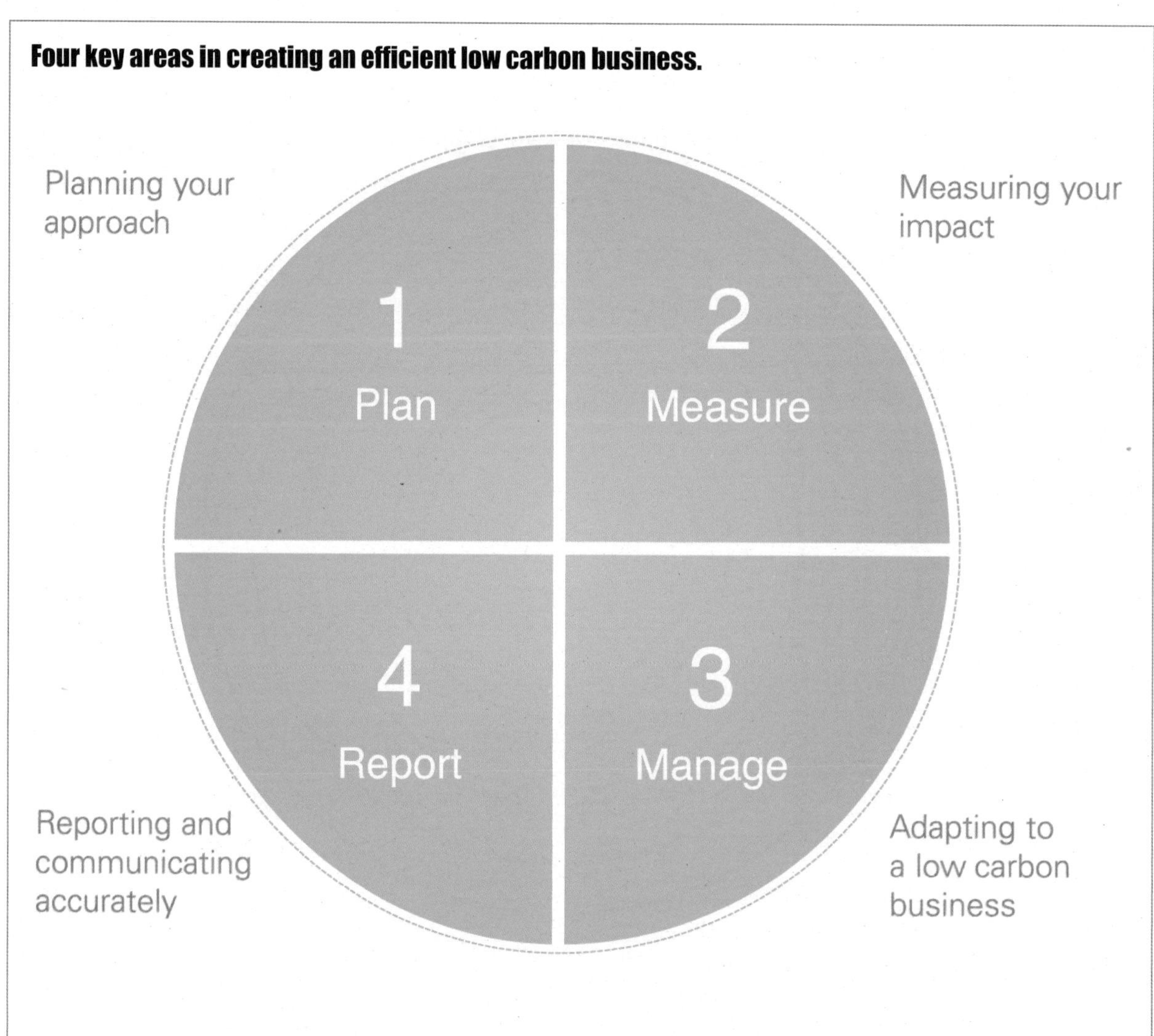

Source: 'Is your business ready for life in the low carbon economy?' KPMG LLP (UK), September 2007

KPMG: SUPPORTING BUSINESS IN MEETING CARBON REDUCTION OBJECTIVES

Carbon trading and offsetting – investigation and dispute

Organisations seeking to reduce their impact on the environment through carbon reduction face a major issue – lack of clear and widely-accepted regulatory regimes. Where allegations of fraud or misconduct, or disputes arise, KPMG can provide support around the following areas:

- Integrity due diligence on potential or existing carbon reduction partners
- Investigations into allegations of fraud or misconduct
- Assessment of the fraud and misconduct risks associated with carbon reduction schemes
- Dispute advisory services including acting as expert witnesses.

Carbon efficient logistics

Today's businesses are under constant pressure from investors, customers and regulatory bodies to adapt their business models to demonstrate a carbon efficient strategy. The carbon efficient supply chain team provide support around the following areas:

- Network mapping
- Scenario building
- Risk and change management
- Benefit tracking and realisation.

Carbon management

Climate change is progressively affecting the business environment and organisations that understand the implications and develop clear strategies aligned with core business activities will benefit. KPMG can help businesses focus on the integration of climate change into the business strategy, develop and implement carbon-reduction activities and provide clarity around the measurement and reporting of carbon.

Environmental tax and incentives

Understanding the tax consequences which may arise following a carbon reduction strategy will be important in allowing the true costs and benefits of such actions to be assessed by businesses. Investors in projects that generate carbon credits and/or other incentives will want to maximise financial return by managing tax leakage and maximising tax relief. Emissions trading can also lead to tax mismatches if not structured carefully. KPMG's Environmental Tax and incentives team can support clients in all of these emerging areas.

Accounting for carbon

The key to accounting for carbon is to establish appropriate policies and to ensure those policies are communicated effectively to key stakeholders. From the strategy and planning through to review, reporting and de-briefing we can help a business understand the impact that activities in the carbon arena will have on the financial results.

Mergers and acquisitions in the carbon market

As businesses look to expand their development pipelines, by acquisition rather than purely by development, we expect to see more consolidation in the carbon market place. Focusing on capital raising, mergers and acquisitions and valuations, KPMG can provide advice around various aspects of corporate finance and transaction services.

Carbon restructuring

Concern about climate change has sparked significant shifts in legislation, regulation and consumer habits.The full impact of these shifts can be hard to predict and often force businesses into rapid and complex programmes of change. This can put great strain on management teams and their employees and can undermine initial investment decisions. KPMG can help businesses, their lenders and investors to deal with such problems.

Organisation

KPMG is the global network of professional services firms who provide audit, tax and advisory services. KPMG LLP operates from 22 offices across the UK with over 10,000 partners and staff. KPMG's Carbon Advisory Group, made up of over 200 professionals, is providing insight and strategies to help organisations understand and manage the many business implications of climate change. The Carbon Advisory Group provides its clients with a range of linked services in order to help create an efficient low carbon business.

Enquiries

Richard Sharman
Lead Partner, Carbon Advisory Group
Tel: +44 (0) 207 311 8228
E-mail: Richard.mas.sharman@kpmg.co.uk
Website: www.kpmg.co.uk/services/ras

Caring for Climate:

THE BUSINESS LEADERSHIP PLATFORM

GEORG KELL
EXECUTIVE DIRECTOR,
UNITED NATIONS GLOBAL COMPACT

The rise of climate change as a fundamental issue for society has emphasised the need for leadership and voluntary action on the environment. Against this background The Global Compact, together with the United Nations Environment Programme (UNEP) and the World Business Council for Sustainable Development (WBCSD), has developed Caring for Climate: The Business Leadership Platform as a way to mobilise business action on climate change. Within months of launching the initiative, 200 companies from around the world, including many of the world's largest transnational corporations, have endorsed the statement and are already taking action.

INTRODUCTION

Through the United Nations Global Compact, thousands of companies in over 100 countries are working to integrate 10 principles in the areas of human rights, labour, environment and anti-corruption into their strategies and day to day operations. The Global Compact's commitment to environmental protection is firmly embedded in our foundational spirit and three environmental principles which are drawn from the 1992 Rio Declaration on Environment and Development:

- Business should support a precautionary approach to environmental challenges.
- Business should undertake initiatives to promote greater environmental responsibility.
- Business should encourage the development and diffusion of environmentally friendly technologies.

The rise of climate change as a fundamental issue for society has emphasised the need for leadership and voluntary action on the environment, including in the areas of biodiversity and responsible management of chemicals, wastes and water. The importance of early action is increasingly recognised, particularly in today's globally integrated world where the ability to compete is more and more linked to a company's capacity to innovate and demonstrate viable solutions for increasing energy efficiency and reducing carbon emissions.

CARING FOR CLIMATE INITIATIVE

Against this background, the Global Compact, together with the United Nations Environment Programme (UNEP) and the World Business Council for Sustainable Development (WBCSD), has developed Caring for Climate: The Business Leadership Platform as a way to mobilise business action on climate change. It is based on a shared belief that seriously addressing the climate challenge requires a preparedness to publicly express a vision and strategy, to make specific commitments, to set examples, to strengthen execution capability, to enhance public disclosure and to conduct pro-climate advocacy.

Launched in July 2007, Caring for Climate is a voluntary and complementary action platform for Global Compact

CARING FOR CLIMATE: THE BUSINESS LEADERSHIP PLATFORM

A Statement by the Business Leaders of the UN Global Compact

We, the business leaders of the UN Global Compact:

Recognise that:

1 Climate Change is an issue requiring urgent and extensive action on the part of governments, business and citizens if the risk of serious damage to global prosperity and security is to be avoided.

2 Climate change poses both risks and opportunities to all parts of the business sector, everywhere. It is in the interest of the business community, as well as responsible behaviour, for companies and their associations to play a full part in increasing energy efficiency and reducing carbon emissions to the atmosphere and, where possible, assisting society to respond to those changes in the climate to which we are already committed.

Commit to:

1 Taking practical actions now to increase the efficiency of energy usage and to reduce the carbon burden of our products, services and processes, to set voluntary targets for doing so, and to report publicly on the achievement of those targets annually in our Communication on Progress.

2 Building significant capacity within our organisations to understand fully the implications of climate change for our business and to develop a coherent business strategy for minimising risks and identifying opportunities. (It is understood that the setting of voluntary targets will be in accordance with different responsibilities and capabilities.)

3 Engaging fully and positively with our own national governments, intergovernmental organisations and civil society organisations to develop policies and measures that will provide an enabling framework for the business sector to contribute effectively to building a low carbon economy.

4 Working collaboratively with other enterprises nationally and sectorally, and along our value-chains, by setting standards and taking joint initiatives aimed at reducing climate risks, assisting with adaptation to climate change and enhancing climate related opportunities. (The term "setting standards" is clearly meant to refer to environmental performance standards, such as energy consumption, environmental impact and emissions. It does not refer to "international standards" whose design is the prerogative of governments.)

5 Becoming an active business champion for rapid and extensive response to climate change with our peers, employees, customers, investors and the broader public.

Expect from governments:

(It is understood that the call to governments to develop frameworks is meant to be framed under the current International framework.)

1 The urgent creation, in close consultation with the business community and civil society, of comprehensive, long term and effective legislative and fiscal frameworks designed to make markets work for the climate, in particular policies and mechanisms intended to create a stable price for carbon.

2 Recognition that building effective public-private partnerships to respond to the climate challenge will require major public investments to catalyse and support business and civil society led initiatives, especially in relation to research, development, deployment and transfer of low carbon energy technologies and practices.

3 Vigorous international cooperation aimed at providing a robust global policy framework within which private investments in building a low carbon economy can be made, as well as providing financial and other support to assist those countries that require help to realise their own climate mitigation and adaptation targets whilst achieving poverty alleviation, energy security and natural resource management.

And will:

1 Work collaboratively on joint initiatives between public and private sectors and through them achieve a comprehensive understanding of how both public and private sectors can best play a pro-active and leading role in meeting the climate challenge in an effective way.

2 Invite the UN Global Compact to promote the public disclosure of actions taken by the signatories to this Statement and, in cooperation with UNEP and the WBCSD, communicate on this on a regular basis, starting July 2008.

participants wishing to demonstrate leadership on the climate issue. A company's decision to join the platform by endorsing the Caring for Climate statement (see sidebar) must follow the Global Compact's established leadership and organisational change model. It requires CEO level support, strategic and operational changes within the organisation, and ongoing public communication on related activities and performance in line with the existing Global Compact 'Communication on Progress' framework. CEOs who endorse the statement will be prepared to set goals, change practices, publicly disclose emissions, and communicate on an annual basis.

Due to the broad geographic spread of Global Compact participants, involving leading actors from both developed and emerging economies, and the multistakeholder composition of the Compact, Caring for Climate transcends national interests and responds to the global nature of the issue. It is our hope, therefore, that the platform will become a leading voice of reason on pragmatic solutions to climate change.

Voluntary initiatives, and more specifically the concept of 'making markets work for climate', need an appropriate regulatory framework in order to function. Therefore, Caring for Climate involves both a commitment to action by business and a call for governmental action. In a world where business interacts with government primarily at the national level, Caring for Climate has the potential to become a neutral platform for exchange between business and governments at the global level. More specifically, the platform will hopefully be used as a unique space for business-government collaboration in the context of major international dialogues, particularly those leading up to COP-15 in Copenhagen, Denmark in 2009.

COMPANIES TAKING ACTION ON CLIMATE

(Material based on "Caring for Climate: Tomorrow's Leadership Today" published by the UN Global Compact, WBCSD and UNEP in July 2007.)

Within months of launching Caring for Climate, over 200 companies from around the world have endorsed the statement, among them are many of the world's largest transnational corporations. (Figures as of October 2007.) Driving this global interest is an understanding that addressing climate change offers space for tangible value creation.

The global climate challenge is affecting every part of society, so thinking inclusive and big is essential. As renewable energy technologies come of age, business is challenged to make greater advances in energy efficiency and the introduction of cleaner technologies. Cleaner products and services show the way in taking on new market opportunities and realities. Examples from the electronics and household goods sector include ecomagination by General Electric, EcoVision by Philips and Eco products by Sony. A number of international business associations, including the WBCSD, have shown leadership and are exercising due pressure via their constituents.

Companies from sectors such as finance, tourism, building and construction, and information and communications technology have also joined UNEP in voluntary initiatives to promote best practice in new technologies and life cycle approaches to advance more efficient resource use and dematerialisation. All of these turn climate risks into opportunities involving alternative business models and cleaner development.

Looking at recent top corporate sustainability reports, one can see examples of leading companies introducing climate change campaigns, CO_2 strategies, climate strategies and climate action plans. This shows that companies are moving from *ad hoc* efforts, to comprehensive strategies and targets and are reporting progress on these. It signals an opportunity for leaders to employ the Global Compact Performance Model and assess how climate can be addressed strategically, creating targeted results and impacts in the value chain and society.

For small enterprises, this could involve basic steps such as more efficient use of resources, recycling, installing energy efficient light bulbs and more efficient electricity use to reduce power bills. For a public institution, it can take the form of reducing air travel, making greater use of teleconferencing and improving building heating and cooling systems, all steps that if executed and monitored systematically help pave the way to climate friendly economy.

CONCLUSION

It is increasingly clear that responsible business can pay. Ignoring climate change implications and delaying proactive responses are not viable options if long term financial success and societal benefits are to be assured. It is true that proactively and comprehensively addressing climate issues will challenge established business practices. However, the alternative of not taking action would certainly be enormously disruptive, through sudden, unpredictable shocks, perhaps to supply and demand, as well as longer term effects.

Voluntary efforts can never be a substitute for government action; however, they can accelerate the process of solution finding and inspire consumers, peers and policy makers to have the courage to face the climate challenge as early as possible. We invite companies that are committed to advancing climate change solutions to endorse Caring for Climate. Together, we will show that responsible business practices can have positive, lasting impacts on our planet.

Author

Georg Kell is the Executive Director of the United Nations Global Compact, the world's largest voluntary corporate citizenship initiative with 4,500 participants and other stakeholders from 115 countries. Following extensive experiences in Africa and Asia as a financial analyst, Kell began his career at the UN in Geneva, where he worked from 1987 to 1990 with the UN Conference on Trade and Development (UNCTAD). In 1990, he joined the New York office of UNCTAD, which he headed from 1993 to 1997. In 1997, Kell became a senior officer in the Executive Office of UN Secretary-General Kofi Annan, responsible for fostering cooperation with the private sector. He has served as head of the UN Global Compact since 2000. A native of Germany, Kell holds advanced degrees in economics and engineering from the Technical University of Berlin.

Organisation

The UN Global Compact brings business together with UN agencies, labour, civil society and governments to advance 10 universal principles in the areas of human rights, labour, environment and anti-corruption. Through the power of collective action, the Global Compact seeks to mainstream these 10 principles in business activities around the world and to catalyse actions in support of broader UN goals. With over 4,500 stakeholders from more than 115 countries, it is the world's largest voluntary corporate citizenship initiative.

Further information on Caring for Climate: A Business Leadership Platform can be found on the Global Compact website.

Enquiries

Website: www.unglobalcompact.org

FSC
By buying products with an FSC-label you are supporting the growth of responsible forest management worldwide.
www.fsc.org
Cert no. SGS-COC-2086 / SGS-COC-1693
©1996 Forest Stewardship Council A.C.
nature
proof

ARCTIC PAPE

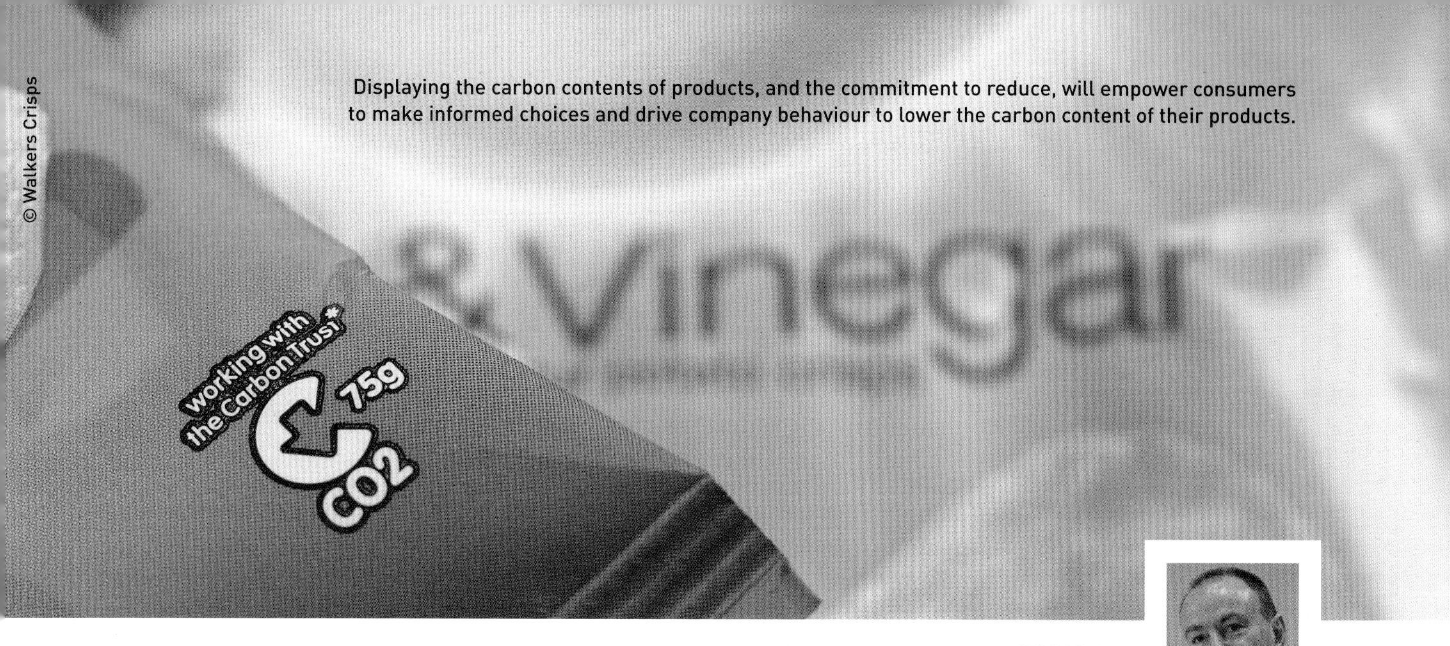
© Walkers Crisps
Displaying the carbon contents of products, and the commitment to reduce, will empower consumers to make informed choices and drive company behaviour to lower the carbon content of their products.
& Vinegar
working with the Carbon Trust*
75g
CO2

TOM DELAY
CHIEF EXECUTIVE,
CARBON TRUST

Carbon footprints in the supply chain:

THE NEXT STEP FOR BUSINESS?

To mitigate the effect of climate change, business needs to develop innovative strategies which challenge the way in which consumer needs are satisfied and which move beyond incremental improvements in process efficiency. By undertaking a carbon investigation of their supply chains, businesses can calculate and manage the carbon footprint of their products and unlock significant emissions reductions.

As climate change has emerged as the greatest environmental challenge we face, so reducing carbon emissions, the main cause of climate change, has become a central issue for business. As we move to a more carbon constrained world, business will have to make fundamental changes to deliver products and services to the consumer in a way that generates fewer carbon emissions.

Many companies are already focussing their attention on reducing direct emissions, with measures ranging from heating and lighting upgrades and staff awareness programmes, to developing low carbon energy sources such as onsite generation.

FOCUSING ON THE SUPPLY CHAIN

To mitigate the effect of climate change, business needs to develop innovative strategies that challenge the way in which consumer needs are satisfied and that move beyond incremental improvements in process efficiency. The important next stage is to tackle indirect emissions and manage the carbon footprint of products across the supply chain. It is the logical next step after traditional carbon management initiatives which analyse the operations of single companies or even single sites.

The carbon footprint, that is the carbon dioxide and other greenhouse gases emitted across the supply chain for a single unit of a product, is calculated by considering all of the raw materials and processes required to get a product to market. For example, the carbon footprint of cola is the total net amount of carbon dioxide emitted to produce, use and dispose of a single can of cola. The total carbon emissions are not just those due to the manufacturing processes or those due to 'food miles' but are based on all the steps in the supply chain to produce, use and dispose of or recycle the can of cola. This information can then be used to identify opportunities to make significant cuts in emissions and energy costs across the lifecycle of the product.

BENEFITS OF THE HOLISTIC VIEW

This approach, often called carbon life cycle assessment, offers a holistic view which can help business understand the reasons why emissions are generated across the economy. Processes, and their emissions, do not occur in isolation but are always part of the supply chains for different products or services. At the individual product level, this supply chain approach has the potential to find significant emissions reduction opportunities and large financial benefits by reducing the carbon footprint of a product.

It can help companies to understand the carbon emissions across their supply chains and allow them to prioritise areas where further reductions in emissions can be achieved. The approach can ultimately help all companies make better informed decisions in product manufacturing, purchasing, distribution and product development by considering the costs and liabilities that exist whenever carbon emissions are generated. As consumer attitudes change, supply chain analysis will also allow forward thinking companies to develop low carbon products to capture new markets and generate higher profits over time.

Consumers are becoming more empowered to take carbon into account when making a purchase, and businesses need to respond to that customer demand by taking carbon out of their supply chain and delivering lower carbon products. In the UK, about half of all personal carbon footprints are related to the 'embodied' carbon in the products that we buy and the amenities we enjoy: from groceries to clothes, from offices to motorways, pretty much everything creates a carbon footprint as it moves from raw material to the point of purchase or use.

Knowing the level of embodied carbon in a product is a reminder that everything we buy or use has a carbon impact and shows us which types of product and amenity have a particularly low or high impact. As consumers become more familiar with carbon as a currency, they'll be able to compare products on the basis of their carbon impact and consumer power could be mobilised as a force for good. This is good for business as well. Many ways to reduce embodied carbon involve energy efficiency that actually reduces costs. So a low carbon product will often be cheaper than the regular version.

ECONOMY-WIDE APPROACH

Managing carbon footprints across the supply chain offers an economy wide approach to tackling carbon emissions. Central to this approach is the view that all emissions across the economy are generated to meet the needs of the end consumer. This turns the traditional view of business carbon emissions on its head and gives a powerful insight into how the daily decisions by individual consumers drive economy wide emissions. So, iron ore is not made into steel because steel bars themselves are useful but because they, in turn, can be made into components for the televisions we all watch and the buildings we all live in. To fully understand the carbon emissions associated with our television sets, we need to consider not only the electricity used to run them but also the energy used to make and deliver all the parts, and the energy to dismantle, dispose and recycle them afterwards.

THE DETAILED METHODOLOGY

To calculate the carbon footprint of a product, the Carbon Trust, the UK Government Department for Environment, Food & Rural Affairs (DEFRA), and the British Standards Institute (BSI), is developing a draft Publicly Available Specification (PAS) standard by which the carbon content of all products and services can be measured. The final PAS standard, which will be made publicly available, will enable businesses of all sizes to identify and analyse the greenhouse gas (GHG) emissions associated with the products and services they provide.

The PAS standard draws heavily on well established life cycle assessment techniques (LCA). The United Nations Environment Programme (UNEP) highlights the key aspects of LCA as "identifying and quantifying the environmental loads involved – the energy and raw materials used and the emissions and wastes consequently released; assessing and evaluating the potential environmental impacts of these loads; and assessing the opportunities available to bring about environmental improvements". These key aspects and other aspects of LCA best practice have been built into the draft PAS standard, making sure it strikes the right balance between being analytically rigorous but simple and practical for business to apply.

In October 2007, Tesco, the world's third largest grocery retailer, partnered with the Carbon Trust to measure the carbon footprint of 30 of its own brand products, from tomatoes to light bulbs, using the draft standard. Nine other companies, including Cadbury Schweppes and Coca-Cola, have also announced that they will be working to test it too.

CARBON REDUCTION LABEL

We have also launched a trial carbon reduction label which not only shows the carbon footprint of a product but also shows the company's commitment to reduce the carbon emissions of the product within two years. The label is being trialled by a number of UK companies including Walkers, UK's largest snack foods manufacturer, Boots The Chemists, the UK's leading Health and Beauty retailer, and Innocent Drinks, the leading fruit smoothie brand.

UK research has indicated that 66 per cent of consumers want to know the carbon footprint of the products they buy, and climate change is increasingly becoming an issue that will impact upon the corporate reputation of business. The Carbon Trust envisages that the label will eventually act as a bridge between carbon conscious companies and their consumers, and not only provide a carbon measure, but also demonstrate a corporate commitment to manage and reduce the carbon emissions of the product. Displaying the carbon contents of products, and the commitment to reduce, will empower consumers to make informed choices and drive company behaviour to lower the carbon content of their products.

SUPPLY CHAIN SUCCESS

Boots The Chemists is the UK's leading Health and Beauty retailer, with approximately 1,500 stores in the UK and the Irish Republic. The company employs over 2,300 people in its factories and around 3,400 people in the group's distribution centres. The company is a member of Alliance Boots, which was created as Europe's leading pharmacy-led health and beauty group through the merger of Alliance UniChem Plc and Boots Group Plc.

For Boots, carbon management activities were aligned with corporate objectives for corporate social responsibility (CSR), energy and transport. Site surveys were used to identify a list of specific carbon abatement projects. From this, Boots developed a five year carbon management strategy to improve its carbon footprint, with initiatives falling into two categories: technical solutions and changing people's behaviours.

We've reduced the
carbon footprint of
Botanics shampoo
by 20%

ou can help too. Using cooler water to wash
our hair cuts CO_2 emissions, reduces your
nergy bills and is actually better for your hair.

working with
e Carbon Trust
148g
CO_2
Trust
Boots
BOTANICS

The Botanics range of shampoos is the first of Boots products to carry the carbon label.

Boots reported cost saving opportunities worth between £1 and £2 million and carbon emission reductions of over 10,000 tonnes.

The supply chain is one area where Boots has discovered new opportunities. As part of this work, Boots looked at a range of shampoos to compare processes and raw materials and can now identify the carbon footprints of a range of its shampoos. Unique to Boots is its Product Journey – a vertically integrated supply chain, with experts at all stages of the journey. This enables Boots to make positive changes that drive efficiency improvements across the entire supply chain.

Boots is currently working with the Carbon Trust on the initial phase of the carbon reduction labelling scheme. As a company displaying the label, Boots has signed up to a 'reduce it or lose it' clause whereby if they fail to reduce the carbon footprint of the product over a two year period they will have the label withdrawn. The Botanics range of shampoos is the first Boots products to carry the label.

Andrew Jenkins, Sustainable Development Manager, Boots the Chemists, said: "Working with the Carbon Trust has enabled Boots to measure and subsequently reduce the carbon footprint of everyday products such as shampoo by as much as 20 per cent. With Boots as the most trusted brand in the UK, providing this information and advice to consumers on reducing personal carbon footprints will raise public awareness about the part we can all play in combating climate change and protecting the environment."

Boots is also committed to developing new products based on sustainable development principles. As well as focusing on reducing its carbon footprint, Boots is engaged in a number of other initiatives focused on the product journey of products, from recycling to responsible packaging. An example is that Boots is designing their packaging to reduce environmental impact and waste. This includes research projects with designers to develop new sustainable packaging formats, and 80 per cent of its goods are now supplied to stores without transit packaging.

WHAT NEXT FOR BUSINESSES?

The supply chain approach has the potential to unlock significant emissions reductions and large financial benefits by reducing the carbon footprint at an individual product level. It can help individual companies to understand the carbon emissions across the supply chains in which they operate and allow them to prioritise areas where further reductions in emissions can be achieved. The benefits extend beyond business. Organisations can engage with their consumers about the carbon impact of their products, enabling them to make more informed buying decisions.

It can also help all of business to make better informed decisions in product manufacturing, purchasing, distribution and product development by considering the costs and liabilities that exist whenever carbon emissions are generated. This is the next step in the evolution of efforts to reduce carbon emissions and mitigate climate change.

Author

Tom Delay was appointed as the first Chief Executive of the Carbon Trust in 2001. A chartered engineer with extensive experience of the energy sector, Tom worked for Shell for 16 years in a variety of commercial and operations roles including four years as General Manager of Pizo Shell – a Shell subsidiary in Gabon, Africa. He moved into management consultancy with McKinsey and Co and then as a Principal with the Global Energy Practice of A T Kearney before joining the Carbon Trust. Tom gained a first class honours degree in mechanical engineering from the University of Southampton in 1981 and completed an MBA at INSEAD, Paris in 1988.

Organisation

The Carbon Trust is a private company set up by the UK Government in response to the threat of climate change. The Carbon Trust works with UK business and the public sector to accelerate the move to a low carbon economy by developing commercial low carbon technologies and helping organisations reduce their carbon emissions. This is delivered through its five complimentary business areas; insights, solutions, innovations, enterprises and investments. Together these help to explain, deliver, develop, create and finance low carbon enterprise.

Enquiries

E-mail: Customercentre@carbontrust.co.uk
Website: www.carbontrust.co.uk

Listening to the planet

For many years we have heard about the ozone hole over Antarctica, the greenhouse effect and other possible effects of man's activities on our planet. For a long time these alarms had little effect outside the world of biologists and other specialists. But today, with an increasing number of indications, the environment and climate issues have finally reached a whole new level of attention among entire populations. The time has come to take what Mother Earth tells us seriously.

Having started in the 1960s, Arctic Paper's contribution to reducing the impact of paper production on nature has contributed to their good reputation. Today, it offers a wide range of environmentally friendly paper, with many FSC certified, and the company's dedicated research continues to develop ways of supporting the surrounding nature. This is the story of Arctic Paper's way to environmental consciousness.

IN 1871 THERE WERE NO WORRIES

At a site where monks once fished for salmon, where the Örekil River joins the Gullmar Fjord, the enterprise today known as Arctic Paper Munkedals AB was founded in 1871. The Örekil River and the Gullmar Fjord run through one of the most beautiful natural areas on the west coast of Sweden, a landscape that gradually changed into a rural terrain.

For many decades, no one worried about the paper production effect on this natural resource with its rich animal life. But eventually, the water became less clear and the fish became scarcer. A similar development could be observed in the lake system where Håfreströms, another Arctic Paper mill, was situated at Åsensbruk in Dalsland.

A LONG TERM COMMITMENT

In the 1960s, it became obvious that the emissions from paper production were a threat to the environment. This strong warning was the beginning of a long term commitment. The mills of Arctic Paper were among the first to take measures to prevent the impact of paper production on nature. In 1966 it stopped producing pulp at Munkedal for this reason. Throughout the years, as new insights grew, knowledge and experience went into developing processes to reduce unwanted environmental effects.

In the course of time, Arctic Paper found ways of reducing its use of water and energy. In the 1980s it introduced biochemical processes to clean the process water. In the 1990s, it was the first to launch chlorine free coated paper.

In the same era, the Arctic Paper Group acquired the Kostrzyn mill, close to the Warta and Oder rivers in Poland. Since then, the Kostrzyn mill's environmental work has been developed on several levels and now has a leading position in Poland. Recently, the new heat and power plant, fuelled by gas, was opened to replace the old coal based plant.

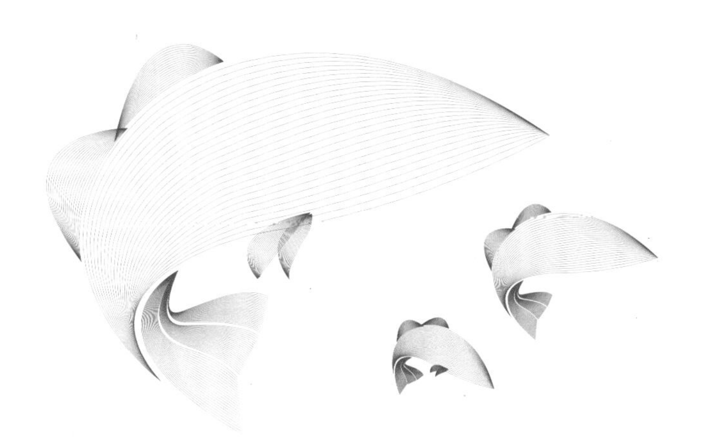

REDUCING AIR EMISSIONS AND PURIFYING WATER

The Håfreström mill also engaged early in developing methods for reducing energy consumption, emissions to air and for purifying the process water. In Håfreström a three step biochemical process was introduced for treating wastewater. The purified wastewater is discharged into Lake Nedre Upperudshöljen where a continuous water recipient analysis is conducted of water samples taken from 25 fixed testing stations in the lake system. Many of the Group's efforts have gone into dealing with cleaning and re-using its processed water, and to ensure that the water put back into nature is fully restored.

THE ENVIRONMENTAL CENTRE IN MUNKEDAL

This process can be studied, for example, at Arctic Paper's Environmental Centre in Munkedal. The exhibition hall is next to the large ponds that are at the final stage of purification. Here they receive customers, school classes and others who want to learn more about environmental care.

The Munkedal mill is now one of the most environmentally friendly paper mills in the world. In addition to low water consumption, the mill has some of the lowest discharge figures in the industry. Where other fine paper mills use 16 litres of water to make one kilo of paper, the Munkedal mill now uses less than three litres. Year by year, emissions into the fjord have been reduced and the water now carries hardly any particles.

As a result of all the measures taken, many qualities of the fjord have regained their previous status. Again the Gullmar Fjord is a place where salmon and seagulls enjoy freedom, clean water and fresh air.

THE ROLE OF FSC CERTIFICATION

Forest certification has the aim of ensuring that the forests are managed in a sustainable way. In Europe and North America this is additional to existing legal frameworks and places further requirements on the industry. One of the most important certification schemes is Forest Stewardship Council, (FSC). FSC is an independent organisation founded in Toronto in 1993 on the initiative of environmental organisations from 25 countries.

Among certifying institutions, FSC has the strictest criteria for ecological forest treatment and the monitoring of these criteria is performed directly in the forests. The FSC approach is global; it also takes responsibility in the developing countries. As the only organisation having social criteria, it recognises that ecological problems also arise due to human problems. Therefore FSC is accepted by environmental organisations all over the world.

A whole range of FSC certified papers

The number of certified products has increased since 2003 when Arctic Paper's first Munken paper quality was certified. Since then, the company has worked further to FSC certify both coated and uncoated papers and offers its customers a complete range of FSC certified products. Today, the whole Munken range from Munkedal is FSC certified as well as part of the Arctic range from Håfreströms. The Kostrzyn mill is using 40 per cent FSC pulp and 60 per cent PEFC (Programme for the Endorsement of Forest Certification schemes) certified pulp for its production of Amber.

> "The Munkedal mill is now one of the most environmentally friendly paper mills in the world."

Organisation

Arctic Paper specialises in producing high quality, graphic fine paper and is one of Europe's leading companies in this field. The company, whose head office is in Gothenburg, Sweden, has factories in Sweden and Poland and sales companies throughout Europe. Arctic Paper's products are marketed under the Amber, Arctic and Munken brand names and are used within the printing industry, including book production, printed matter and advertising products, office materials and other specialist areas. The Group has a turnover of about 3.3 billion Swedish kronor and has approximately 1,400 employees.

Enquiries

Arctic Paper AB
Box 383, 401 26 Göteborg
Sweden
Tel: +46 770 110 120
E-mail: info@arcticpaper.com
Website: www.arcticpaper.com

Responding to climate change –

THE INSURANCE INDUSTRY PERSPECTIVE

DR EVAN MILLS
STAFF SCIENTIST,
LAWRENCE BERKELEY NATIONAL LABORATORY

With core competencies in risk management and finance, the insurance industry is uniquely positioned to further society's understanding of climate change and advance creative solutions to minimise its impacts. Insurers have now begun to embrace this huge opportunity, which will enable them to prosper while reducing the claims from climate change.

INTRODUCTION

There is growing acknowledgement among insurers that the impact of climate change on future insured losses is likely to be profound. The chairman of Lloyd's of London has said that climate change is the number-one issue for that massive insurance group. And Europe's largest insurer, Allianz, stated that climate change stands to increase insured losses from extreme events in an average year by 37 per cent within just a decade. Losses in a bad year could top US$1 trillion. Insurers increasingly recognise that it is the lack of action to combat climate change that is the true threat to their industry and the broader economy; engaging with the problem and mounting solutions represents not only a duty to shareholders but also a boon for economic growth.

The insurance sector thus finds itself on the front lines of climate change. The response of many, particularly in the United States, has been to focus on financial means for limiting their exposure to high-risk areas along the coastlines and areas prone to wildfires. Allstate, for instance, has said that climate change has prompted it to cancel or not renew policies in many Gulf Coast states, with recent hurricanes wiping out all of the profits it had garnered in 75 years of selling homeowners insurance. The company has cut the number of homeowners' policies in Florida from 1.2 million to 400,000 with an ultimate target of no more than 100,000. The company has curtailed activity in nearly a dozen other states.

Figure 1: Range of insurer activities documented in this article.

* For these three categories, a maximum of one is tallied, as there is too much subjectivity in assigning weights to each individual activity.

** Multiple-year responses to a given disclosure initiative (eg Carbon Disclosure Project) are counted once.

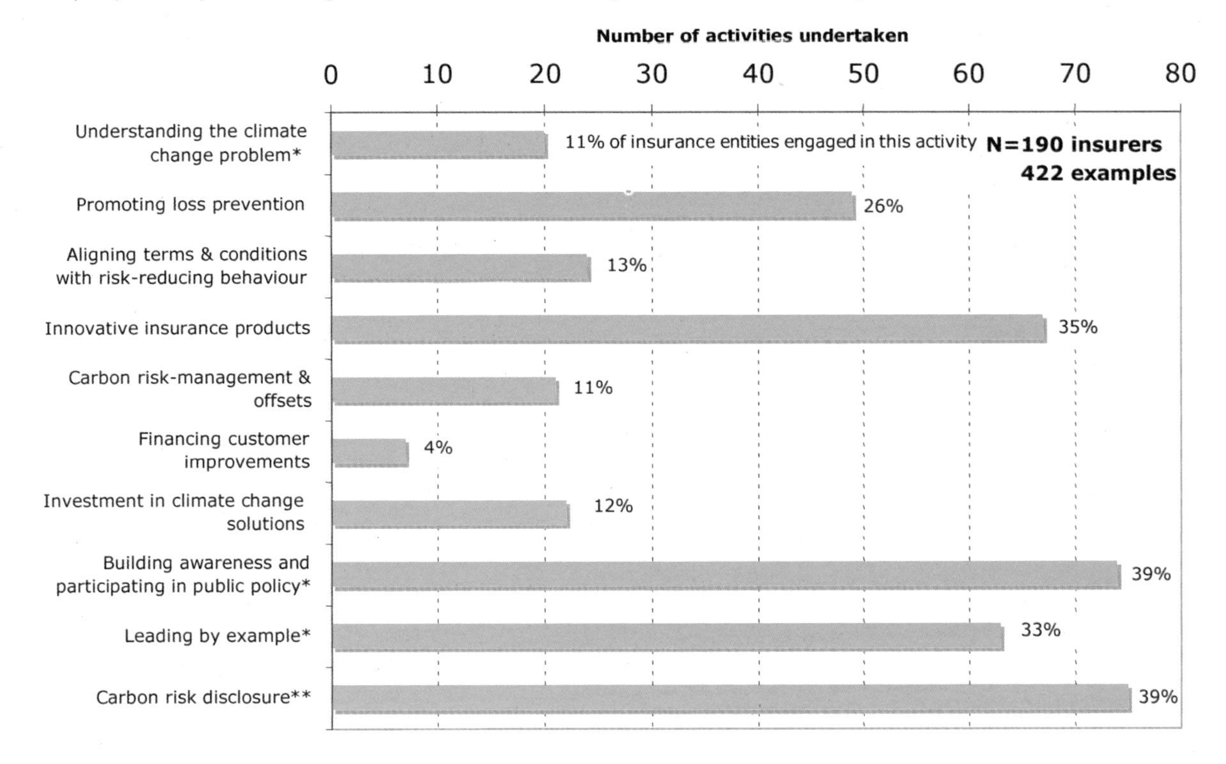

More difficult to detect than formal withdrawals or price spikes is the 'hollowing out' of coverage through increased deductibles, reduced limits, and new exclusions. A similar crisis in availability is occurring in many commercial insurance markets such as hotels and the energy sector, despite the absence of price regulation for non-household insurance.

A PROACTIVE APPROACH TO CLIMATE CHANGE

While many insurers continue to focus chiefly on financial risk management in response to climate change, others are realising that a more proactive, holistic approach to the issue presents significant opportunities to grow revenues, reduce risk and improve brand value. The Association of British Insurers and the European Insurance and Reinsurance Federation have recently called on insurers to actively pursue climate change solutions to ensure the preservation of private insurance markets.

Regarding the business opportunities presented by climate change, hundreds of billions of dollars will be spent on clean energy technologies and other responses, which represents an enormous new capital base with associated business operations requiring insurance. Several large insurers have already established special practices dedicated to the diversity of customers participating in this new market. Examples include AIG's Global Alternative Energy Practice, Allianz's Climate Solutions, Aon's Agri-Fuels Group and Chubb's Green Energy Team.

ADVANCING SOLUTIONS

Just as the insurance industry has historically asserted its leadership to minimise risks from building fires and earthquakes, insurers have a huge opportunity today to develop creative loss prevention solutions and products that will reduce climate change related losses for consumers, government and insurers.

Tokio Marine Nichido home page illustrates its mangrove restoration project designed to offset the company's carbon footprint and improve protection of insured infrastructure to storm surge.

Figure 1 identifies a wide spectrum of insurance opportunities, with 422 real world examples from 190 insurers, reinsurers, brokers and insurance organisations from 26 countries. These are grouped into 10 broad categories described below. An additional 23 non-insurance organisations, ranging from energy utilities to foundations to governmental agencies, have collaborated with insurers or otherwise supported their initiatives. These embody a wide range of activities that help improve disaster resilience and adaptation to climate change, while reducing climate-related risks through strategies, such as energy efficiency programmes, green building design, sustainable driving practices, and carbon emissions trading.

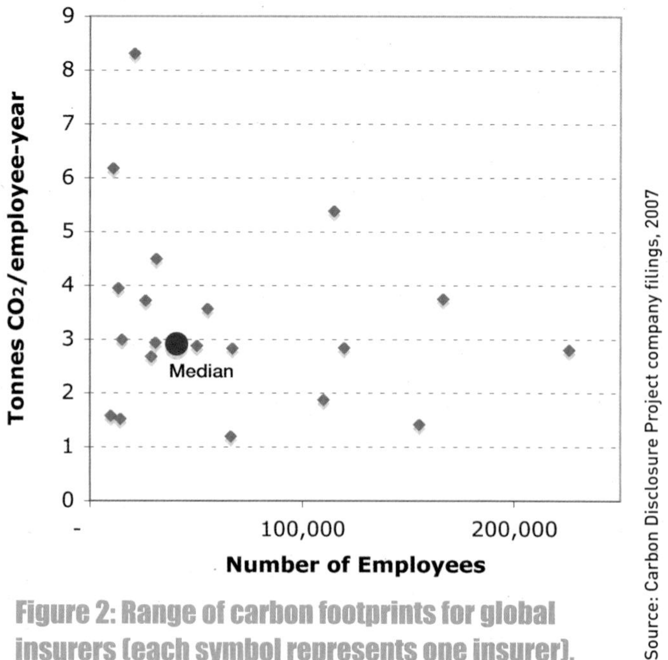

Figure 2: Range of carbon footprints for global insurers (each symbol represents one insurer).

1. Understanding the climate change problem

Insurers, including AIG, Allianz, Lloyds of London, Marsh, Munich Re and Swiss Re, are beginning to apply their expertise in data collection, catastrophe modeling, and risk analysis to track trends and define problems posed by climate change, pointing towards solutions for both the industry and society. Insurers are also looking to the scientific community to help it build forward-looking risk models that take climate change into account, with profound results.

2. Promoting loss prevention

Managing risks and controlling losses is central to the insurance business, and is evident in the industry's history as founders of fire departments and advocates for building codes. While the primary focus in recent years has been on financially managing risks, physical risk management is receiving renewed attention and could play a large role in helping to preserve the insurability of coastal and other high risk areas. Improved building codes and land use management are important starting points. Beyond that, innovations include a whole genre of energy efficient and renewable energy technologies that also make infrastructure less vulnerable to insured losses. Improved management of forests, agriculture and wetlands also offers dual benefits: withdrawal of carbon from the atmosphere and storage in biomass and soils coupled with increased resilience to drought, coastal erosion and other products of weather extremes. Tokio Marine Nichido is a leader in this area.

> "Insurers have an enormous opportunity to develop new profit centres by providing innovative insurance products for energy users or providers of clean energy services."

3. Aligning terms and conditions with risk reducing behaviour

New kinds of insurance terms and policy exclusions designed to instill behaviours that reduce greenhouse gas (GHG) emissions, as well as appropriate efforts to prepare for the impacts are beginning to emerge. Pay-as-you-drive insurance products have now been offered by 19 insurers, recognising the link between accident risk and distance driven and co-benefits for the environment. Among the most discussed possibilities is the liability of corporate directors and officers for their actions regarding climate change risks. Conversely, customers with a tendency to reduce climate vulnerabilities, eg drivers of hybrid cars, are being seen by companies like Farmers, Sompo Japan and Travelers as 'good risks', and rewarded accordingly through premium discounts.

4. Crafting innovative insurance products

Insurers have an enormous opportunity to develop new profit centres by providing innovative insurance products for energy users or providers of clean energy services. Examples include 'Green buildings' insurance and products from AIG, Fireman's Fund and Sompo Japan that cover risks associated with energy efficiency or renewable energy projects. London based Willis Holdings has launched a new product to cover potential underproduction of power from wind farms.
Insurers can also tap their core competencies to offer new services to assess and mitigate climate risks. Such activities would naturally develop into new business lines in energy auditing, retrofit evaluation, installation and management, as well as a host of quality assurance services that manage the performance risks of energy saving and carbon offset projects. New products such as micro-insurance are being introduced by companies such as Munich Re for small farmers and others in the developing world currently lacking access to traditional insurance.

5. Offering carbon risk management and carbon reduction services

Combined expertise in risk analysis and finance makes insurers natural participants in emerging markets for carbon offsets and trading. Some companies, eg Allianz and Insurance Australia Group, are also bundling carbon offsets with their products, particularly automobile and travel insurance. Insurers are also involved in providing property and liability insurance for carbon reduction capital projects, as well as consultative services in designing and managing such projects so as to maximise their technical and financial upside.

6. Financing climate protection improvements

Insurers, especially those associated with banking operations, are in a position to engage in financing customer-side projects that either improve resilience to the impacts of climate change or contribute to reducing emissions. One example is Fortis's preferential mortgage rate for energy efficient appliance and home upgrades. The company also offers Clean Car Credit, ie financing for low-emission vehicles. ING car leasing, which operates 300,000 cars across Europe, offers its customers in the Netherlands fuel efficient options, selected by 70 per cent of customers.

7. Investment in climate change solutions

Insurers are among the most significant players in financial markets, with US$16.6 trillion in financial assets as of 2005. There has been at least US$6 billion in green investment from 10 of the leading companies (total investment is not known), as well as significant examples of 'green' real estate asset management.

8. Public policy

It is in the business interest of insurers to support public policies that reduce and make risks more predictable. Insurers are now beginning to add their voices to the national and international discussion regarding climate change. Thirty eight insurers and insurance organisations from around the world have joined in the ClimateWise programme to promote a policy and market agenda for proactive responses to climate change risks. AIG and Marsh joined companies like ConocoPhillips and Duke Energy in the US Climate Action Partnership, which calls on the US to establish mandatory targets to reduce GHG emissions 60 to 80 per cent over several decades.

9. Leading by example

While insurers are not major emitters of GHGs, the energy used by their vast real estate holdings and employee travel is significant. Some insurers have pledged to become carbon neutral through various combinations of reducing energy intensity and the purchase of carbon offsets. Others prepare annual sustainability progress reports. It is notable that the median emissions by insurers, about three tonnes of CO_2 per employee per year, see Figure 2, is equivalent to the global average emissions per capita for transportation energy, and greater than that for housing.

10. Carbon risk disclosure

The process of assessing and disclosing climate risks contributes to insurers' ability to evaluate the impacts of climate change on their business. Disclosure also enables consumers and investors to gauge whether to purchase a policy from, or invest in, a particular insurance company, and it helps regulators to monitor the financial condition of insurance companies and the progress they are making towards addressing climate change risks. Such disclosures have been made in documents submitted to federal regulatory agencies or in response to formal requests from institutional investor groups, the largest of which is the annual call by the Carbon Disclosure Project (CDP), representing global investors with US$41 trillion in assets.

TOWARDS BEST PRACTICES

The concrete opportunities described here have in common the potential for improving the business position of insurers while addressing the risks posed by climate change. To be successful, insurers will need to partner with other actors, such as energy providers and governments.

However, most insurers remain behind the curve in developing forward-looking products and services in response to climate change. As shown in Figure 1, only about one in 10 of the insurers in this compilation are working and contributing to understanding the mechanics and implications of climate change, with a similarly small proportion incorporating these considerations into asset management. About a third are offering innovative products and services, and only four in 10 have disclosed climate risks to shareholders. Insurers engaging in the policy discussion of climate change, or leading by example through energy and carbon management in their own operations, remain in the minority. One challenge will be to ensure that responses are brought to scale in time to have a material impact on what is likely to be the biggest challenge facing the industry in its history.

It is also important to anticipate and avoid inadvertent adverse side effects of carbon-reduction strategies. For example, questions have arisen about unquantified liabilities associated with the rising popularity of carbon capture and storage. The insurance sector may also be unwilling to insure a rebirth of nuclear power, argued by some to be an important climate mitigation strategy.

The prospect for insurers' involvement in the development and promotion of climate change mitigation and adaptation strategies stands as an immense but as yet largely untapped opportunity.

Author

Dr Evan Mills is scientist at the US Department of Energy's Lawrence Berkeley National Laboratory. Dr Mills works in the areas of energy management and the impacts of climate change on economic systems, particularly with respect to the insurance sector. He has published over 200 technical articles and reports and has contributed to nine books. He contributed to the Intergovernmental Panel on Climate Change (IPCC) assessments in 2001 and 2007. IPCC scientists shared the 2007 Nobel Peace Prize with former US Vice President Al Gore.

Organisation

Founded more than 70 years ago, the Lawrence Berkeley National Laboraory is the oldest of the US Department of Energy's National Laboratories. The Lab is managed by the University of California, with an annual budget of more than US$500 million and a staff of about 3,800 employees, including more than 500 students. Today there are 11 Nobel Laureates associated with Berkeley Lab. The Lab conducts unclassified research across a wide range of scientific disciplines with key efforts in fundamental studies of the universe, quantitative biology, nanoscience, new energy systems and environmental solutions, and the use of integrated computing as a tool for discovery.

Enquiries

Evan Mills, PhD
Lawrence Berkeley National Laboratory
US Department of Energy, University of California
MS 90-4000, Berkeley, California 94720 USA
Tel: +1 510 486 6784 | Fax: +1 510 486 6996
E-mail: emills@lbl.gov
Website: http://eetd.lbl.gov/emills

The Inuit people used human rights as the focus to bring a petition asking for relief from violations as a result of climate change caused by acts of the US.

PRUE TAYLOR
DEPUTY DIRECTOR,
NEW ZEALAND CENTRE FOR ENVIRONMENTAL LAW

Climate change litigation:

A CATALYST FOR CORPORATE RESPONSES

This overview considers a variety of legal actions against both governments and government agencies (public actions), and against corporate entities (private actions). It is generally considered that governments and corporations in non-Kyoto Protocol ratifying nations are the most vulnerable to litigation, eg USA and Australia. However, nations that are dramatically failing to meet Kyoto obligations will also be at risk, eg Canada. As the gap between Kyoto obligations and the stringent cuts recommended by the Intergovernmental Panel on Climate Change (IPCC) grow, no nation and no corporation can consider themselves safe.

Over the past decade, the rate of climate change litigation has been rising slowly. By 2006 there were around ten significant cases. In 2007, increasing scientific certainty, emerging legal precedent and growing public awareness of climate change, are combining to create potent motivators for legal action.

AN OVERVIEW OF CLIMATE CHANGE LITIGATION

Climate change litigation began a decade ago but, until recently, few cases have delivered significant success. Progress has, however, been made in resolving some of the difficult legal issues inherent in climate change litigation. These include legal standing, causation, present harm, and duties of care. Parallels have been drawn between climate change and large environmental pollution cases, such as acid rain, and with early tobacco litigation. While there are similarities, there are also significant differences. These differences include complexities in proving that anthropogenic greenhouse gases caused the harm that is claimed, widespread generalised harm from global to local and the fact that contributing causes are the result of common economic and social activity.

Plaintiffs in climate change litigation are likely to be individuals including group actions, environmental and social responsibility groups and governments. In the case of governments, actions can be between state and federal governments or between sovereign nation states. Defendants include large GHG emitters, suppliers of fossil fuels, producers of products using fossil fuels, eg cars, and governments and government agencies with regulatory or permitting authority. Recent scientific evidence is beginning to lead to the door of less obvious parties including foresters, in the case of deforestation, and biofuel producers in the case of significant land use changes.

> "In 2007, increasing scientific certainty, emerging legal precedent and growing public awareness of climate change, are combining to create potent motivators for legal action."

Motivations for litigation include compensation for present or future anticipated harm, prevention of continuing high emissions and the focusing of public attention. The most likely motivation will be to use the threat of litigation to influence larger changes to government and corporate policy.

PUBLIC LAW ACTIONS (INTERNATIONAL)

There have been few cases drawing on international environmental legal obligations, partly due to the difficulties inherent in establishing a cause of action and in finding a forum with jurisdiction. But the biggest problem is in finding nations with sufficient political will to pursue other nation states. The most likely litigants are those of small island nation states vulnerable to inundation. In 2002, for example, the island state of Tuvalu threatened legal action before the International Court of Justice against the USA and Australia but, to date, no action has occurred.

International human rights actions are generally considered to present better opportunities for success, based on either express rights to a decent environment, or on violation of fundamental rights such as those to life, liberty and property, as a result of environmental degradation. A petition to the International American Commission on Human Rights, by the Inuit people, asks for relief from human rights violations as a result of climate change caused by acts and omissions of the US.

Public international law obligations under the World Cultural and Natural Heritage Convention have also been the subject of climate change related petitions. Petitioners are currently asking for specified World Heritage Sites to be placed on the List of World Heritage in Danger, due to the impacts of climate change.

PUBLIC LAW ACTIONS (DOMESTIC)

Domestic environmental laws often provide a basis for action against governments and their agencies. Such action is usually intended to ensure statutory obligations to protect the environment are met, including those in relation to GHGs and the impacts of climate change. Where a government agency is required to use environmental assessment procedures as part of decision making on development projects or on the scope of regulatory measures, corporate interests are likely to be directly involved. For these reasons, corporate bodies often join government agencies in proceedings.

New Zealand

A number of cases for the permitting of gas fired power stations, have taken GHG emissions into account. These emissions have been treated as contributing to climate change and to resultant ecological harm. NZ courts have recognised international obligations under the Kyoto Protocol and the UN Framework Convention and the need for united compliance efforts to avoid Hardin's tragedy of the commons.

In 2005, Greenpeace NZ challenged a decision to permit a coal-fired power station. A 2004 law change meant that GHG emissions could no longer be treated as an air discharge. However, the plaintiff successfully argued that climate change was still a relevant consideration, requiring decision makers to compare GHG intensive energy projects with those using renewable energy, such as wind power. This case is currently on appeal.

Australia

An interesting line of cases challenging the permitting of Australian coal mines began in 2004. The issue raised was whether environmental impact assessment (EIA) for coal mines should extend to include GHG emissions from coal subsequently purchased and used by third parties. Courts in a number of states generally accepted arguments from environmental groups, opening the way to challenge administrative decisions with respect to GHG producing developments and industries. A particularly important aspect of the 2004 decision was judicial comment that requirements to reduce GHG emissions from coal-fired power stations was important for ecological reasons and as a means of balancing present interests with future interests. This has been read as an indication that current economic benefits should no longer be traded off against the interests of future generations.

USA

One of the most significant cases on climate change so far has been the US Supreme Court ruling in *Commonwealth of Massachusetts, et al v EPA*. This litigation was initiated by 13 US States and territories, 13 NGOs and a number of municipalities. The respondent was the federal Environmental Protection Authority (EPA), supported by automobile manufacturers and associations, 10 US States, fossil fuel suppliers and power generators. The petitioners sought a review of EPA's decision not to regulate GHG emissions from new motor vehicles under the Clean Air Act (CAA). The EPA maintained that the Act did not authorise them to

Defendants in climate change litigation are not always those directly associated with fossil fuels.

regulate emissions due to an absence of Congressional intent that the CAA be used to regulate GHGs as a means of responding to climate change.

The US Court of Appeals for the District of Columbia Circuit dismissed the petition in 2005. This Court found that the EPA did have authority to regulate under the CCA, but that the EPA's decision not to exercise that authority was based on proper grounds. This decision left difficult issues of legal standing unresolved. These included issues of harm, actual and potential, and the viability of regulation to redress harm.

In 2006 the US Supreme Court agreed to hear an appeal and delivered its decision in April 2007. It found in favour of the petitioners, namely that the EPA has authority to regulate under the CAA as carbon dioxide is an air pollutant. Further that there were no legal grounds supporting the EPA's refusal to use authority under the CAA to regulate emissions.

Legal commentary is of the view that this decision will both significantly embolden potential litigants and fundamentally alter US political discourse on climate change. This is partly due to judicial findings in relation to legal standing. The Court determined that there was now sufficient scientific consensus on the link between anthropogenic GHG emissions and climate change and associated harms, the combined effect of which was to create sufficient standing to sue in courts to address climate change. Further, that states and individuals were suffering harm now and that those harms were both serious and well recognised. The shared nature of the injuries did not preclude individual petitioners right to sue. The EPA must now reconsider its decision and decide what steps it will take to address emissions from both motor vehicle and stationary sources. These sources currently account for around 60 per cent of US carbon dioxide emissions.

Nigeria

Human rights based claims centre around the argument that acts or omissions contributing to the release of GHG emissions and climate change create environmental harms that infringe on fundamental human rights, such as those to human life, health and property. In jurisdictions where there are constitutionally protected rights to a decent or healthy environment, the potential for successful action is enhanced. One landmark case under this cause of action is the 2005 decision of the federal High Court of Nigeria, ordering Shell Petroleum of Nigeria to stop gas flaring in the Niger Delta. It was found that the GHG emissions and toxins released violated the constitutional rights to life and dignity, of the people of the Delta State. Contempt of court proceedings were laid for failure to comply with the order to stop flaring, later in 2005.

PRIVATE LAW ACTIONS

There are a number of other private law causes of action available including product liability, negligence and the enforcement of general environmental duties. Two further causes of action are of particular interest to corporations.

Public nuisance

The potential of public nuisance is illustrated by the 2004 case *Connecticut et al v American Electric Power Company Inc et al*. Eight US States filed suit against six large power companies considered responsible for approximately 10 per cent of US anthropogenic carbon dioxide emissions. The plaintiffs claimed that harms to their citizens included injury to health and property, including ecological harm such as loss of biodiversity. They also claimed that the defendants could significantly reduce emissions through a range of cost effective measures. The claims were ultimately dismissed on the grounds that the case raised political questions for political branches of government to resolve. However, it cannot be presumed that this type of case would be decided the same way today. Judicial views change over time and will be influenced by rapidly emerging cost effective technology and scientific consensus regarding actual and imminent harm.

Director liability

Generally speaking, corporate law places legal duties on directors and officers to make money in the interests of the corporation and its shareholders. There are few legal duties in corporate law to consider the environment. In some circumstances, directors putting environmental protection ahead of profits could be sued for failing in their fiduciary duties to shareholders. However, there are three important developments that argue against directors' duties being considered a shield.

> "Climate change litigation now presents a real and present financial risk for corporations. Actual costs of litigation are likely to be far outweighed by associated commercial risks."

First, a growing number of commentators are arguing that corporations and their officers have social and environmental responsibilities, beyond returning profits to shareholders. A number of reforms have been proposed including those contained in the UK Company Law Reform Bill 2005. If passed as originally proposed, this Bill would have imposed a statutory duty of care upon directors, in respect of the environment and communities. Such a duty of care would have prevented directors and officers from acting in the best interests of the corporation, in a manner that also caused harm to the environment and communities. While this particular law reform failed, proponents will likely pursue it again in the future.

Second, commentators and reports have suggested that the financial liability arising out of climate change litigation could extend to billions of dollars. In these circumstances directors of corporations supplying fossil fuels, producing GHGs and financing harmful projects, could be in breach of a duty to act in the best interests of the corporation by failing to consider this increased financial liability and associated financial risks including those to shareholder value, corporate reputation, insurance and other associated commercial risks.

Third, there are now a range of mechanisms and studies that attempt to build environmental and social responsibility considerations into investment policy, taking these decisions beyond the traditional criteria of maximisation of financial returns. These developments are likely, in time, to extend beyond investment institutions to influence fiduciary law generally.

CORPORATE RESPONSES

Climate change litigation now presents a real and present financial risk for corporations. Actual costs of litigation are likely to be far outweighed by associated commercial risks. Traditional responses to this situation would include comprehensive risk assessment and measures to manage that risk including inhouse GHG mitigation measures. However, climate change demands responses that are much more comprehensive and structural in nature, ones that go beyond business as usual.

Appropriate responses require emission targets and timetables that are determined by ecological limits and not present economic efficiency or business as usual. Efforts to achieve anything less will expose businesses, people, property and living systems to devastating injury and harm. Governments will be reluctant to commit to stringent targets and timetables, fearing lack of support. Corporations have the influence and economic power to lead governments down the path of the ecologically determined regulation. Beneficial outcomes include regulatory certainty but also the achievement of a level of regulation that will be effective in the long term to protect and restore ecological systems upon which humanity depends.

Finally, climate change provides an important opportunity for corporations to seriously consider expanding their own understanding of corporate social and environmental responsibility, and the ethical and ecological arguments upon which this responsibility is based. Voluntary exercises of responsibility will only ever achieve limited outcomes. To be effective and fair, governments need to legislate.

The federal High Court of Nigeria ordered Shell Petroleum to stop gas flaring in the Niger Delta on the basis that emissions released violated the constitutional rights to life and dignity of the people of the Delta State.

Author

Prue Taylor currently teaches environmental and planning law to graduate and undergraduate students at the University of Auckland, NZ and is the Deputy Director of the New Zealand Centre for Environmental Law. She is also a member of the IUCN Commission of Environmental Law and its Ethics Specialist Group. Prue's specialist interests are in the areas of climate change, human rights, biotechnology, environmental governance, ocean law and policy, and environmental ethics. In 2007 Prue received an outstanding achievement award from the IUCN in recognition of her contribution as a world pioneer on law, ethics and climate change.

Organisation

The New Zealand Centre for Environmental Law (NZCEL) was established in 1998 as a specialist centre hosted by the Faculty of Law at the University of Auckland to provide expertise in the area of environmental law and resource management law.

Enquiries

Prue Taylor
School of Architecture and Planning
University of Auckland
Private Bag 92019
Auckland
New Zealand
Tel: +64 9 373 7599, ext 88649
Fax: +64 9 373 7652
E-mail: prue.taylor@auckland.ac.nz
Website: www.nzcel.auckland.ac.nz/

HANS JÜRGEN STEHR
CHAIR OF THE EXECUTIVE BOARD OF THE CLEAN DEVELOPMENT MECHANISM

The Clean Development Mechanism –

EVOLVING TO MEET CLIMATE AND DEVELOPMENT CHALLENGES

The Clean Development Mechanism is addressing both climate and development issues. It is up and running but the challenge now is to increase efficiency, and scale up the mechanism so that CDM can fulfil its potential.

Let projects in developing countries earn saleable credits for reducing greenhouse gas emissions, and let developed countries buy those credits to help meet their emission targets. It's a simple idea, so simple in fact that it takes up less than two pages in the Kyoto Protocol. Those few paragraphs, Article 12, which describe the mechanism's objectives of emission reduction, sustainable development and support for adaptation to climate change, have given rise to the first globally traded environmental commodity: the certified emission reduction (CER).

The rules, the so-called modalities and procedures, and the growing body of guidance, tools and clarifications on how projects can actually create CERs, are more complex. Yet the mechanism is up and running, and it works. CDM is stimulating green investment flow to developing countries and giving industrial countries some flexibility in how they meet their emission reduction commitments. The next challenge is how to make the CDM fulfil its potential.

NUTS AND BOLTS OF CDM

Despite its good start, there is still a great deal of confusion about CDM, indeed about emissions trading in general, and its role in combating climate change and contributing to sustainable development. The central feature of the Kyoto Protocol is a requirement that developed countries accept commitments to limit or reduce emissions, bound by international law – in a word, 'caps'. The adoption of caps creates a commodity – emission credits – which can be traded and sold. Countries or companies that emit below their cap have credits to sell, and those that go over their cap must look to the market.

The ability to trade emission credits:

- Gives countries/companies some flexibility in how they plan for and meet their emission targets.
- Offers lower cost opportunities to meet emission targets.
- Provides a market-based financial incentive to reduce emissions.

What about developing countries? How might they be involved in efforts to reduce emissions? More importantly, how might they secure the investment necessary to carry them down a green path to development, and spurn the well-worn path that has led the planet to its present crisis? What about adaptation to climate change? Where will the most vulnerable countries find resources to counter the effects of climate change? These were questions that occupied the negotiators of the Kyoto Protocol. An idea was floated that a fund be created for adaptation, financed through fines for non-compliance.

This idea evolved into a plan to generate revenue for adaptation as a by-product of some flexible means of assisting countries to comply with their emission targets. Thus was born the CDM. Reasoning that an emission reduction in one part of the world is as good for the atmosphere as an emission reduction achieved anywhere else, the negotiators provided for a market-based mechanism that would allow projects in developing countries to earn saleable credits for emission reductions that are real, measurable, verifiable and additional to what would have occurred without the project. A portion of the CERs generated would be placed in a fund for adaptation to climate change. It is somewhat ironic that a country which played a big part in the birth of the CDM, the United States of America, is not a Party to the Protocol.

Interest in CDM is strong in developed and developing countries alike. In a few short years, more than 840 projects have been registered, and about another 1,800 are in the validation/registration pipeline. The projects could generate more than 2.5 billion CERs, each equivalent to one tonne of carbon dioxide, by the close of the Kyoto Protocol's first commitment period in 2012. The numbers, still on the rise, tell a story of success, even if the estimated volume of credits is reduced because not all projects might fully live up to expectations.

“In a few short years, more than 840 projects have been registered, and about another 1,800 are in the validation/registration pipeline.”

The other achievements of the CDM are less obvious, but perhaps far more important. Consider how it has engaged the private sector, for without the private sector's participation and imagination, little can be achieved. Consider the partnerships it has forged between developing and developed countries. Forgetting for a moment the billions of dollars in investment flow that CDM will generate, emissions trading and the CDM have given virtually all the countries of the world a common, cooperative means by which to take action against climate change.

Is CDM the answer to global climate change? No. CDM is only one of the answers to climate change. The objective now is to consolidate, increase efficiency and scale up the mechanism to take advantage of its great potential.

ACHIEVING POTENTIAL

A new, promising way of scaling up CDM will be through registration of programmes of activities. Until now, the mechanism has focussed on single projects at single sites. It was a necessary first step in the 'learning by doing' process, akin to learning to walk before you run. The Executive Board issued procedures and guidance on programmes of activities in June 2007, which means the way is now clear for registration of programmes with activities operating at multiple sites, and activities can be added to a programme over time. It's the difference between installing energy-efficient light bulbs in a single home and successively installing energy efficient bulbs in an entire town or state. That's just an example, but it is quite helpful in illustrating the potential of programmes of activities to leverage the considerable work that necessarily goes into setting up a CDM project, not least the development of methodologies for determining baseline emissions and then monitoring resulting emission reductions.

“A number of variables impacting on the ability of sub-Saharan African countries to attract CDM investment are the same constraints that limit FDI in the region.”

Exactly what such a programme would look like is up to project developers, because the CDM is basically a bottom up process. The modalities and procedures for the CDM were agreed upon in Marrakesh in 2001. Since then, project developers have built the mechanism, one project at a time. The Executive Board's role, aided by expert groups and panels, has been to ensure that what developers submit meet the criteria: emissions must be real, measurable, verifiable and additional to what would occur without the project.

Along the way, the Board, assisted by the secretariat of the United Nations Framework Convention on Climate Change (UNFCCC), has crafted a wide array of guidance and clarification. Also along the way, the number of meeting days per year has more than doubled. Against a backdrop of dynamic growth and development, the Board provides the professional and regulatory competence needed to supervise the CDM, a mechanism of substantial size, global spread and sectoral diversity. To help the Board keep pace with activities, the resources in the UNFCCC secretariat have had to be increased. This has allowed for, among other things, improved communication with CDM stakeholders, for example through the holding of information forums for the national authorities that handle CDM. Nevertheless, the Board sees the need to increase dialogue with project participants and other stakeholders in its endeavour to strike the right balance between environmental integrity and flexibility and practicability.

© GLIBBERY-UNEP/Still Pictures

This is the kind of challenge that the Board is pleased to face. It means that the mechanism is working and is expanding. Where might it go from here? The programme of activities offers one route to expansion. Might there be other ways to leverage the mechanism? Surely there are, and these will no doubt be proposed and pursued by the many stakeholders that engage in the mechanism: project developers, investors, financiers, multilateral development agencies and banks, brokers, analysts, consultants and national authorities.

THE CHALLENGE OF AFRICA

As at November 2007, sub-Saharan Africa has hosted 15 registered CDM projects. This equals 1.8 per cent of all projects. Twelve of the projects are hosted by South Africa. Forty of the 45 sub-Saharan African countries have ratified the Kyoto Protocol, but many have limited ability to implement CDM projects. Twenty-six have established Designated National Authorities to administer CDM activities. Several of these countries, however, have not invested sufficient resources in establishing the required institutional structure to approve CDM projects due to the number of other development priorities they face, such as poverty alleviation, food security, health and education. The limited capacity is reinforced by the lack of designated operational entities (DOEs) from these countries. Project verification and validation are undertaken by these third party entities, which are generally large international consulting firms whose rates are expensive for project developers in poor countries.

What is true for African countries, is also true for many countries in other less developed regions. Many countries have suffered from a lack of capacity building. The largest amount of official development assistance for CDM capacity building has gone to middle income countries that are better positioned to capitalise on carbon funding due to their better investment climates and greater potential for emission reductions. Important lessons can be learned from these efforts, leading to greater efficiency and improved results in activities in sub-Saharan Africa and elsewhere.

South-South cooperation and transfer of learning could prove effective for sub-Saharan Africa and developing countries in other regions. The market-based nature of the CDM means that most investment will naturally gravitate to low-risk, high-opportunity projects and locations, similar to foreign direct investment (FDI). In this respect, a number of variables affecting the ability of sub-Saharan African countries to attract CDM investment are the same constraints that limit FDI in the region. Studies indicate that multilateral finance predominates in the region, and trying to do carbon investment projects relying on private finance is exceptionally difficult. The continent has a low savings rate and few indigenous financial institutions. This indicates the need for innovative financing arrangements and risk reduction schemes.

MICROFINANCE AND CDM

In the context of promoting a more equal regional distribution of CDM projects, the CDM Executive Board has identified potential synergy between CDM and microfinance activities in least developed countries. This could develop into a major initiative, in combination with a programme-of-activities approach to expand CDM's usefulness to the poor. Poor people rarely access services through the formal financial sector. They address their need for financial services through a variety of financial relationships, mostly informal. Credit is available from informal commercial and non-commercial money lenders, but usually at a very high cost to borrowers. Savings services are available through a variety of informal relationships, such as savings clubs, rotating savings and credit associations and other mutual savings societies that have a tendency to be erratic and insecure. Provision of financial services for the poor has proved to be a powerful instrument for poverty reduction that enables local populations to build assets, increase incomes and reduce their vulnerability to economic stress.

> **"Credit is available from informal commercial and non-commercial money lenders, but usually at a very high cost to borrowers."**

However, nearly one billion people, especially the very poor, still lack access to basic financial services. As such, the Executive Board is naturally interested in exploring the complementarities of microfinance and programmatic CDM, in order to further expand the reach of microfinance and enhance the quality of life of people in developing countries.

Microfinance as a development strategy combines massive outreach, far-reaching impact and financial sustainability, which makes it unique among development interventions. However, it is usually not considered to be a key strategy in addressing climate change. After all, the poor contribute relatively little to the build up of greenhouse gases. Then again,

considering that developing countries, and particularly their poorest residents, are among those most likely to be adversely affected by climate change, it makes sense that the Executive Board explore ways to leverage a proven development approach. CDM, a market-based mechanism, might be a perfect match for microfinance, a finance-based sustainable development approach.

RESPONDING AND EVOLVING

The Kyoto Protocol's first commitment period runs to 2012. Already countries must begin working diligently to craft what will come next. Whatever that is, it should be guided by a long-term emission goal and be clearly anchored in the sustainable development aspirations of Parties. It will need to address adaptation to current and foreseeable impacts, while delivering aggressive mitigation action to minimize further climate change. It will need to establish effective enabling mechanisms to allow emission reductions to be cost effective, while allowing countries with varying national circumstances to actively contribute to the effort. And any agreement will need to stimulate the diffusion, transfer and deployment of existing technological solutions and investment in additional technological solutions.

> " Microfinance as a development strategy combines massive outreach, far-reaching impact and financial sustainability, which makes it unique among development interventions. "

The clean development mechanism has shown that it can respond and evolve. Its success has led people negotiating the international response to climate change to suggest that market-based mechanisms, such as CDM, could be central to an agreement after 2012. Meanwhile, CDM is up and running and already doing its part to spur investment and reduce greenhouse gas emissions.

Author

Hans Jürgen Stehr is currently serving his second term as Chair of the Executive Board of the Clean Development Mechanism. He is former Director for Research and Development at the Danish Energy Authority, where he was responsible for the implementation of the Danish Energy Research Programme and for developing Danish energy research and innovation strategies. Prior to his assignment on energy research and development, Mr Stehr supervised Danish preparations for the UNFCCC Kyoto conference in 1997 and legal and political follow up activities nationally and in relation to EU regulations. He holds an MA in Law from the University of Copenhagen, and was for many years Assistant Professor of Constitutional Law.

Organisation

The Clean Development Mechanism was established in Marrakesh in 2001, following negotiations at the Kyoto Protocol in 1997 and comes under the remit of the UN Framework Convention on Climate Change (UNFCCC). It was established as a market-based mechanism that would allow projects in developing countries to earn saleable credits for emission reductions that are real, measurable, verifiable and additional to what would have occurred without the project. These credits can be bought by developed countries to help meet their emission targets.

Enquiries

Website: http://cdm.unfccc.int

The use of targets or market mechanisms for emissions reductions in developing countries

© Kati Molin/Fotolia

RAE KWON CHUNG
UN ESCAP

There is a strong perception among developed countries that net global greenhouse gas emissions (GHG) reductions are only possible by imposing targets on developing countries. But to do so is neither politically feasible nor technically practicable. We should not waste time designing targets into any post-Kyoto climate change regime. Instead, we should explore other options, such as market mechanisms. This article looks at the arguments for doing so.

Target obsession and political deadlock

Climate change negotiations are facing serious political deadlock regarding the reduction of GHG emissions from developing countries in designing the climate change regime after 2012. Most developed countries share the view that imposing targets on developing countries is the only available option to ensure emission reductions from developing countries. But the technical difficulties and uncertainties involved in projecting a future emission trajectory are making many developing countries uneasy and reluctant to accept any form of target. Misunderstandings arising from this issue are beginning to sour mutual trust and threaten to block any positive progress in the negotiations.

Targets in the form of equal global quota

Calculating an equal global *per capita* carbon quota can easily be achieved by dividing the target volume of GHG gases by the number of people in the world. However, there is little chance that this idea will gain political acceptance and be agreed on as an immediate option for the climate change regime after 2012.

TARGETS AGAINST BUSINESS AS USUAL

Most targets now being proposed tend to look at reducing expected emissions based on business as usual (BAU) scenarios, eg intensity, sectoral and even no-lose targets. There are, however, considerable technical difficulties for developing countries in setting up such targets where the task is less onerous for developed countries whose economic growth rate and industrial structure are stable and predictable unlike developing countries.

TARGET SETTING FOR DEVELOPING COUNTRIES

Even the best economists often fail to predict the exact economic growth rate of developed economies for a given year. How then can we expect reliable long term projections for GHG emissions for developing countries whose economies are far more volatile and vulnerable? If actual emissions unexpectedly drop below projected BAU, the imposed target will mean very little. If actual economic growth and GHG emissions unexpectedly shoot up, the target will in effect impose growth capping, which is detrimental to the economy of any developing country.

There are three major reasons why BAU emission trajectories for developing countries are difficult to predict: vulnerability to external factors; transformation of industrial structure; and high impact of non-climate policies.

Vulnerability to external factors

Many developing countries are far more vulnerable to external factors such as unexpected sudden economic crises than developed countries, eg the financial crisis experienced by many Asian countries during the late 1990s. Those countries were enjoying high growth rates immediately before the crisis that reduced their growth rate for several years.

"The technical difficulties and uncertainties involved in projecting a future emission trajectory are making many developing countries uneasy and reluctant to accept any form of target."

If China and India set targets based on the current high growth rates as is suggested by many developed countries, there is a higher risk that growth rates may slow in the middle of the commitment period due to unforeseen reasons such as high oil price (external factor) or domestic political turmoil such as democratisation and labour strikes (internal factors).

Some might argue that the economies of developed countries are also vulnerable to external factors. However, developed countries, such as the EU member states, the US and Japan, have the power to lead or drive the world economy. They set the basic parameters of the world economy while developing countries follow, having to live by the rules set by developed countries. The degree of vulnerability and fluctuation is much more unpredictable for developing countries.

Transformation of industrial structure

The economies of developing countries are undergoing rapid structural change as globalisation opens the door to world trade. Many countries are changing their industrial structure from primary industry to manufacturing and a service economy. However, it is difficult to predict this rate of change. It is even more difficult to estimate the impact of such structural changes on GHG emissions.

While developed countries also experience industrial structural changes, these are generally far more subtle and predictable given the mature nature of their economies.

HIGH IMPACT OF NON-CLIMATE POLICIES

There are many socio-political factors that impact the GHG emissions of developing countries, such as the democratisation process, workers rights, urbanisation, land planning, infrastructure, housing, transportation or even population growth rate. China's one child policy, for example, which has been credited with saving 1.3 billion tonnes of CO_2 in 2005 may well change for social reasons as *per capita* income rises along with the inevitable CO_2 increase. Most governments will give higher priority to such social issues over climate change.

Agreeing on specific targets

Even if we do have a reliable projection on BAU for the emission trajectory, we still face the far more difficult political problem of haggling about the absolute level of an acceptable target for each developing country. We have all observed the difficulties of setting up targets even among developed countries (Annex 1) during the negotiation process of the Kyoto Protocol. Agreeing on specific targets for so many developing countries will be far more difficult. We will need a miracle to arrive at such a consensus.

Agreeing on target level of per capita emissions

If we were to set *per capita* emissions, the question is how would we set the most appropriate level for developing countries. How can we ask China or India to freeze their *per capita* emissions by, for example five tonnes of CO_2, which is higher than the current *per capita* level but much lower than that of Japan, EU and the US?

Difficulty of verification

Even if we could set up a perfect target thanks to a super computer or modern day Nostradamus, we still have the difficulty of imposing and verifying compliance. Given the low reliability of data in many developing countries, it will not be easy to do so.

DIVERSE MODIFIED TARGETS

In view of these difficulties, experts have come up with modified target ideas, such as intensity, sectoral or no-lose targets. However, none of these are free from the difficulties mentioned above.

Intensity targets

Intensity targets have merit as they are relative, not absolute, giving more flexibility. However changes of GHG emission intensity are not easy to predict or to curb and control as they are directly related to industrial structure. As economic development progresses, carbon intensity generally goes down. But if energy intensive heavy industry continues to grow, the intensity would increase.

Sectoral targets

Sectoral targets are more easy to project, but they face the more difficult problem of industrial competitiveness. Energy intensive industries, such as steel, cement and chemical are usually the backbone of industrial competitiveness of any country. It is highly unlikely that these countries will submit their key strategic industries under an international target.

No-lose targets

No-lose targets attempt to provide incentives. But they are not free from the difficulty of projection and face the added problem of negotiating an acceptable target between the buyers and sellers of carbon credit expected to be generated by developing countries who successfully reduce their emissions target.

Pledge and review targets

Imposing capping and forcing compliance on developing countries, necessary for pledge and review targets, will be difficult to achieve; any legally binding target seems technically impracticable and realistically unfeasible.

> "If targets are not workable, market mechanisms could provide incentives for developing countries to act on climate change."

A voluntary pledge and review target could, however, function as a driver for best possible efforts to reduce GHG emissions. Developing countries have a strong incentive to improve their energy efficiency as the price of oil rises.

Compatibility of climate action and economic goals

The rising oil price makes energy efficiency and major climate action increasingly compatible with economic interests. China, for example, has set up its own target of improving energy efficiency by 20 per cent during its current economic development plan, not because of climate change, but because of the economic need to save energy. The Chinese government is vigorously pursuing this goal and the 20 per cent energy efficiency target is a good example of a voluntary pledge and review target for climate change action.

WHY NOT MARKET MECHANISMS

If targets are not workable, market mechanisms could provide incentives for developing countries to act on climate change. Economic instruments and market mechanisms are generally regarded as more efficient than command and control measures in environmental governance. Why then, are we so obsessed with imposing targets which represent a typical command and control mechanism?

The Clean Development Mechanism

There are already market mechanisms available. The Clean Development Mechanism (CDM), developed under

© Olivier Blondeau/iStockphoto

the Kyoto protocol, is one such international mechanism and the only one that engages developing countries. Designed to promote sustainable development in developing countries and to assist Annex I countries in meeting their GHG emission commitments, the CDM can function as an incentive mechanism for emission reduction projects in developing countries.

There are other forms of market mechanisms that could be developed, such as a global carbon tax but time is running out to negotiate and while the CDM, in its current form, has limitations, reforms could reap results.

> "Originally designed as an Annex 1 compliance mechanism, the CDM can function as an incentive mechanism for emission reduction projects in developing countries."

Currently, the CDM has limited scope with very complex procedures. To expand the scope of CDM, among the three criteria for financial, technical and project additionality, project additionality has to be relaxed while technical additionality should remain strict in order to ensure the quality of Certified Emission Reduction (CER), the unit of carbon credit generated by CDM projects.

There is no reason to discriminate among projects that reduce emissions based on arbitrary criteria. Project additionality criteria was introduced from an ideological point of view that CDM projects should be purely designed for climate change that have not been tainted by commercial interests. In the real world, however, most projects have certain commercial components which would very likely make them ineligible under the CDM.

In order to make CDM function as a market mechanism, such ideological purity should be abolished. CDM should be more practical and not isolated from the dynamism of the private sector. Any project that reduces GHG emissions according to a strict technical baseline should be eligible to become a CDM project without going through the hassle of proving that it is not an 'anyway' project. Once the project additionality criteria is removed, CDM can function as a genuine market mechanism and would provide incentives for many more meaningful climate action projects.

> **"The emission reductions expected from unsold CER represent a real, tangible and measurable option, while emissions reductions based on a fictitious BAU target can never be verified or certified."**

NET GLOBAL EMISSIONS REDUCTIONS

A reformed CDM can function as a market mechanism to provide incentives for emission reduction of developing countries. Nevertheless, CDM itself does not generate net global emission reductions, since the CER generated by CDM is sold to developed countries (Annex 1) and will increase their own emissions. Currently CDM simply shifts emissions from developing countries to developed countries but if CERs were discounted this could change.

A CER discounting scheme

If only certain portions of CER were sold to Annex 1 buyers, the unsold portion could contribute to net global emissions reductions. If for example, out of two million tonnes of CO_2 reduced by a CDM project, only one million tonnes were sold, this would result in a net global emissions reduction. Once we introduce such a rule, CDM will be able to generate net global emission reductions.

CONCLUSION

Developing countries are prepared to take meaningful climate action as part of their economic development plan and are in fact already doing so. They simply do not accept that fixed targets are workable and realistic for them. The changes to the CDM discussed here represent a real, tangible and measurable option to reduce GHG emissions, while emissions reductions based on a fictitious BAU target can never be verified or certified.

If we could reform and redesign CDM, the current climate crisis can be turned into an economic opportunity for developing countries. Redesigned CDM would enable developing countries to contribute to net global emission reductions, moreover without imposing any binding target. In the UNFCCC climate meeting in Bali, climate change negotiators need to focus on reforming and redesigning CDM rules.

Of course, the basic assumption of this idea is that developed countries would accept deep targets after Kyoto, so that a strong demand for CER from CDM could be sustained. This is one of the basic requirements for any kind of climate change regime after Kyoto Protocol.

This article is the work of the author in his personal capacity and does not represent the views of his organisation.

Author

Rae Kwon Chung is Director of the Environment and Sustainable Development Division of the United Nations Economic and Social Commission for Asia and the Pacific. He is a former Korean career diplomat and negotiator for many global environment conferences, inserting 'compulsory licensing' and 'transfer of publicly-owned technologies' in Agenda 21 of Rio Earth Summit in 1992, and proposing unilateral CDM for the Kyoto Protocol in 2000.

Mr Rae Kwon Chung served as Director-General for International Economic Affairs Bureau at the Korean Foreign Ministry until July 2004 before joining the UNESCAP.

Organisation

The United Nations Economic and Social Commission for Asia and the Pacific (ESCAP) is the regional development arm of the United Nations for the Asia-Pacific region. With a membership of 62 governments and a geographical scope that stretches from Turkey in the west, to the Pacific island nation of Kiribati in the east, and from the Russian Federation in the north, to New Zealand in the south, ESCAP is the most comprehensive of the United Nations five regional commissions.

Enquiries

E-mail: chung1@un.org

Purchasing carbon offsets

A HOW-TO GUIDE

MARK KENBER
POLICY DIRECTOR,
THE CLIMATE GROUP

Purchasing carbon offsets can set an internal price for carbon, allowing a company to assess the cost of other emissions-reducing initiatives. It can also help companies gain a greater understanding of carbon markets and support existing corporate responsibility policies. But offsetting has also been seen as an easy 'get-out' from tougher internal choices with some media casting doubts over the extent to which offsets represent real and additional emissions reductions. This simple 10 step guide is designed to help companies avoid the pitfalls and use carbon offsetting as a useful part of overall greenhouse gas (GHG) reduction strategy.

STEP 1: UNDERSTAND WHAT OFFSETS ARE

An offset is a carbon dioxide equivalent (CO_2e) emissions reduction or removal that is used to counterbalance or compensate for emissions from other activities. Offsets can be purchased by countries, companies or individuals to meet their own reduction requirements.

Offsets can either be bought from within the compliance system (under the Kyoto Protocol emissions reductions are bought and sold through the Clean Development Mechanism and Joint Implementation) or in the voluntary market (these are labeled VERs).

The key criterion for an offset is that the CO_2e reduction or removal used as an offset would not have otherwise happened, ie is additional to business-as-usual activity because there would be no net reduction in emissions if this were not so. The seller of the offset should be able to prove that the emission reduction would not have occurred if the offsetting payment (in aggregate across the project) had not taken place.

STEP 2: DECIDE HOW MANY OFFSETS NEEDED

The first step is to calculate your organisation's carbon footprint. The GHG Protocol, developed by the World Resources Institute and the World Business Council for Sustainable Development, and ISO 14064 developed by the International Standards Organisation provide globally recognised standards for measuring and reporting the carbon footprint of an organisation. All emissions necessary for operating a business and those emissions assumed to be under the company's control should be looked at.

Company targets then need to be set, both for internal reductions and offsetting. These targets should take into account the potential for cutting emissions at source, eg decreasing energy consumption, improving the efficiency of operations, and purchasing renewable energy electricity, and whether the company wants to be carbon neutral overall (net zero emissions including reductions and offsets).

> **"The GHG Protocol and ISO 14064 provide globally recognised standards for measuring and reporting the carbon footprint of an organisation."**

It is also important to consider that, for some organisations, the largest footprint impacts come from working with the supply chain, employees or consumers. Many firms have found that there are a number of emissions-reducing opportunities that also allow for significant cost cuts and/or productivity gains.

STEP 3: DEVELOP A TAILORED OFFSET STRATEGY

Companies need a robust strategy for purchasing the residual emissions that require offsetting.

Questions to consider during the process

- What are the drivers for offsetting?
- What are the key audiences to reach with the offset programme?
- How important is alignment with government and other standards?
- Are there particular areas of geographic focus or sectors the company prefers?
- Are there particular types of technologies that the company does (or does not) wish to support?
- How important are sustainable development attributes of projects?
- What is the budget?

OFFSETTING CASE STUDIES

Sky In May 2006, Sky became carbon neutral through measuring, reducing and offsetting its CO_2e emissions. After reducing what it could at source through the purchase of renewable energy, reductions in energy use and finding more innovative ways to operate, Sky purchased offsets to compensate for the remainder of its emissions, deciding to focus on voluntary market projects. Carbon credits were purchased in two renewable energy offset projects: a gold standard wind power project in New Zealand and a micro-scale hydro-electricity scheme in Bulgaria.

HSBC HSBC became carbon neutral in 2005, reducing the Group's contribution to climate change through three different approaches: energy efficiency; buying 'green' electricity where possible; and offsetting remaining CO_2e emissions. A Carbon Management Task Force directly oversaw the process of finding credible, genuinely-incremental and cost effective offsets. Four offset projects were used to buy a total of 170,000 tonnes of carbon offset credits: a wind farm in New Zealand; an organic waste composting project in Australia; an agricultural methane capture project in Germany; and a biomass cogeneration project in India. The average price per tonne of offset amounted to US$4.43.

One further important issue to consider is whether to buy offsets from the regulatory (Kyoto-compliant) market, delivered under the Clean Development Mechanism, or from the unregulated voluntary market. A recent report by consultancy ICF International suggests that the voluntary carbon market could expand from 10-25 million tonnes (Mt) of carbon dioxide equivalent (CO_2e) in 2005 to a mid-range estimate of 400 Mt CO2e by 2010. This is equivalent in size of about a third of the Kyoto Protocol compliance market in 2006, or the total emissions of Spain or the Ukraine.

Despite the growing enthusiasm, its unregulated nature is particularly susceptible to criticisms regarding the quality of offsets being traded. Of most concern to those wishing to purchase offsets is the environmental and social integrity of the carbon reductions they are purchasing.

Two aspects, in particular, are crucial here:

- Voluntary offset quality, ie that voluntary offsets represent genuine additional emissions reductions
- Delivery quality, ie mechanisms that ensure offset buyers get what they are paying for.

Being aware of the key standards in the voluntary offset market (explained further below) will help companies take advantage of the flexibility and competitive prices offered, while ensuring offsets are really delivering the reductions in emissions they claim.

© Ben Heys/iStockphoto

STEP 4: DECIDE HOW TO BUY OFFSETS

Companies should think about whether they want to choose a preferred provider or buy from the market, and whether they want a long term agreement or to buy year-on-year. In addition, companies should also decide who will pay for the offsets. Making individual units pay can provide an incentive for reductions at source.

STEP 5: CHOOSE THE OFFSET STANDARD THAT BEST SUITS YOUR COMPANY

There are a number of offsetting standards that suit different needs. If choosing voluntary offsets, The Climate Group recommends offsets approved under the two main international standards: the Voluntary Carbon Standard (VCS) and the Voluntary Gold Standard. Both deal directly with VER quality, establishing objective verification criteria and procedures and procedures developed through expert panels and public consultation, against which emission reduction projects and the credits they generate can be assessed.

The relationship between the VCS and the Voluntary Gold Standard can be likened to that between standard commodity certification and attribution of premium quality characteristics. The VCS, for example, seeks to assure users internationally that Voluntary Carbon Units represent real, additional, permanent and independently-verified emission reductions. Voluntary market offsets meeting the Gold Standard must do this and meet other goals, including the use of technologies that will be part of the long term solution to climate change (renewable and energy efficiency) and make a positive contribution to sustainable development in the host country, as assessed by local stakeholders.

In addition, the Climate, Community and Biodiversity Alliance (CCBA), a group of NGOs, research institutions and businesses, have created a climate, community and biodiversity (CCB) standard to guide investments in land-based projects, including forestry, range management and agriculture. This standard seeks to ensure that such investments achieve broader ecological and social goals as well as reducing GHG emissions.

> **“Being aware of the key standards in the voluntary offset market will help companies take advantage of the flexibility and competitive prices offered, while ensuring offsets are really delivering the reductions in emissions they claim.”**

If widely adopted, these standards should go a long way to providing much needed clarity and credibility in the voluntary carbon market and help the market grow and contribute to mitigating climate change. The voluntary carbon market should be able to achieve this if it:

- Has the Voluntary Carbon Standard as its basic benchmark;
- Provides, through the Voluntary Gold Standard and the CCB, an opportunity for buyers to invest in high-quality projects that contribute to other social and environmental goals.

STEP 6: UNDERTAKE DUE DILIGENCE ON PROJECTS

Before and after purchasing offsets, companies should complete their own project assessment to ensure they are delivering what they say. Consultants and NGOs can help deal with this. This is particularly important if the offsets are not verified under either of the standards outlined above.

STEP 7: TALK TO DIFFERENT SELLERS

Offset prices generally increase as they go up the offset supply value chain. Average prices charged by offset retailers are US $8.04/t$CO_2$e compared with US$6.03 for brokers, US$5.31 for wholesalers and aggregators, and US$3.88 for project developers. Ask sellers to disclose how much of the company's money goes directly to the project and how much covers their administrative costs. Companies may want to consider directly developing offset projects. While this increases a company's exposure to project risk, it should also decrease offset costs and ensure the project characteristics are the ones wanted.

STEP 8: ENSURE OFFSETS ARE RETIRED ON A CREDIBLE GHG REGISTRY

To ensure offsets are not used more than once, companies should require that they are retired on a credible GHG registry. There are several registries that currently offer these services and many offset retailers also operate internal registries that are independently audited.

STEP 9: BE TRANSPARENT ABOUT OFFSET PURCHASES

Any offset strategy should be transparent and publicly communicated. Information to be disclosed should include carbon footprint calculations, emissions reduction activities, the type of offset being used, where offsets have been retired and any uncertainties related to these issues.

STEP 10: REVIEW THE APPROACH TO OFFSETTING ON A REGULAR BASIS

An organisation's carbon footprint and approach to carbon management will no doubt change over time. The carbon market is also likely to change over time. With this in mind, ensure the company's approach to carbon offsetting is reviewed on a regular basis to ensure it is still in line with best practice.

Author

Mark Kenber is an economist who has worked on environmental issues for over a decade in non-governmental organisations and the public and private sectors. Immediately prior to joining The Climate Group, Mark was Senior Policy Officer at WWF's International Climate Change Programme, focusing on carbon market and finance issues and coordinating the Programme's economics-related work. His other experience includes being Director of Planning at Fundacion Natura, Ecuador's largest environmental organisation, acting as climate change advisor to the Ecuadorian government and a wide range of consultancies. Mark is an occasional lecturer at Sussex University's Institute for Development Studies and also serves on the advisory boards of a number of environmental organisations.

Organisation

The Climate Group is an independent, non-profit organisation, founded in 2004 and dedicated to advancing business and government leadership on climate change. We are based in the UK, the USA and Australia and operate internationally. Proactive companies, states and cities around the world are demonstrating that cuts in GHGs required to stop climate change can be achieved while growing the bottom line. Using the work of these leaders as a catalyst, The Climate Group works to accelerate international action on global warming with a new, strong focus on practical solutions. The organisation promotes the development and sharing of expertise on how business and government can lead the way towards a low carbon economy while boosting profitability and competitiveness. Focused on solutions and positive collaboration across the government, business and non-profit sectors, The Climate Group acts independently of special interests and political affiliations.

Enquiries

Mark Kenber, Policy Director
The Climate Group, The Tower Building, 3rd Floor,
York Road, London SE1 7NX UK
Tel: +44 (0)20 7960 2970 | Fax: +44 (0)20 7960 2971
Website: www.theclimategroup.org

Contraction and Convergence:

THE PROPORTIONATE RESPONSE TO CLIMATE CHANGE

© Deb Kushal/Still Pictures

AUBREY MEYER
DIRECTOR,
GLOBAL COMMONS INSTITUTE

The United Nations Framework Convention on Climate Change (UNFCCC) was agreed in 1992 with the objective to halt the rising concentration of greenhouse gas (GHG) in the atmosphere. In 2007, efforts to this end remain insufficient and the danger of 'runaway' rates of global climate change taking hold is increasing. The science-based, global climate policy framework of Contraction and Convergence (C&C) offers an equitable solution to cutting carbon emissions in the hope that global collective efforts to reduce emissions can be successful. Three elements are at the core of the C&C campaign: the constitutional concept of Contraction and Convergence (C&C); the techniques and processes developed to focus the debate on rates of C&C that are relevant; the sustained effort to present C&C as the basis of the proportionate response to climate change.

THE BASIS OF C&C

Technically, the C&C model is a coherent and mathematically-stable framework. It holds the science-policy content together as a unity; science-based on the contraction side of the argument and rights-based or 'constitutional' on the 'political' side of the argument. C&C is in effect a bill of rights; it simply plots a full term event for achieving equal *per capita* emissions rights globally (Convergence) but governed by the overall emissions limit over time that stabilises the atmosphere concentration of GHG at a 'safe' value (Contraction).

> "It becomes possible to go beyond the merely aspirational character of the current debate around the UNFCCC, to communicating the rationale and constitutional calculus of C&C."

The UNFCCC makes C&C generically true, but C&C specifically embraces a calculus built on this truth that strategically focuses the negotiations at the Climate Convention on two necessarily finite, global assumptions:

- A trajectory to a safe and stable atmospheric GHG concentration limit, allowing for a range of calculations of the global emissions contraction limit to carbon consumption consistent with that.
- The calculation of equal rights to the global total of emissions permits to the global total of people consuming within that limit, again allowing for different rates of convergence and even a population base-year to be considered. This is in preference to the irresolvable complexity of assuming any inequality of rights.

With this calculus, C&C captures the goal focus of the UNFCCC process in a structure of reconciliation. It is a universal first order numeraire. From this it becomes possible to go beyond the merely aspirational character of the current debate around the UNFCCC, to communicating the rationale and constitutional calculus of C&C.

THE LONG TERM PAST

Figure 1 shows data from ice cores for half a million years before industrialisation. Throughout this period, with natural sinks for CO_2, such as the oceans and the forests in balance with the natural sources, the level of atmospheric CO_2 concentration varied between 180 and 280 parts per million by volume (ppmv) averaging at 230 ppmv.

Since 1800 with the onset of industrialisation and fossil fuel burning, human emissions have caused the concentration of CO_2 to increase by over 40 per cent to 380 ppmv. The rise in ppmv CO_2 is higher and faster than anywhere in the historical record. This rise is because CO_2 emissions from human sources, particularly CO_2 from fossil fuel burning, are going to the atmosphere and accumulating. Furthermore, for the past 200 years, on average 50 per cent of any year's human emissions has remained in the atmosphere while the remaining 50 per cent has returned to the natural sinks.

> "Instead of 100 years, we now realise that to reduce human CO_2 emissions and other GHGs in the atmosphere to zero globally, we have only the next 50 years."

A slowly increasing fraction of these emissions in the atmosphere remain there, accelerating the rise in concentrations even more. Column one in Figure 2 (see overleaf) demonstrates that the average retention over the past decade has increased from 50 per cent to 60 per cent. This recognises that the capacity of the natural sinks for CO_2 capture is now gradually declining. If this continues unchecked as the graphics suggest, the rise in the concentration of atmosphere GHG will accelerate towards the level at which dangerous rates of rise translate to a climate change crisis that becomes unavoidable. To be UNFCCC-compliant, we need to enact C&C now to prevent the chaos that is otherwise inevitable.

THE SHORT TERM PAST AND FULL TERM FUTURE LIMITS

The UNFCCC objective is to avoid dangerous rates of climate change by stabilising concentrations and we are all both circumstantially and legally bound by this. Compliance is governed by the need for a finite answer to the questions: 'what is a safe GHG concentration value for the atmosphere?' and 'what is the scale of the full term emissions contraction event required to achieve it'?

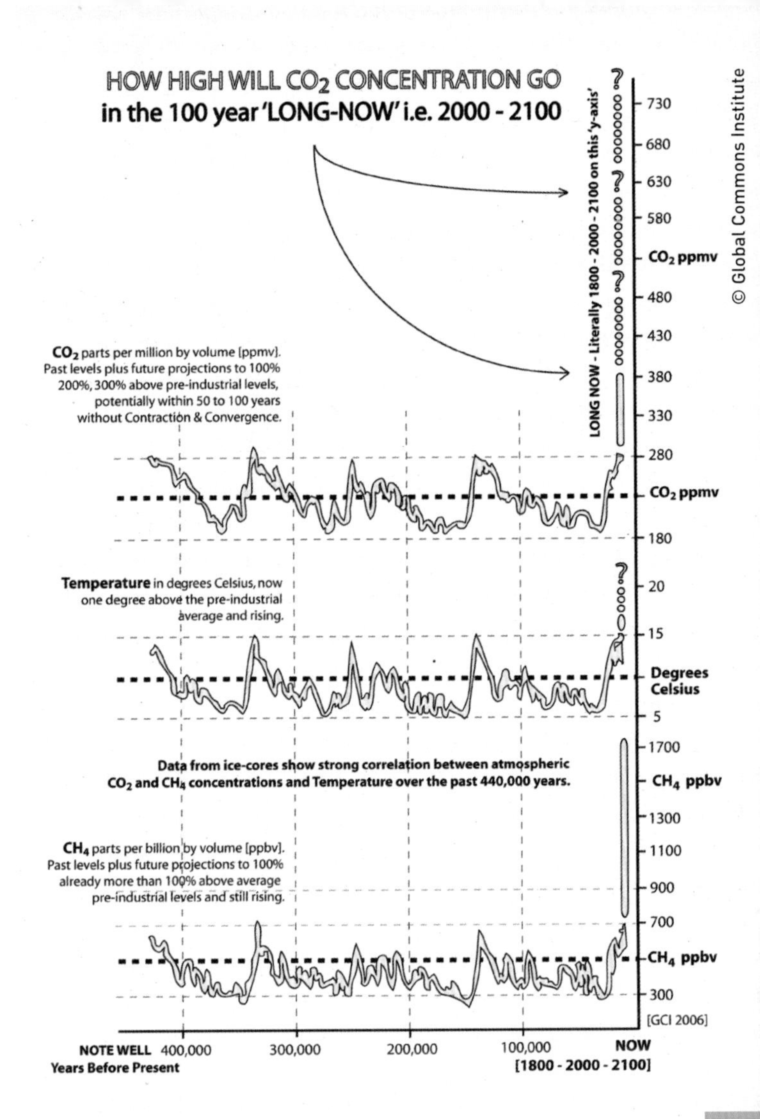

Figure 1: Data from ice cores 500,000 years ago to present day and beyond.

Without answers, traditional evaluation of the economics of abatement and the social consequences is not possible. Because of weakening sinks, analysis now shows that to stabilise GHG concentration in the atmosphere below the level that prevents dangerous rates of climate change taking hold, requires a rate of overall emissions control that is faster than was previously assessed. Instead of 100 years, we now realise that to reduce human CO_2 emissions and other GHGs in the atmosphere to zero globally, we have only the next 50 years [IPCC AR4 and Hadley Centre, 2007].

As activities under the Kyoto Protocol show, unless we are visibly organising globally by a shared commitment not to exceed that safe concentration number, the probability increases that our collective efforts to avoid dangerous rates of climate change will be too little too late.

Already under Kyoto, the slight gain of CO_2 emissions avoided has been more than negated by more carbon accumulating in the atmosphere at an accelerating rate as the result of changes in the climate system as a whole. Consequently, a global arrangement for emissions control in future that is sufficient in the light of this is *sine qua non* for success. As the original authors of the UNFCCC understood at the outset, embracing this primary question of the sufficient, and indeed the proportionate response, is fundamental to the whole global engagement.

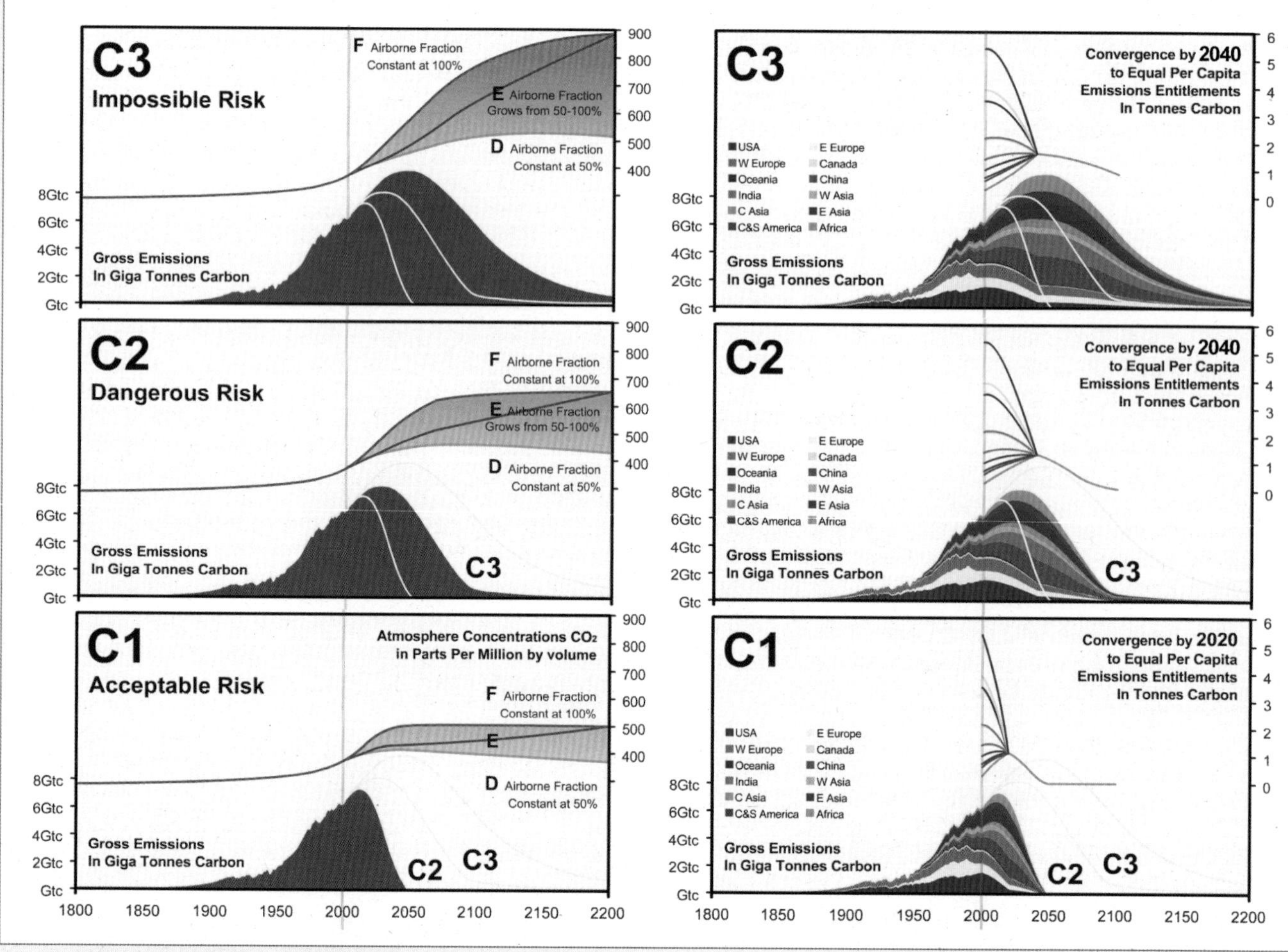

Figure 2: Charting the UNFCCC Objective & Principles, the Development Benefits of Growth versus the growth of Climate Change Related Damage Costs. (http://www.gci.org.uk/images/Proportionate_Response.pdf)

Columns one and two address the objective and principles of the UNFCCC. Columns three and four compare the development benefit of growth with the growth of climate damage and costs. The left hand side of each graph shows:

- Expanding fossil fuel emissions of CO_2, measured in billions of tonnes of carbon between 1800 – 2000.
- Rising concentration of atmospheric CO_2 as parts per million by volume (ppmv) between 1800 – 2000.

The key questions for integration are in four columns:

Column 1: Contraction and Concentration: what is a safe level of concentrations and, in the light of sink failure, how rapid must contraction be to avoid GHG concentration going too high in future?

Column 2: Contraction and Convergence: what is the internationally equitable agreement necessary to ensure this level is not exceeded?

Column 3: Contraction and conversion: what is the rate at which we must convert the economy away from fossil fuel dependency?

Column 4: Damage costs and insecurity: what is the environmental and economic damages trend associated with this analysis?

Each Row has a different level of Risk projected across the four columns:

- **C1 (bottom row) Acceptable risk**: global GHG emissions contraction complete by 2050 so concentrations end up around 400/450 ppmv with damages potentially still under control.
- **C2 (middle row) Dangerous risk**: global GHG emissions contraction complete by 2100 so concentrations keep going up through 550/750 ppmv with the illusion of progress maintained, while damages are going out of control.
- **C3 (top row) Impossible risk**: global GHG emissions contraction complete by 2200 so concentrations keep going up through 550/950 ppmv while the illusion of progress is being destroyed, damages costs are destroying the benefits of growth very quickly and all efforts at mitigating emissions become futile.

In each graph, different futures are projected on the right-hand side as scenarios or rates of change that are linked to the objective of the UNFCCC where three levels of risk for stabilising the rising concentration of CO_2 are understood in the light of the rising fraction of emissions that stays airborne.

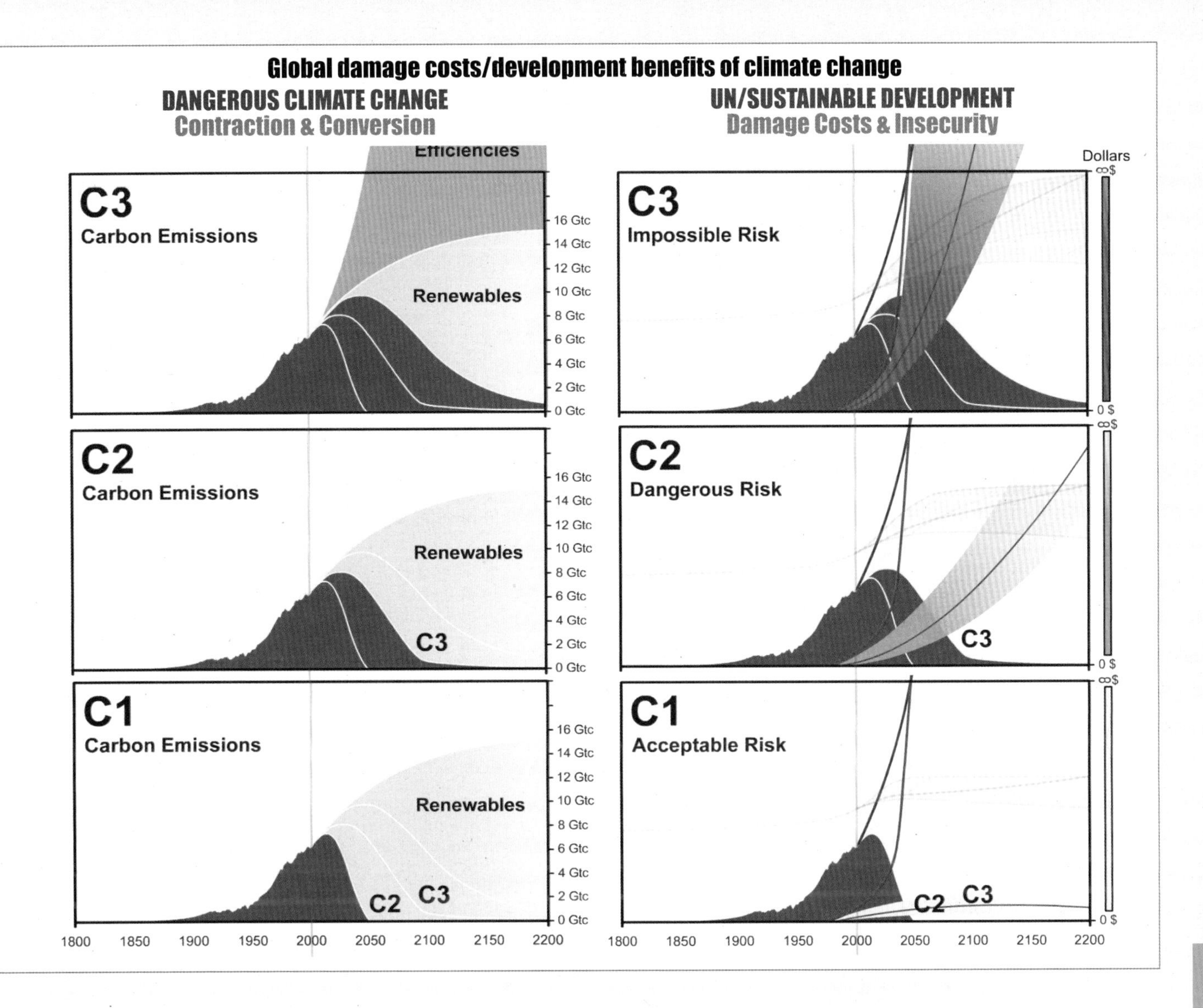

DAMAGES

We are still locked into causing global climate change much faster than we are mitigating it. Treating climate change as a global emergency is now long overdue and responding proportionately is vital. Unless the risk analysis is focused by this understanding, our best efforts will be in vain.

According to the reinsurers, the weather-related damages trend is growing at twice the rate of the global economy, see Figure 2, column four. To prevent this damage trend from running out of control, emissions need to contract to zero globally by 2050 if it is to be fast enough to stabilise atmosphere GHG concentrations at a level that prevents change accelerating uncontrollably. This is corroborated by the latest coupled climate modelling results from the UK Government's Hadley Centre, published in the IPCC Fourth Assessment. While the notion of global emissions control is certainly heroic, the only vector of the problem over which we can still posit direct control, is our GHG emissions and thereby the level to which GHG concentrations will rise in the future.

With this integrated approach we can more clearly visualise the challenge within a finite calculus of collective responsibility, and so keep focused on the imperative of solving the problem faster than we are creating it. Communicating and implementing this remains the primary challenge.

A FRAMEWORK-BASED MARKET

With the C&C operational framework, we can compare how much must be achieved globally to avoid dangerous climate change, with the widening margins of error in which we are becoming trapped.

> "Treating climate change as a global emergency is now long overdue and responding proportionately is vital."

There are more complicated 'alternatives to' and 'derivatives from' C&C. While defending the evolutionary nature of the politics, these have also attempted to be non-chaotic. They include for example the Kyoto Protocol, which seeks to interpose a partial and random market-based framework in support of the Convention. But such an evolutionary response to its objective and principles is guesswork by definition, and there is no evidence

Satellite image of Hurricane Katrina, which has cost the south-eastern US billions of dollars. Damages from extreme weather events are increasing with climate change.

supporting claims that merely incremental activity at the margins will collectively generate a sufficient response fast enough to be effective. Until recently, the unguided inertia of evolutionary process under the Kyoto Protocol has been projected as *ne plus ultra*.

“C&C overcomes the stand-off where a one sided agreement is not an agreement and where half an argument is not, nor will ever become, a whole solution. It recognises that separate development is not sustainable development.”

The fact is that this is a lottery where everybody loses. This approach has obscured the global objective of safe and stable concentrations and the obviously urgent need for a trajectory to this objective by design.
C&C starts with an integral response to the Convention's objective and allowing a full term framework-based market to result, where:

- Equity as collateral is the 100 per cent entirety of the emissions contraction event necessary for concentration stability.
- The social equity as the equal per person claim on the same 100 per cent throughout that event but softened by convergence.
- The commercial equity is the shares pre-distributed this way sum to the same 100 per cent and are tradable so as to accelerate the positive sum game for the emissions-free economy that must emerge if we are to prosper in the future.

In a nutshell, this integration puts rational principle ahead of stochastic practice in order that the former guides the latter. In practice this arrangement is flexible and will create a lucrative framework-based market for the zero emissions industries within a future structure that corrects and compensates for the asymmetric consumption patterns of the past while saving us all from dangerous rates of climate change.
In this context C&C overcomes the stand-off where a one sided agreement is not an agreement and where half an argument is not, nor will ever become, a whole solution. It recognises that separate development is not sustainable development.
In September 2007, the German Government recognised this when mediating between supporters and opponents of the Kyoto Protocol with C&C as the basis of the post-Kyoto agreement. Their urgent call for a whole and proportionate solution should be supported vigorously.

Author

Aubrey Meyer is the Director of the Global Commons Institute [GCI] responsible for the formulation of Contraction and Convergence [C&C] framework. For his work he has won several prestigious awards including the Andrew Lees Memorial Award, 1998, the Schumacher Award in 2000, the Findhorn Fellowship in 2004, a City of London Lifetime's Achievement award in 2005 and was made an Honorary Fellow of the Royal Institute of British Architects in 2007. In a recent edition of the *New Statesman*, he was listed as one of the 10 people in the world most likely to affect climate change.

Organisation

The Global Commons Institute [GCI] is an independent body based in the UK, concerned with the protection of the global commons. GCI was founded after the UN's Second World Climate Conference in 1990 and since then has contributed to the work of the United Nations Framework Convention of Climate Change and the Intergovernmental Panel on Climate Change.

Enquiries

E-mail: aubrey.meyer@btinternet.com
Website: www.gci.org.uk

© apg/iStockphoto

Carbon rationing: the only realistic strategy

MAYER HILLMAN
SENIOR FELLOW EMERITUS,
POLICY STUDIES INSTITUTE

The considerable reduction of carbon emissions needed to prevent climate catastrophe cannot be achieved on a voluntary basis. Instead, the urgent transition to very low energy lifestyles requires government to set mandatory targets based on per capita carbon rationing. Such a strategy will lead to a far greater and faster take-up of energy saving practices and energy efficient measures than would otherwise occur and an essential process that industry will be increasingly motivated to facilitate.

INTRODUCTION

The time is over for denial that catastrophe is inevitable without drastic reduction in the world's use of fossil fuels and in deforestation. The problem largely stems from burning the reserves of the sun's energy accumulated over millions of years to raise material standards during the past 200 years, dramatically so in affluent countries.

Atmospheric concentrations of greenhouse gases (GHGs) are dangerously close to exceeding a safe 'ceiling'. The consequences are already apparent in the recent acceleration in the melting of the Arctic and Antarctic ice shelves, growing desertification in Africa and China, flooding in Bangladesh and heat waves in Australia. There is the distinct prospect of sea level rises resulting in a shrinking habitable land mass on which a burgeoning world population will have to live. The latest IPCC report includes the dire calculation that a drastic curtailment to zero carbon emissions must be achieved: not the widely accepted 60 per cent reduction on 1990 levels by 2050. The fact that the calculation did not take account of the implications of more alarming recent data on feedback mechanisms, such as the effects of methane release in tundra regions of the Northern hemisphere due to increasing temperatures, should add to our concern.

SLEEPWALKING

The focus of efforts at all levels of society, including government, to limit damage from climate change appears to be very narrow. It is aimed at the actions individuals and industry can take in switching to lower carbon lifestyles and the barriers that are seen to be in the way of adoption. Against a background of considerable opportunities for reducing carbon emissions, most attention is in promoting energy efficiency, exploiting energy renewable sources of supply for generating electricity, and identifying the most effective policies that government can adopt to encourage these actions. Implicit in this approach is the view that, in time, they will lead to sufficient emissions reduction and that the public, industry and commerce can be motivated to deliver this voluntarily, encouraged by better information, offers of grants and the setting by government of higher standards.

With a personal carbon allowance, those who lead a more carbon-light life (see image) will not only spend less on fuel but will be able to sell their surplus carbon units.

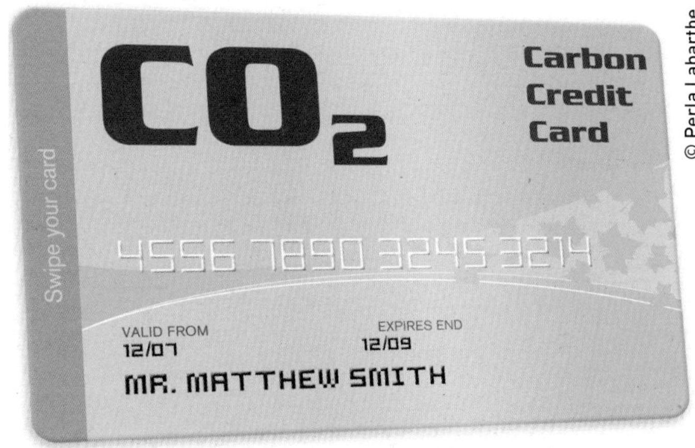

The public has been led to believe that it has a right to ever rising improvements in its fossil fuel-dependent material standards and life choices. Statements of all main political parties give a strong impression that such a future is possible without the need for major behavioural changes. There is a failure to alert the public to the awesome prospects for life on Earth later this century and beyond if businesses dependent on high use of fossil fuels continue on a path of growth.

> "There is a failure to alert the public to the awesome prospects for life on Earth later this century and beyond if businesses dependent on high use of fossil fuels continue on a path of growth."

FALLACIOUS ASSUMPTIONS

Behind this lack of recognition of the need for drastic action lies a judgement that the primary way of improving the public's welfare and quality of life is through the medium of economic growth; its slow down cannot under any circumstances be contemplated. Indeed, it is asserted that growth can be reconciled with protection of the global environment from the ravages of climate change and implied that it can be maintained in perpetuity.

Allied to this judgement are many beliefs which have wide support such as:

- 'Green' taxation, combined with market forces and regulation to limit any adverse effects, can allow the economy to grow without leading to accelerating diseconomies
- People have an inalienable right to engage in environmentally damaging activities if there are no acceptable alternative means of doing so
- Science and technology can be relied on to make a major contribution in reversing the process of climate change and major behavioural changes will be unnecessary
- Offsetting carbon emissions by paying for their reduction in developing countries, thereby enabling people in affluent parts of the world to maintain their energy intensive lifestyles, should be interpreted both as a welcome contribution to their overall reduction and as morally defensible
- Modest reductions in GHG emissions are indicative of a process that should be supported as it will eventually lead to sufficient reductions
- There is sufficient time left for this process to prove effective and the costs will be found to achieve the transfer to a sufficiently low carbon economy.

QUIS CUSTODIET?

Government is rightly seen as having both the prime responsibility and the authority to reach decisions to protect the public interest in light of how grave it sees any situation to be. Given the public's increasing addiction to energy intensive lifestyles, and given little indication of its preparedness to give them up, it is evident that government must be far more effective than it has demonstrated to date. It is unrealistic to expect many individuals or communities to act unilaterally when others are not doing so. Nor is it realistic to expect sufficient success to come in the wake of businesses 'going it alone' in adopting green practices, even though more and more of them are doing so.

At the heart of the problem is the need, yet to be fulfilled, to educate the public about the gravity of the situation. Only in this way can there be the realistic prospect of the draconian measures that must be taken not being dismissed as unacceptable *diktat* in a democracy.

THE NEED FOR A COMPREHENSIVE STRATEGY

Climate change is a global problem requiring a global solution with no country immune from its depredations. Any effective strategy must be based on reaching an internationally binding agreement. Only in this way can the world community be adequately engaged in sharing the atmospheric sink for the safe absorption of GHG emissions. The framework for such an agreement, Contraction & Convergence (C&C), first put forward by the Global Commons Institute 12 years ago, is explained in an article by Aubrey Meyer, p122. It is founded on the fundamental principles of justice and equity. It is obvious that the issue of the fair distribution of a basic component of everyday life to which everyone has an equal claim cannot be side-stepped.

TRADABLE PERSONAL CARBON ALLOWANCE

As people cannot be relied upon to meet their ecological responsibilities, governments must require them to do so and therefore the national manifestation of C&C has to be personal carbon rationing. Based on the equity principle, everyone would be given the same annual Personal Carbon Allowance (PCA). They would then be strongly motivated to reduce activity dependent on fossil fuel use, improve energy efficiency and use renewable energy in order to live more easily within their carbon allowance. The allowance will have to decrease steadily year on year in line with the negotiated international reductions agreed on the basis of the most up to date scientific knowledge of the safe level of concentrations of GHGs. By giving due warning of the annual reduction in the future allowance, people will be able to alter their homes, transport arrangements and general lifestyles at the least cost to themselves and in the way they prefer.

> "The PCA will act as a parallel currency to real money with a key feature being the buying and selling of its units."

PROMOTING CARBON THRIFTY BEHAVIOUR

The PCA will act as a parallel currency to real money with a key feature being the buying and selling of its units. In this way, it will create an ecologically-virtuous circle. Those who lead less energy intensive lives and who invest in energy efficiency and renewable energy are unlikely to use all their allowance. Thus the energy thrifty will not only be spending less on fuel but will add to their income by being able to sell their surplus units. Those who maintain high energy activity patterns will have to buy these units. But the cost of doing so will not be determined by economists attempting the near impossible task of attaching a monetary value to the effects of adding to the concentration of carbon dioxide in the atmosphere for over 100 years, such as resettling ecological refugees fleeing their homes owing to climate change. Instead the value will simply be a function of the availability of the surplus set against the demand for it on the open market, inevitably rising in line with the annual reduction in the allowance. In effect, this 'conserver gains principle' will complement the conventional 'polluter pays principle' and act as a driver towards minimising the awesome impact of climate change far more effectively than by attempting to encourage individuals to follow green practices.

THE BUSINESS COMMUNITY

What are the implications for the business community of the adoption of the international framework of C&C and the application of PCAs? In the light of the previous comments, that is easy to answer. Business will logically react to the changes in public demand for the goods and services it provides. The more it is able to anticipate what these changes will be, the more successful it will be.

As the annual allowance is reduced year on year down to the level that must be reached urgently, so will the demand fall away for such energy intensive practices and activities as using energy inefficient appliances, living and working in poorly insulated buildings, taking up jobs and shopping in locations which entail much motorised travel. By marked contrast, a sharply rising demand for heavily insulated homes and the most efficient equipment available, for local patterns of activity and holidays that can be reached without surrendering an unrealistic number of units of the annual PCA for transport, will be the logical outcome. The business community will be able to foresee and cater for these changes following the take up of this inevitable political strategy.

CONCLUSIONS

We all have a crucial role in tackling climate change. But it is wishful thinking to believe that this will happen at the level necessary unless everyone is subject to a mandatory requirement to do so. In the UK, for example, this necessary level translates to a reduction in CO_2 from the current annual average of about 10 tonnes to just one tonne. No one must be allowed to continue to pass the buck between individuals, industry and government in an attempt to evade responsibility for contributing their fair share of a radical shift to a drastic reduction of carbon emissions. Government is the only body that can achieve this, by taking the immediate steps to reach an international agreement on the massive switch to very low carbon lifestyles and to introduce carbon rationing. Responding to climate change is ultimately a moral choice. Given the urgency of the situation, the implications of failure to limit our emissions to a safe level are dire. We can no longer proceed as if we have a right to turn a blind eye to the damage we are causing. The distinct likelihood if we do, is to bequeath a dying planet to the generations succeeding us on the planet.

Author

Dr Mayer Hillman, a social scientist, is Senior Fellow Emeritus of Policy Studies Institute in London. His studies since 1970 have been concerned with transport, urban planning, energy conservation, health promotion, road safety and environment policies, notably those relating to climate change. He is the author or co-author of more than 40 books on these subjects. Many of the themes in this article are drawn, for example, from the contents of his books: *How we can save the planet*, published by Penguin and *The suicidal planet: how to prevent global climate catastrophe*, published by Thomas Dunne Books.

Organisation

Policy Studies Institute is run on a not-for-profit basis and carries out studies on environmental, employment and social issues. It uses the most advanced methods and professional expertise to aid the formulation of policy. It was formed in 1978 through the merger of Political and Economic Planning (established in 1931) and the Centre for Studies in Social Policy (established in 1972) and in 1998 became an independent subsidiary of the University of Westminster.

Enquiries

Dr Mayer Hillman
Policy Studies Institute
50 Hanson Street
London, W1W 6UP
Tel: 020 7911 7500 | Fax: 020 7911 7501
E-mail: mayer.hillman@blueyonder.co.uk
website@psi.org.uk
Website: www.psi.org.uk

PV plant of 8.4 MW installed in 2007 in Saarland, Germany by City Solar AG. In the background, the facilities of a coal mine.

Sustainable energy:

A NEW PARTNERSHIP WITH THE SUN

WOLFGANG PALZ
CHAIRMAN,
WORLD COUNCIL RENEWABLE ENERGIES

Energy is a basic need, yet the world relies primarily on polluting and exhaustible fossil and atomic energies. Consequently, we are faced with three essential geopolitical challenges: local security of energy supply; global depletion of conventional energy resources; and climate change. A solution for these problems is the speedy development of renewable energies. Renewable energies have been a shining star in some national economies and global investments are now exceeding US$50 billion/year with the creation of hundreds of thousands of new jobs. An overview of the current positioning of renewable energies is given here.

THE CASE FOR RENEWABLE ENERGY PROMOTION

The potential for renewable energies (RE) to provide the world with clean, inexhaustible, and decentralised energy that is accessible to all is now generally accepted. The UN/IPCC report of early 2007 called for immediate action to alleviate climate change before it becomes irreversible. Renewable energies are perfectly placed to help in this action as they emit no greenhouse gases or atomic waste. And with most experts agreeing that fossil fuel resources, with the exception of coal, will be used by the middle of this century, REs offer an inexhaustible replacement.

The challenge is tremendous, requiring in not more than 50 years the complete restructuring of the current energy system that stands for a global business of trillions of dollars. And in the case of global change, we are running out of time. The limitation of fossil resources is already affecting today's price markets. The oil price that leads the market price of all fossil energies, for example, was over US$90 in October 2007 – many times more than just a few years ago. For atomic energy the trend is similar with the price of fissile Uranium increasing in the past six years by a factor of 10. Prices for RE, however, are decreasing thanks to the economy of scale facilitated by a globalisation of the energy markets.

Many renewable energies are already competitive in today's markets, eg solar water heaters in China, biofuels for transport in Brazil, passive solar gains in buildings and wind energy in Germany. In a few years, predictions are that market based competition will swing in favour of REs across the board and those political incentives, so important now, can be phased out.

Energy security is another area where REs come out on top. The sun shines everywhere and wind resources are widespread, whereas most of the world's oil and gas reserves are in the ground of the Gulf States.

THE RE SHARE IN TODAY'S ENERGY MARKETS

Power generation

In 2007, 20 per cent of all investments in the global power sector went to RE power. Large hydro and the 'new' REs (eg in order of market share: wind power, small hydro, bio power, PV) together slightly exceed the overall capacity of all nuclear plants. Renewable energies also develop much faster than atomic power plants. In 2006, six atomic plants were taken from the grid and no new ones added, whereas 15 gigawatts (GW) of new wind capacity were installed worldwide, a €20 billion business.

Wind power

Global wind capacity reached 74.3 GW at the end of 2006, a tenfold increase since 1996. Two-thirds of the world's wind capacity (48.6 GW) was installed in Europe, 28 per cent in Germany. During 2006 15 GW of new wind capacity was installed and, with 2.4 GW, the US took over the leadership as the largest wind market for new installations from Germany. The total German wind capacity in July 2007, however, covered almost seven per cent of the country's net electricity consumption.

Today's wind technology is highly mature with extraordinary progress being made in recent years. Market prices vary in the range of €1,000/kW and are unlikely to come down in the future. On top of the turbine price, there is around 30 per cent for installation costs and 4.8 per cent for operations and management. The optimal economic size for wind turbines corresponds to between 1.5 to 3 MW for physical reasons, with the majority of machines sold on the market falling in this range. Rotor diameters will be between 60 and 90 metres with tower heights ranging from 60 metres to the world record of 200 metres.

Virtually all wind turbines are currently deployed on land with offshore wind farms still in their infancy. By the end of 2007, globally installed wind power capacity will reach between 90 and 100 GW. Order books of the wind industry are currently full with virtually all wind turbines sold out. By 2011, observers expect a world market of over 200 GW.

Denmark was a pioneer of modern wind technology and, although, for political reasons, the Danish market has almost disappeared, the Danish company Vestas still leads the World's industry for wind.

© Peben Maegaard, Folkecenter, DK

Small hydro (units of less than 10 MW)

The world leader for small hydro is China with 38 GW in operation. (In 2006, global operations for small hydro stood at 60 GW.) In the European Union the installed capacity is 11.7 GW producing 'normally' 44 terawatt hours (TWh). Italy has the European leadership in this field.

Compared to the wind sector, market growth rates in the EU are small with capacity by 2015 possibly reaching 13 GW. China foresees at least 100 GW production by 2020.

Bio power

Bio power today represents over 40 GW of global capacity. It comprises different sectors, such as co-firing of wood in coal plants or dedicated bio power plants, small cogeneration plants employing vegetable oils and biogas plants feeding into power generators with Germany the world market leader in both vegetable oil cogeneration and biogas. In early 2007, 1,660 plant oil micro CHPs were in operation in Germany along with 3,260 biogas units.

Finland, followed by Sweden and Germany, is the lead country for solid biomass (logs, wood chips, pellets, sawdust, straw and crop residues). In 2005, the EU generated 44 TWh of electricity from solid biomass with three-quarters of this energy generated in cogeneration CHP plants. This sector shows an impressive growth rate of 16 per cent between 2004 and 2005. To these 44 TWh, a further 10.7 TWh can be added from the renewable part of municipal solid waste.

Photovoltaics

Solar photovoltaics (PV) are semiconductor plates that convert radiation directly into electricity without any moving parts. Market PV expansion has been so explosive that precise figures are not available, only approximations. Currently, seven GW of PV are in operation, a seven fold increase with respect to 2004. Germany has the largest share of that capacity (a third), followed by Japan and California.

In 2006 the global market accounted for around 2.5 GW with half of that installed in the EU. The global PV business exceeds €10 billion annually. Approximately 300,000 new jobs have been created in the past few years in this new sector with the industry developed by billions of Euro in investments.

Today's global PV market is booming with modules sold out currently in 2007. A bottleneck is the availability of the silicon feedstock, the same that is used in semiconductor chips. Silicon is currently being sold at 10 times its production cost, resulting in alternative thin-film PV cells aggressively entering the world market. By 2010 when the quantity of silicon feedstock on the market will have quadrupled, silicon solar cells may be less embattled than now.

PV modules currently cost a little less than 3€/Watt on today's world markets. With the technologies available to us now, costs of 1€/Watt are around the corner. PV will then be able to achieve market competitiveness for electricity production from building integrated PV generators. Predictions see hundred of GW of PV installed around the globe in the long term.

Combined with battery storage, PV is ideally suited for independent power generation, such as in remote villages for lighting purposes and communication. Fifteen per cent of world production goes to these kind of applications in developing countries.

© Qin Haiyan, cgc, Beijing

PV plant for village electrification in Tibet; Chinese Township Electrification Program.

Concentrator solar power plants

Most solar power plants employ parabolic mirror troughs for solar concentration; other types are solar tower plants. Recently a number of 50MW solar plants have been built, with a few of these being built in areas with bright sunlight, such as the deserts near Las Vegas, Andalusia, and the Algerian Sahara.

Geothermal power

In 2007 9.7 GW of global capacity were in operation, an increase of 10 per cent over 2005. In the EU there is 863 MW capacity which generates 5.7 TWh. Almost all of this is in Italy.

Renewable biofuels for transport

The world relies on liquid biofuels for three per cent of its petrol consumption. More than three-quarters of global bio-alcohol for transport is produced and used in Brazil and the US. In Brazil, its share in petrol consumption is 44 per cent where it is used pure or blended. The United States currently produces 16 million tonnes of ethanol per year, slightly more than Brazil. The US development is breathtaking, with production tripled over the last three years and predicted to double in a year's time. The US invested US$3 billion in 2005 and created 150,000 new jobs in the sector. Ethanol in the US is corn based and subsidised; Brazil employs sugar cane as a feedstock that is more economic and needs no subsidy. There are disagreements within the US agricultural sector on an over-reliance on alcohol with important US players seeing a risk to food and feed supply and inconsistencies with market demand.

The EU has fixed a target of 5.75 per cent biofuels contribution in the transport market by 2010 and 10 per cent by 2020. For years the Common Agricultural Policy of the EU was plagued by overproduction of food and set aside regimes had to be introduced. At that time, solar advocates had proposed farming energy crops as an additional activity without success. In 2007 things have changed dramatically. Prices for cereals and sugar have soared on the global markets and farmers see biofuels as a new opportunity.

At current world market prices, cereals and sugar beet have become expensive feedstocks for bioethanol production. Now, second generation bioethanol produced from wood via gasification and Fischer-Tropsch synthesis is being developed, but will be expensive. Another hope will be the development of alternative crops, such as the highly productive sweet sorghum. The other biofuels market segments are the vegetable oils and the derived biodiesel. These are grown with highest productivity as palm oil in South-east Asia leading to rising concern that forests are being cleared to give way for new plantations.

Solar heat collectors

In 2007 the global park of solar heat collectors amounted to 200 million m². Half of this surface area is installed in China with 10 per cent in the European Union and five per cent in Germany. In China, 200 million people benefit from solar heat; it will be 300 million by 2010. Europe expects to double its surface to 40 million m² by that time. Unlike in Europe, China's solar hot water systems are not subsidised.

Geothermal heat

Under the leadership of Hungary, Italy and France, the EU had a thermal capacity of 2.5 GW in 2006. Market growth rates are low. At the same time, Europe counted 600,000 low enthalpy geothermal heat pumps with an overall capacity of 7.3 GW (thermal). Unlike direct geothermal heating systems, the market for heat pump based geothermal grew an impressive 20 per cent between 2005-2006.

Bioenergy

In 2005, Europe consumed 58.7 million tonnes of oil equivalent (TOE) of bioenergy in the form of logs, pellets, chips, straw, residues and black liquor, a 5.6 per cent growth over 2004. France, Sweden, and Germany are the lead European countries. The market is expected to grow to 100 million TOE by 2010.

Passive solar buildings

Low energy houses and passive solar architecture, coupled with solar collector integration, has one of the highest potentials for renewable energy utilisation. In October 2007, California energy regulators adopted the target that all homes built after 2020 must produce at least as much energy as they consume through advanced

A bioethanol production plant in Brazil.

Source: EUBIA European Biomass Industry Ass, Brussels

Typical Chinese solar hot water system on a domestic building.

Solar building Burri in Zurich, Switzerland.

insulation and solar power, one of the many examples currently being promoted around the globe.
Beyond individual buildings, whole communities and cities must become sustainable in terms of energy consumption by deploying renewable energy. The new 'Masdar' city currently being built by Lord Norman Foster in Abu Dhabi is such an example.

OUTLOOK

Renewable energies have benefited from significant market development over the past few years, which is encouraging for the future. New RE mass markets will rapidly accelerate their competitiveness with conventional energies. But the challenges ahead are tremendous, with China aggressively developing the new coal plants for power generation, making it the world's largest CO_2 emitting country in two years. There is also every chance that the world will endeavour to consume all remaining oil and natural gas resources to the last droplet.
The European Union has adopted the ambitious goal of tripling its RE share over the next 13 years to 20 per cent. As far as the US is concerned, new initiatives for more RE implementation are coming forward. Beyond the current administration's 20 per cent biofuel share target in transport by 2017, the US Senate has unanimously adopted a resolution establishing a 25 per cent RE share by 2025 as a national energy goal.
What is needed most now is raising awareness about RE facts and figures and their unique role in controlling climate change. The alternative is simply a world in jeopardy that we leave to the next generation.

Author

Wolfgang Palz holds a PhD in physics and was for 20 years (1977-97) an EU official in charge of the European Union's research and development of all REs. In 1997 he developed the target figures for RE implementation in Europe by 2010 and is currently Chairman of the World Council for Renewable Energy (WCRE). He is based in Paris. Wolfgang Palz has won a number of awards including the European Awards for merits in Photovoltaics, Wind power and Biomass and an Order of Merit of the Federal Republic of Germany.

Organisation

The WCRE is a not for profit, non-governmental international organisation with the goal to promote renewable energies. Its secretariat is in Bonn.

Enquiries

Wolfgang Palz
E-mail: w@palz.be
info@wcre.org
Website: www.wcre.org

The wind is always blowing somewhere in the world

In recent years, the debate about global warming and its consequences has reached new heights. What started as a grassroots movement has become a fully integrated issue in the global public debate and the hunt for sustainable solutions has intensified. One of these solutions is wind power.

Today there is broad consensus in the scientific community that climate change is related to the increasing volume of CO_2 emissions generated by human activities. Today, we need to think along new lines if we are to accommodate the constantly rising demand for energy.

MODERN ENERGY

In just a few years, wind power has developed from being an alternative form of energy to its status as a large scale reliable source of energy integrated in the energy mix along with oil and gas. Multinational energy companies are now seriously making their entrance into the wind energy business.

Wind is an inexhaustible, free and clean source of energy, and a fully operational wind power plant can be established in as little as 18 months. There is every reason, therefore, to refer to wind power as Modern Energy.

There are as many as eight persuasive arguments to see wind power as a great contributor in curbing CO_2 emissions:

1 Wind power is increasingly competitive on price.

2 Wind power generates no emissions of CO_2, SO_x or NO_x.

3 Wind is free.

4 Wind is an inexhaustible source of energy.

5 Wind is not limited by geographical borders.

6 Wind power generates no waste.

7 Wind power plants can be established quickly.

8 Wind power does not consume water for energy production.

FIGURES THAT SPEAK FOR THEMSELVES

When conventional sources of energy are to be supplemented or replaced in a country's energy mix, it is necessary to take additional factors, such as cost, into consideration. And in this area, wind power has reached a level of technological maturity where it can compete on price with conventional forms of energy, if the comparison is made on equal terms.

Moreover, it is worth remembering that the 'fuel' for a wind turbine is free, and not subject to the uncertainty of future energy prices. This means that wind power hits the supply curve for a mix of energy technologies – the curve that shows the relationship between volume and price – at the bottom. A large supply of wind would therefore push the entire curve to the right, resulting in falling market prices. Wind power provides financial savings, even though the directly related costs may be slightly higher than for the conventional technologies.

MORE THAN A DROP IN THE OCEAN

The climate challenge does not simply involve reducing CO_2 emissions, but doing so in a sustainable manner, so that the problem is not just palmed off onto other areas. In this context, it is particularly relevant to focus on global supplies of clean drinking water, which, within a relatively short space of time, may become very scarce with serious consequences. This focus on water is very important, because a recent report by the Intergovernmental Panel on Climate Change (IPCC) in 2007 concluded that "global warming will hit through water", and that some of the major challenges to adaptation are related to water resources development and management.

In 2006, an analysis performed by the consultancy company Emerging Energy Research on behalf of Vestas Wind Systems revealed that even without including the costs of CO_2 emissions, land based wind power is only marginally more expensive than electricity from a new coal-fired power station. Its cost is equal to that of the cost of electricity from a new natural gas-fired plant. If a realistic quota price of €30 per tonne of emitted CO_2 is included in the calculations, land based wind power, with an average price of €0.050/kWh is more competitively priced than electricity from coal (€0.065/kWh), Intensified gasification combined cycle (IGCC) (€0.058/kWh) and natural gas (€0.055/kWh).

> **"It is worth remembering that the 'fuel' for a wind turbine is free, and not subject to the uncertainty of future energy prices."**

> **"Wind power has developed from being an alternative form of energy to its status as a large scale reliable source of energy integrated in the energy mix along with oil and gas."**

Both nuclear power and conventional energy production (oil, coal and natural gas) consume gigantic volumes of clean drinking water. A report from The US Geological Survey has established that in 2000, electricity generation in the United States was the second largest consumption of clean water, exceeded only by agriculture. In fact, every single day, almost 515 billion litres of clean water were pumped up for thermoelectric cooling at American power plants.

So let us make it absolutely clear: wind turbines do not use even one drop of water when generating electricity.

NO WASTE FOR FUTURE GENERATIONS

Another significant consideration is the problem of waste. In this area, too, wind power is a natural choice for the simple reason that wind turbines generate no waste at all when producing electricity. In other words, there are no toxic or otherwise hazardous substances that need to be dealt with.

Taking a Vestas V90-3.0 MW onshore wind turbine as an example, as much as 80 per cent of the materials used can be recycled, when the turbine is scrapped at the end

of its projected 20 year service life. During that period, this turbine will, from the perspective of a complete life cycle analysis, have 'covered' its own environmental impact more than 35 times.

> **"Wind turbines do not use even one drop of water when generating electricity."**

ENERGY WITHOUT BORDERS OR LIMITATIONS

The argument that wind is a limitless resource should also carry a lot of weight today among both industrial and political decision makers. Taking into consideration the geopolitical events of recent years and the fact that most fossil energy resources are located in unstable regions of the world, it is obvious that greater independence of energy imports is a necessity for many countries, particularly in the western world. It is worth noting that the European Wind Energy Association (EWEA) has calculated that wind power has the potential to cover one-fifth of Europe's electricity requirements as early as 2030 – and this calculation even takes into account a 50 per cent rise in consumption.

That this is more than just figures and theory is proved by countries including Denmark, which already covers a fifth of its total energy requirements through wind power. What is more, Denmark does so by using less than 20 per cent of its actual wind resources. On very windy days, Denmark receives all its electricity from the country's 5,000 wind turbines. It should interest every responsible politician to know that in its *Survey of Energy Resources* from 2004, the World Energy Council concluded that the world's total consumption of electricity could be covered several times over by exploiting just one-tenth of the technical potential of the Earth's wind resources.

> **"Wind turbines generate no waste at all when producing electricity."**

WIND POWER IN A LONG TERM PERSPECTIVE

With the rapidly accelerating pace that has distinguished the development of wind technology, there is every reason to have great expectations for its role in the medium to long term. The EU, for example has a binding ambition to source 20 per cent of electricity from sustainable energy sources in 2020. In addition, China has also stated its intent to invest a tenth of its GDP in energy, so that 15 per cent of China's energy supply can come from sustainable sources in 2020.

These ambitious visions mean that the total installed wind power capacity worldwide will need to increase from 75,000 MW in 2006 to 1,000,000 MW in 2020. This corresponds to average growth of 20 per cent per year; it is no exaggeration to claim that all indicators point to an industry that is still growing very strongly, if the political preconditions remain in place.

WHAT IS NEEDED FROM A CLIMATE POLITICAL PERSPECTIVE?

Continued expansion of wind power makes demands, not only on the wind power industry, but also on the politicians who are to help pave the way for an energy mix involving the lowest possible CO_2 emissions and the greatest possible supply reliability.

Through the political implementation of the Kyoto Protocol, two means have already been adopted. Firstly, the EU has introduced a CO_2 quota trading system intended to make it more expensive to generate energy that involves the emission of high volumes of CO_2. Secondly, climate credits (Joint Implementation (JI) and Clean Development Mechanism (CDM)) have been introduced, so that companies and countries, on the basis of the principle that CO_2 savings are good for the climate, no matter where in the world they are made, can choose to implement CO_2 saving projects in countries where the associated costs are lower.

> **In its Survey of Energy Resources from 2004, the World Energy Council concluded that the world's total consumption of electricity could be covered several times over by exploiting just one-tenth of the technical potential of the Earth's wind resources.**

The leading player in the sector, Vestas Wind Systems, identifies three key areas that require the full attention of politicians in the important discussions and negotiations to come:

1. Energy prices should reflect the actual costs of a given form of energy. As long as they do not, supplementary regulation will be required to promote sustainable energy. A well functioning CO_2 quota system could constitute part of the solution, but is unlikely to be sufficient on its own.
2. Regarding JI/CDM, the current system is costly and complicated and removing these barriers would enable more countries to host projects. The challenges that lie ahead are the following: to increase cooperation and communication between project developers, governments and UN bodies (CDM Executive Board, Methodology Panel and JI Supervisory Committee); to increase governmental support to these bodies; to ensure better criteria for methodology approval in the CDM; and to improve the harmonisation of national project approval procedures, in order to minimise the administrative burden – varying national requirements create barriers and prevent a well functioning CDM and JI system to develop. Moreover, not all countries have given sufficient financial backing to the numerous different institutions in the system, which has further limited the activities of the UN bodies and, therefore, hampered the success of the mechanisms.
3. Consideration should be given to financing options over and above the existing climate-political instruments, particularly in regions expected to undergo both large and energy intensive economic growth in the immediate future. Both the Stern Report and UN studies indicate that investment requirements over the coming 20–30 years will exceed the options that exist in the current flexible mechanisms in the area of climate issues. Banks and other financing institutions are reluctant to provide the necessary supplementary loan financing for climate projects.

A wider system allowing the active participation of more sectors in more countries should also be one of the top considerations in the design of a post-2012 framework. A more efficient system is needed to ensure the uptake of new, clean, and renewable power generating capacity – this is what will make a difference in the long term.

For references and further information about the article, please contact Group Communications, tel: + 45 97 30 51 62, or e-mail: mjvvi@vestas.com.

Organisation

Vestas installed its first wind turbine in 1979 and has since played an active role in the fast moving wind power industry. From being a pioneer in the industry with a staff of approximately 60 in 1987, it is today a global, market leading group with over 14,600 people employed. Vestas is the leading producer of high technological wind power solutions. The company's core business includes development, manufacturing, sales, marketing and maintenance of wind power systems that use wind energy to produce electricity.

Enquiries

Vestas Wind Systems A/S
Alsvej 21, 8900 Randers, Denmark
Tel: +45 9730 0000 | Fax: +45 9730 0001
E-mail: vestas@vestas.com
Website: www.vestas.com

© timur1970

Back to the future:

INVESTING IN CLEAN ENERGY

SYLVIE LEMMET
DIRECTOR, DIVISION OF TECHNOLOGY, INDUSTRY AND ECONOMICS, UNEP

Investments in clean energy reached new heights in 2006 with over US$100 billion financial transactions in the global sustainable energy sector. Looking back from 2050, researchers might see this investment-high as the tipping point for the future energy market; the point where renewable energy and energy efficiency technologies not only moved to the mainstream, but became the fundamental components of the global energy system.

Historians in 2050 may see the year 2007 as a watershed period; the time where leaders and citizens accepted the imminent threat of climate change. This acceptance would come after several concurrent events. First, the Intergovernmental Panel on Climate Change released its 4th Assessment Report confirming there was a 90 per cent probability humans were changing the climate. Then along came a man named Stern from Britain who found the costs of addressing climate change were less than the costs of not doing so. And a former US Vice-President by the name of Al Gore won the Nobel Peace Prize *and* an Academy Award for his documentary on climate change called *An Inconvenient Truth*.
In a year of somewhat gloomy predictions, however, they would also see something very positive: a signal that an alternative future was possible to today's fossil fuel dominated energy markets and the threat of severe and unpredictable climate change. That signal was financial transactions of more than US$100 billion in the global sustainable energy sector, setting a new record that was more than three times the amount just three years earlier. Those researchers might conclude that this was a 'tipping point' where renewable energy and energy efficiency technologies started to become the fundamental components of the global energy system.

Although such a scenario is just that, the information is real and comes from a report by the United Nations Environment Programme (UNEP) called *Global Trends in Sustainable Energy Investment, 2007*. In addition to a rapidly increasing US$70.9 billion of new investment and US$29.5 billion of mergers and acquisitions in the sustainable energy industry in 2006, the report found investments occurring in sectors and regions previously considered too risky to merit the attention of the institutional investment community.

SIGNALS

Signals move markets and these are important signals, particularly for governments beginning the next round of climate negotiations to replace the Kyoto Protocol. Although climate change issues generate a lot of discussion about the technologies of tomorrow, the data in *Global Trends* clearly show that the finance sector believes the technologies of today can and will change the energy mix and begin to eliminate 'fossil carbon'.

> **"We are beginning to experience full scale industrial development, not just a tweaking of the energy system."**

Sure, some of this investment is being driven by the price of oil. The 43 per cent increase in investment in sustainable energy from 2005 to 2006 came in a year when oil prices averaged more than US$60 a barrel in

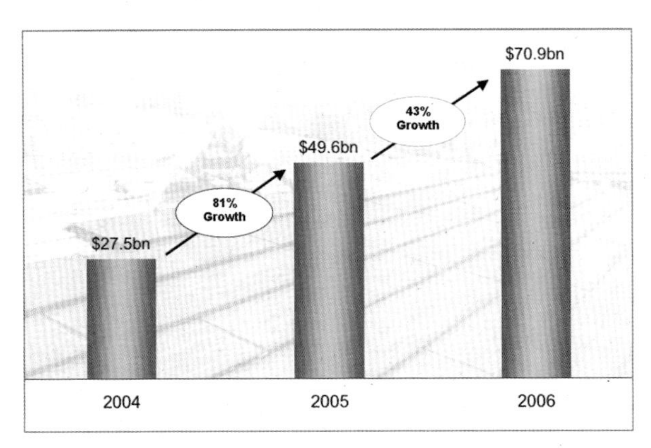

Direct investment in sustainable energy has grown rapidly to nearly US$80 billion. Total financial transactions including mergers and acquisitions reached more than US$100 billion in 2006.

2006 and in October 2007 the price is up over US$90 a barrel. But oil price is only part of the picture and really only has a significant impact on the biofuels industry, the renewables sector that competes most directly with gasoline and diesel. Looking beyond the price of oil, the capital mobilisation now underway indicates that we are beginning to experience full scale industrial development, not just a tweaking of the energy system. Underpinning this development are climate change concerns, energy security, the price of oil and a widening array of clean energy and climate policies at the federal, state and municipal levels. The challenge for governments, energy planners and policy makers is to build off this positive market dynamic, turning near term advances into long term frameworks and continued sector growth.

Many financiers see investments in current renewable energy and energy efficient technologies as key opportunities to profitably address climate change. The general greening of industry is another related driver, with more companies than ever reporting their environmental performance. Growing public cynicism about greenwash, lip service paid by companies to their sometimes non-existent environmental credentials, is fostering an underlying shift towards companies practicing what they preach.

If the current market signal remains and is strengthened through policies to lower carbon emissions, this sector will far surpass the predictions of conventional energy analysts, which have mostly assumed a very minor role for these technologies.

BEATING PREDICTIONS

Today, analysts say new renewable energy sources (excluding hydro) only account for 0.5 per cent of the global energy sector and two per cent of the power sector. Since the capital stock turnover is very slow – most generating facilities having 40 to 60 year operating lives – these figures say little about today's technology choices and even less about the future energy mix. Mostly, they give a picture of the technology options that were available in the 1950s through 1970s, when most of today's plants were built.

A better perspective on the current and future role of sustainable energy technologies in the energy mix,

TRENDS

- Sustainable energy investment was US$70.9 billion in 2006, an increase of 43 per cent over 2005, with total turnover toping US$100 billion when mergers and acquisitions are valued in.
- Investment in sustainable energy is widely spread over the leading technology sectors of wind, solar, biofuels, biomass and waste. Overall the wind sector attracted the most investment, 38 per cent of the total, followed by biofuels, 26 per cent, and solar, 16 per cent.
- Asset finance – the capital going into energy generation projects – dominated the funding mix although the US also attracted significant venture capital and private equity investment, which was equivalent to half its asset finance investment. This was a far higher proportion than in either the EU-27, 11 per cent, or other OECD countries, 28 per cent.
- Solar dominated the list of initial public offerings on stock exchanges in 2006. It experienced US$5.7 billion of this sort of investment activity, putting biofuels in second place with US$3.1 billion, and wind in third place with a relatively modest US$1.4 billion. The solar IPO boom has continued into 2007, including several Chinese companies listing on the Nasdaq exchange in New York.
- The Wilder Hill New Energy Global Innovation Index (NEX) of clean energy stocks was up 33 per cent in 2006 and a further 56 per cent by end of October 2007.
- OECD dominance is being challenged by Chinese, Indian and Brazilian companies. Overall investment in developing countries was 21 per cent (US$15 billion) of global sustainable energy investment in 2006 as compared with 15 per cent (US$4.2 billion) in 2004.
- Chinese companies were the second largest recipient of venture capital in 2006 after the United States and China took a healthy nine per cent of global investment, helped by significant asset financing activity in the wind, biomass and waste sectors. In the same year, India was the largest net buyer of companies abroad, mostly in the more established European markets.

> "The US$21.5 billion of financing in renewable energy plants represents about 18 per cent of total power sector investment in 2006. Such numbers are hardly trivial."

however, is today's investment trends. In 2006, US$110 billion – US$125 billion was invested globally in about 120 GW of new power generation capacity. Of this investment, US$30.8 billion was in new renewable energy power generation, which includes US$21.5 billion of asset finance in new generating plants, and the remainder in small scale systems, such as rooftop solar arrays. The US$21.5 billion of financing in renewable energy plants represents about 18 per cent of total power sector investment in 2006. Such numbers are hardly trivial.

On top of the US$30.8 billion invested in 2006 in new generating capacity, a further US$25.2 billion was invested in new renewable and efficiency technologies and manufacturing facilities, essentially investments in the future prospects for the industry. This significant capital influx would not have happened if investors had been expecting growth to drop from over 20 per cent last year down to the single digits that conventional analysts are continuously predicting for the sector.

> "Some may simply regard circular economy as waste recycling, yet the fundamental goal is to systematically prevent and reduce wastes in the industrial process."

DOT COM BUBBLE OR REAL BOOM?

The surge in sustainable energy investment activity has led some commentators to compare it with the technology boom of the late 1990s and early 2000s. It is true that some specific renewable energy sub sectors might be somewhat overvalued, but overall the fundamentals of the sustainable energy sector are sound. In fact, the sustainable energy sector shed its similarities with the dot com bubble in 2004, when wind and solar companies in Europe and Japan began to generate a lot of revenues and be perceived by investors not as technology players for the future, but profitable energy sector investments for today. Further, renewable energy and energy efficiency are underpinned by real demand and growing regulatory support, which the dot com boom did not have, as well as considerable tangible backing of assets by manufacturers and project developers. Banks formerly interested only in mainstream financing for renewable energy, such as the wind sector in Germany, have begun to innovate in the forms of financing they provide and broaden their exposure to other renewables and energy efficiency.

Even so, there are still significant challenges ahead. Investment is unevenly distributed, not all markets are efficient and many regions and technologies are still fighting to get noticed. Many parts of the developing world have significant sustainable energy resources to exploit, but lack the investment and technology to convert these resources into the energy that can power their development. In all countries, opportunities to improve energy efficiency are being missed, and new investment approaches are needed to upgrade and reengineer how and where we use energy.

The shift from a global economy based on limited and polluting fossil fuels to one based on limitless renewable energy and increased energy efficiency, requires a transformation in the attitude of investors and governments. Clearly that trend is underway and there will substantial profits to be made. Financiers, however, will not be the only ones to profit. Communities will take their returns in cleaner air and water, and new cleantech jobs.

In 2050, historians will see this as a very good thing.

Author

Ms Sylvie Lemmet has been appointed as the new Director of the United Nations Environment Programme (UNEP) Division of Technology, Industry and Economics (DTIE) located in Paris.

Ms Lemmet, a French national, has had a distinguished career in the field of management and environment. She has held a series of senior management positions and has led large teams to success. In leadership roles, Ms Lemmet's responsibilities have ranged from general management to project, budget and financial management. She has been the Chief Financial Officer and, later, a member of the Executive Board of Médecins sans Frontières working in the field of humanitarian emergency assistance.

Organisation

The UNEP Division of Technology, Industry and Economics (DTIE) strategy is to influence informed decision making through partnerships with other international organisations, governmental authorities, business and industry and non-governmental organisations. It also supports implementation of conventions and builds capacity in developing countries. It encourages decision makers to develop policies and practices that are cleaner and safer, make efficient use of natural resources, ensure environmentally sound management of chemicals, reduce pollution and risks for humans and the environment, enable implementation of conventions and international agreements and incorporate environmental costs.

Enquiries

Sylvie Lemmet
15 rue de Milan
75441 Paris Cedex 09
France
Tel: +33 1 4437 1450
Fax: +33 1 4437 1474
E-mail: unep.tie@unep.fr
Website: www.unep.fr

The EDF Climate Change Policy

EDF has a long-standing commitment to sustainable development. As one of the world's premier energy firms, EDF wants to be part of the solution to the climate/energy challenge the world is facing. One of the objectives for EDF is to remain the company releasing the least amount of CO2 per kWh of the seven major European electricity companies through the optimisation of its energy mix, through its investment choices and through the creation and promotion of commercial offers and advice for energy efficiency for all its customers. EDF aims to replace current end of life power plants with competitive generating units releasing little or no CO2 in line with national policies, through R&D focusing on generation technologies and promoting in the same time the high-efficiency use of electricity in households, industry and transport.

EDF Group in Europe : learning the carbon market

The corporate leadership of the Group designs its overall strategy around the requirements of particular national circumstances. Considerations of carbon constraint are integrated within the risk policy of each entity of the Group in order to optimise the use of generation units. Synergies are activated appropriately. A Carbon Fund managed by EDF-T, our trading company, has been set up in order to cover a part of the CER's needs. Also, carbon constraints are included in analysis of investment portfolios. In Europe, our new investments will be strongly driven by the existing and emerging regulation in place. Accordingly, CO2 externalities will be taken into account in a straightforward manner.

Some insight in our present investments

EDF France has decided on an ambitious programme of €40 billion of investments over the next five years. Its generation investment choices are directed by the objective to continue emitting the least of the seven leading European electricity companies. The CO2 emissions of the generation fleet are less than 0,05 t/MWh in France and 0,125 t/MWh for the Group as a whole. In comparison, the other leading European electricity companies average over 0,400 t/MWh. A new 1650 MW nuclear reactor, the Flamanville EPR, will be operational in Normandy in 2012. EDF is also involved in the recent international nuclear revival. New investments in France, Italy and the UK, will be made for fossil-fired generation. The Group is developing the next generation in fossil fired technologies, notably by developing clean technologies, such as clean supercritical pulverized coal, highly efficient combined cycle systems. EDF is developing a major renewable energy program to create 3300 MW of mostly wind power by 2010, mainly in Europe. The project represents an investment of about €3 billion. EDF is also considering wind projects in China. Tenesol, of which EDF owns 50 per cent, is investing in solar energy in combination with EDF's plans to grow these technologies. Finally EDF is investing in a 1000 MW dam at Nam Theun in Lao, which will be operational by the end of 2009, providing electricity to the people of Lao and Thaïland.

Recent commitments and future investments

Last July EDF SA signed a joint statement within the Global Compact ("Caring for climate") that highlights the importance of climate commitments. EDF is presently committing to reduce its CO2 absolute emissions in France by at least 20% by 2020 compared to 1990 (averaged climatic years). Moreover, EDF Energy took its "Climate Commitment" which stipulates that reducing specific CO2 emissions by 60 per cent in 2020 compared to 2006. Outside Europe new investments will be scrutinized through different criteria such as contribution to the climate policy of the country, efficiency, clean coal technologies, and the ease of upgrading to CCS technologies.

Preparing the future with R&D

EDF's R&D teams also work on energy issues, such as the depletion of natural resources, climate change and nature conservation. Efforts focus on planning ahead for the new energy landscape, by developing energy efficient technologies and services in houses and buildings, cooperating to grow renewable infrastructure and to make these technologies more efficient and cheaper. EDF devotes almost €400 million each year to research and development. Two-thirds of this budget is assigned to carbon free technologies research. EDF is further investing in CCS technologies to reduce the impact of the existing world fossil fuel dependence.

A reappraisal of infrastructure to drive the renewables revolution

© Ken Ng/Fotolia

The solutions to climate change already exist. The science is looking bleaker by the day. So how, asks the UK's largest renewable energy trade association, can technological solutions to climate change be better mobilised – and quickly?

LEONIE GREENE
HEAD OF EXTERNAL RELATIONS,
THE RENEWABLE ENERGY ASSOCIATION

In the disorienting field of climate change it can be easy to forget we face a crisis not of energy, but of fossil fuel use. There is no shortage of renewable energy but a shortage of infrastructure to harness it. Societies remain locked into physical and institutional fossil fuel infrastructure that constrains their ability to accelerate renewable solutions. Just as dangerously, societies are also locked into outdated mindsets allowing precious resources and vested interests to replicate the outdated infrastructure that has already pushed humankind to the precipice.

In 2007, PV industry analysts anticipate price cuts of 40 per cent in just three years as technology and polysilicon supplies improve. The potential of concentrated solar became clear with California's Pacific Gas & Electric committing to one of the largest ever purchases of solar power to be generated by mirrors in the Mojave Desert. Off the coast of the UK, planning permission was granted for the world's largest 1,000MW wind farm. But if these advances in renewables are to be captured and accelerated, it is essential that our energy systems stand at the ready to incorporate them. Only a handful of countries, including Iceland or Sweden, can claim to be in that position.

“If these advances in renewables are to be captured and accelerated, it is essential that our energy systems stand at the ready to incorporate them.”

A NEW INFRASTRUCTURE FOR THE OLDEST ENERGY

Admittedly, gas pipe lines, electricity networks, fuelling stations, wires, network interchanges, grid architecture, system planning and the rules and regulations that determine who builds and uses what, where and when, don't hold quite the same immediate appeal as the latest renewables breakthroughs. But if we are to unleash the promise of a renewables revolution, we need to focus far greater political and professional interest on the infrastructure beyond new generation technologies. At its heart, climate change is an infrastructure issue. How do we create an infrastructure that will underpin the rapid technology shift towards a low carbon economy? What does this new infrastructure look like? How do we best get there from where we are and how much will it cost?

From Europe to India, from the remote African community to the powerful multinational, awareness is growing that a renewables revolution means something much more profound, challenging and ultimately liberating than simply replacing, say, the world's coal-fired power plants with wind farms. The infrastructure on which we

Figure 1: Centralised energy – yesterday's technologies.

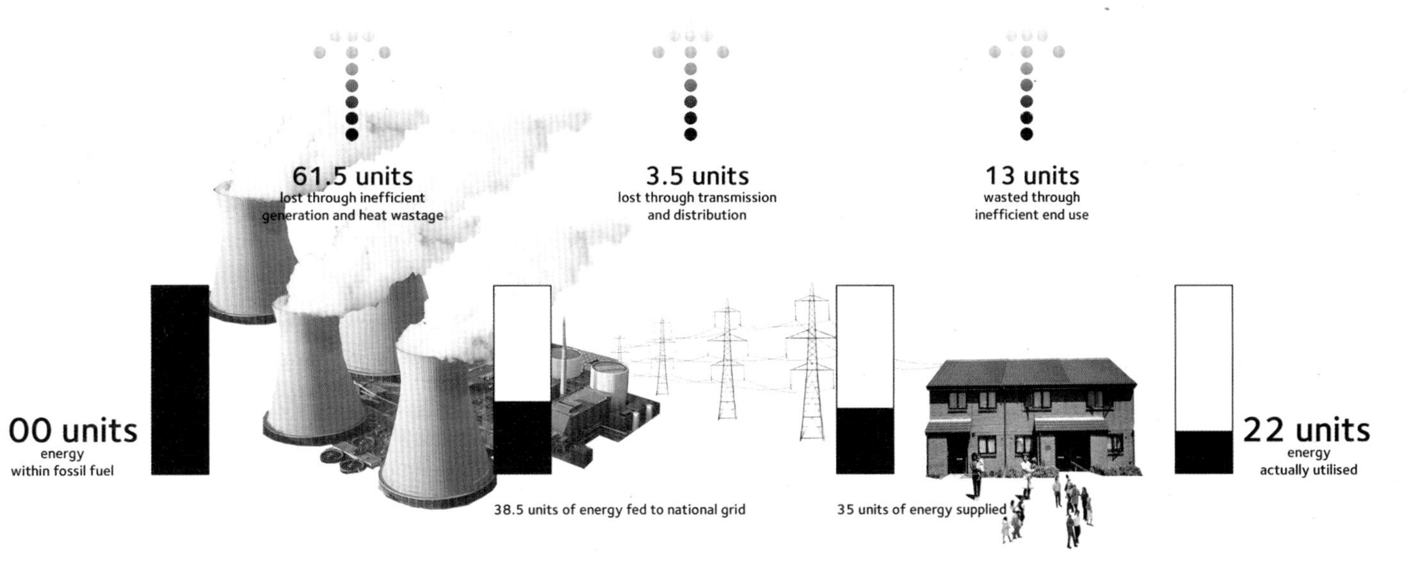

all depend is itself an extension of fossil fuel technology. Found in relatively few places in the world, the fossil fuel industry has developed vast global infrastructure to distribute to almost every corner of the Earth the fuels that underpin people's daily lives.

But renewables don't need inputs shipped in from other continents. Their ambient and decentralised nature makes possible a very different physical infrastructure indeed – and one with huge economic advantages.

> "If we are to unleash the promise of a renewables revolution we need to focus far greater political and professional interest on the infrastructure beyond new generation technologies."

A MOVE AWAY FROM CENTRALISED POWER SYSTEMS

Centralised power systems supply 93 per cent of the world's electricity and are the biggest single source of global CO_2 emissions. The UK is a particularly meaningful model of infrastructure to critique, for it was here, in the 1930s, that the first centralised grid model was constructed in order to bring electricity generated next to coal mines to users all over the country. While today's UK system is unquestionably well managed, losing just eight per cent of power in transmission and distribution, the centralised fossil fuel grid model itself remains staggeringly inefficient, wasting a total of two-thirds of primary energy inputs. Designed in a very different era, when fossil fuels were abundant and climate change the concern of a very few farsighted scientists, no sensible engineer would design a 'coal-by-wire' system like this today.

Worse, in many 'developing' countries the centralised power model can suffer transmission and distribution losses of nearer 40 per cent which, on top of poor generation efficiency, can end up delivering less than 10 per cent of primary energy inputs as usable power.

A CHANGE TO THE BUILT ENVIRONMENT

Buildings are another critical dimension of energy infrastructure increasingly talked about in the context of climate change, but not always recognised as energy technology. Buildings represent humankind's first attempts to master energy in their surroundings to ensure the sustenance and comfort of warmth, cooking, cooling and light. In many ways our built environment has changed remarkably little over centuries, arguably millennia. That must now change. The beauty of renewable technologies is they can work in practical and aesthetic harmony with building materials to deliver a low carbon built environment. A combination of good design, including high levels of insulation and passive solar alongside PV and solar thermal, can offer the 21st century citizen a comfortable and benign building that runs itself. These buildings can stand alone. Or, embedded in a responsive network infrastructure, they are power stations as well as homes or offices, offering clean electricity to neighbouring buildings. The role of buildings in renewable energy infrastructure needs very close attention given the majority of the world's people now live in cities.

While power generation enjoys political attention, heat is often overlooked. Yet in cooler climates, heat is the primary use for energy. Here again infrastructure is key. A highly energy efficient built environment can reduce the need for space heating dramatically. The staggering wastage in the electricity system is mostly in the form of heat; this wastage is visible in the vast white plumes of vapour from the cooling towers of generation plant. Decentralised combustion generation plants, brought closer to users, enables the capture of waste heat for distribution via heat networks to commercial, industrial or domestic buildings. This is the tried and tested technology of Combined Heat and Power.

Figure 2: Decentralised energy future – today's technologies.

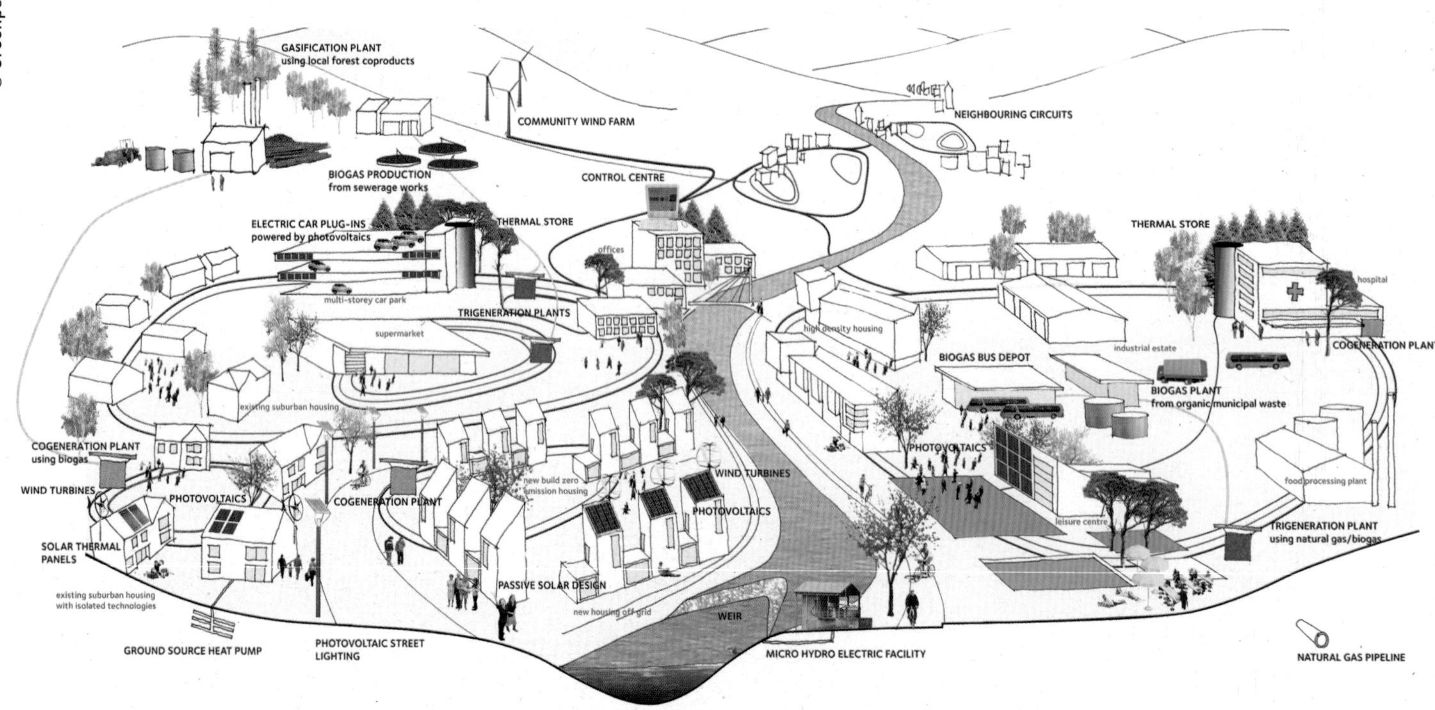

Energy solutions in this town come from local opportunities at both the small and community scale, making use of a wide range of energy resources.
The decentralised system is flexible and able to adapt to future circumstances. More importantly it can be constructed relatively swiftly using technologies available today.

When we start to think holisitically about energy and how to create the most efficient and renewable energy system possible, it becomes clear how much of a role infrastructure has to play in delivering the solutions. The example of heat also shows how much more widely and imaginatively we need to think about infrastructure.

REGULATORY BARRIERS

Infrastructure is a long term investment and cannot be overhauled overnight, but it is vital that today's infrastructure is as receptive as possible to new technologies. Unfortunately the rules over which technologies have access to centralised infrastructure all too often prohibit renewables. For example, it is technically viable to add biogas produced from organic agricultural or food wastes to gas networks ensuring a greener fuel for cooking and heating but no incentives exist to pursue this in the UK. Likewise, under European Law, it is legal to allow renewable power priority access to transmission networks, but some European countries, including the UK, have failed to adopt this measure. With its insensitive regulatory barriers to infrastructure use, it is no accident that the UK is one of the worst performers on renewables in Europe while countries like Denmark, that pursue bold regulatory inventions to support the penetration of renewable energy, excel.

SMART GRIDS

So decision makers need to turn their attention to infrastructure; the way in which its ongoing physical development and day to day management must change to underpin renewables expansion. The European Union has developed a 'smart grids' programme to look at exactly these issues in relation to power supply. The driver for this rethink of electricity network infrastructure is clear; smart grids is a necessary response to the environmental, social and political demands placed on energy supply. The diagram above is a Greenpeace version similar to the EU Smart Grids Initiative and demonstrates a low carbon power system adapted for far higher levels of decentralised and intermittent renewable energy, integrated with the built environment and with an emphasis on active operational management of networks. Modern network infrastructure can be highly sophisticated, using computerised automation, demand-side management technologies and other renewable interface technologies, like storage, to balance any intermittency.

DECENTRALISED AND OFF-GRID

India has recognised the benefits of a network infrastructure rethink, passing laws in 2004 to support greater use of decentralised systems. And in Africa, where many people lack basic access to electricity inhibiting human development, some countries are now questioning the merits of a centralised approach. Renewables can dispose of the heavy fossil fuel infrastructure we are accustomed to – a vital consideration in severely economically constrained countries where renewables allow a piecemeal system development as the skills and resources available permit. Over the past decade, nearly 7,000 homes in South Africa have been provided with solar home systems. Namibia is masterminding a 20 year electricity infrastructure plan setting out which parts of the country will be grid connected, and which will instead be powered by stand alone renewable systems. As the coordinator of South Africa's Renewable Energy and Energy Efficiency Programme says, "The old fashioned idea that everyone should be connected to the grid is finally being recognised as unrealistic and undesirable."

Influential mainstream organisations are also recognising that low carbon means a radical rethink of infrastructure. For example, the International Energy Agency's 'Alternative Scenario', commissioned by the

G8 + 5 Gleneagles Dialogue, underlines the value of 'smart investment' – the replacement of ageing network and generation assets with an efficient and responsive infrastructure model fit for 21st Century technologies.

A RENEWABLES INFRASTRUCTURE

But what is the cost of an energy system explicitly oriented towards renewables? Unusually in the climate change field, there are some very pleasant surprises when it comes to the economics. All too often, analyses of generation options look exclusively at generation costs while ignoring the delivery infrastructure in the mistaken, and often subconscious, assumption that centralised networks are a given. One of the major economic advantages of a more efficient and renewable system is that net system costs can actually be reduced, as the International Energy Agency found out its 2006 Alternative Scenario. The massive expense of profligate, generally drastically over-specified centralised grid networks is not commonly appreciated. In fact, of the US$20 trillion the IEA anticipate the world will spend on its energy systems under a Business as usual scenario, over half of that is on the power sector, of which half again is for network infrastructure alone. Under the IEA's Alternative Scenario, savings of 40 per cent in transmission infrastructure costs and 36 per cent in distribution costs are anticipated. That means liberating greater resources to spend on renewable generation technologies themselves.

Just as importantly, renewables offer a very different pattern of system ownership. While renewables can operate at the largest imaginable scale allowing multinationals their grand projects, they can also work at the very smallest. A more renewable and decentralised system therefore lowers barriers to sector entry, opening up the system for the widest possible array of new actors to invest in the solutions. Sectors particularly well placed to contribute are the construction, farming and waste industries, as well as PR sensitive commerce, and concerned communities and individuals. While fossil fuels subscribe purely to an economic logic of economies of scale, renewables can also offer economies of mass production. Indeed, this opening up of the, hitherto, exclusive energy sector to a flood of new small investors offers exactly the competitive dynamic needed to push the traditional centralised fossil fuel sector towards radical action on renewables.

By making our infrastructure as enabling as possible to renewables at all scales, a virtuous pathway is established of greater renewables penetration and further technological stimulus. The challenge for politicians concerned with climate change and energy security is now clear. It is to ensure the trillions of dollars of public and private resources that energy systems will absorb over coming decades are now robustly redirected to develop network infrastructure tailored to the decentralised and ambient nature of renewable energy. Likewise, R&D must be prioritised away from fossil fuels towards both renewables generation technologies and also interface and network technologies. Barriers to access for all infrastructure, however imperfect, must be removed. Those who want to act to mitigate climate change must be enabled to do so, and very fast indeed.

Only national governments can direct such a bold strategic reorientation of our energy systems and set out the long term frameworks public and private investors need. If we are to emerge from the dark era of climate change, the world must spotlight OECD governments. It is these most wealthy of countries that, as the IEA firmly points out, are now at the critical point in their investment cycles to pursue 'smart investment'. A colossal amount of generation and network assets now require immediate replacement in OECD countries. If these affluent countries get it wrong and merely replace like with like they will frustrate renewables for decades and damage any credible case for urging emerging giants, such as India and China, to do things differently. If they get it right, they may yet unleash the renewables revolution in time. They will offer the most necessary and practical of global leadership imaginable.

> **" Decentralised combustion generation plants, brought closer to users, enables the capture of waste heat for distribution via heat networks to commercial, industrial or domestic buildings. "**

Author

Leonie Greene is the Head of External Relations for the Renewable Energy Association. She was previously Political Adviser for Greenpeace UK and policy officer to the Deputy Mayor of London on sustainable development. She holds an MSc in Environmental Change and authored Greenpeace's report *Decentralising Power; An Energy Revolution for the 21st Century*.

Organisation

The Renewable Energy Association was established in 2001 to represent British renewable energy producers and promote the use of sustainable energy in the UK. REA's main objective is to secure the best legislative and regulatory framework for expanding renewable energy production in the UK. We undertake policy development and provide input to government departments, agencies, regulators, NGOs and others. With over 500 members, the REA is the UK's largest renewables trade association covering renewable power, heat and transport fuels.

Enquiries

Leonie Green
REA
17 Waterloo Place
London SE1Y 4AR
Tel: +44 207 747 1830
E-mail: lgreene@r-e-a.net
Website: www.r-p-a.org.uk

BARD

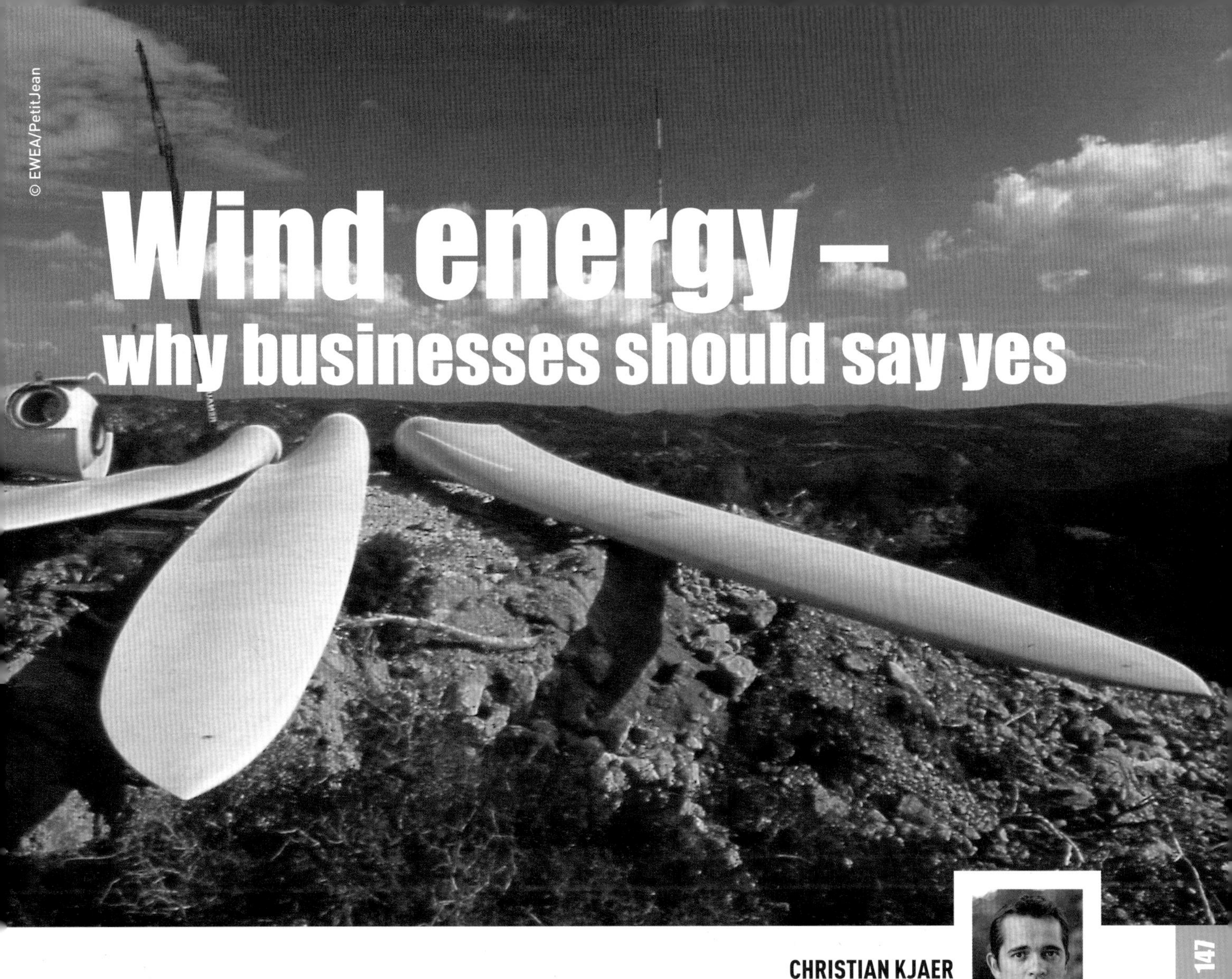

Wind energy – why businesses should say yes

CHRISTIAN KJAER
CHIEF EXECUTIVE OFFICER,
EUROPEAN WIND ENERGY ASSOCIATION

Wind energy presents environmental and economic advantages over other energy alternatives. The picture of current global energy shows a year on year increase in global demand for wind power, with the potential, through effective legislation, to meet 12-14 per cent of Europe's electricity demand. It is not only to the advantage of businesses and governments to embrace and support renewable energies such as wind power, but imperative in order to mitigate the impacts of climate change.

WIND ENERGY TODAY

The wind energy sector is flourishing. In 2006, global demand for wind power capacity grew by 32 per cent, following an increase in the market of more than 40 per cent in 2005. The value of wind turbines sold last year was €18 billion globally. Wind currently supplies three per cent of EU electricity and could, with effective legislation, meet 12 to 14 per cent of Europe's power demand. Thanks to the now widespread awareness of the environmental effects of traditional energy sources, targets have been put in place to encourage the deployment of renewable energies instead. The EU Heads of States agreed last spring to a 20 per cent binding target for renewable energies by 2020, while current average use in Europe stands at just under seven per cent. Wind energy could be one of the main contributors to meet this ambitious objective.

In the US in 2006, wind power was second only to gas in terms of new capacity, and the same was true for the seventh year running in the EU, where new wind power installations in 2006 amounted to 7,588 MW, while gas stood at approximately 8,500 MW of electricity-generating capacity. Between 2001 and 2005, 30 per cent of all new capacity installed in the EU was wind power. In that same period, 53 per cent of all new installed electricity-generating capacity in the EU was gas.

AN ENVIRONMENTALLY FRIENDLY SOURCE OF ENERGY

The two technologies of gas and wind, making up over 80 per cent of new electricity-generating capacity in the EU between 2001-2005, complement each other well, both technically and environmentally. Gas is flexible; it can be ramped up and down much faster than other technologies, apart from large hydro. Gas is also far less aggressive to the environment than other conventional power sources. There is a 10 year window to change habits and systems in order to avoid irreversible damage to the world as a result of human-induced climate change. The use of gas can help the EU meet its Kyoto obligations on greenhouse gases (GHGs) and climate change in the medium term. Gas is an important stepping stone on the way to a future in which energy will come primarily from renewable sources.

Given the climate and energy supply situation, wind energy in the future may well provide the same amount of power as conventional sources do today. By 2010, wind power will save the EU CO_2 emissions equivalent to one-third of its Kyoto commitment.

> "In 2006, global demand for wind power capacity grew by 32 per cent, following an increase in the market of more than 40 per cent in 2005."

FUEL EXPORTERS AND IMPORTERS

The importance of using wind power is not just linked to its green credentials. We are living in a period of energy change. Globally, net exporters of fuel are few, and major new market players, such as India and China, are demanding increasing amounts of energy for their expanding economies. According to the European Commission, dependence on imported oil and gas within the EU, which is currently at 50 per cent, could rise to 70 per cent by 2030.

This reliance on fuel-exporting countries makes importers vulnerable to potential supply cuts or higher prices resulting from international crises. According to the European Commission, every time the oil price goes up by US$20, the price of Europe's gas imports goes up by US$15 billion annually, and a quadrupling of oil prices from US$20 to US$80 (as we have experienced in the past years) adds €30 billion annually to the gas import bill of the EU. In comparison, the value of wind turbines installed in Europe in 2007 was approximately €9 billion.

As a result of our fuel dependence, vast amounts of money are continually being transferred from the many net importing countries to the few net exporting countries. It would require dramatic decreases in the price of oil, and thereby gas, to limit the effect of this transfer of wealth, and such a decrease is highly unlikely. On the contrary, the financial transfer is likely to increase in the future as the EU production of gas, currently meeting approximately 50 per cent of demand, is expected to fall by 50 per cent over the next 20 years.

It is clear that countries will soon have to reduce their import dependence in order to limit exposure to unpredictable fuel import prices and to curb GHG emissions. The impact of such elements on their economies and the global environment can be limited, in the short, medium and long term, through much needed energy efficiency measures and the deployment of renewables.

© EWEA/PetitJean

A FINANCIAL ADVANTAGE

While oil and gas prices fluctuate, wind energy, being an indigenous power source, can guarantee security of supply at a predictable cost because the fuel is free. Money spent on wind energy is an investment: it takes a wind turbine two to three months to produce the amount of energy that goes into its manufacture, installation, operation, maintenance and decommissioning after its 20 year lifetime. If the environmental and social costs of power generation were included in electricity prices, wind power would already be cheaper than any other electricity-generating technology. Importers of energy would reap the benefits of putting money to work in their own economies rather than bearing the economic burden of transferring increasing amounts of their citizens' wealth abroad.

> "While oil and gas prices fluctuate, wind energy, being an indigenous power source, can guarantee security of supply at a predictable cost."

EMPLOYMENT OPPORTUNITIES

Wind, unlike oil reserves, for example, is indigenous to all countries and, when wind power is deployed in a region or area, employment is boosted. The wind power sector currently employs around 64,000 people in Germany, around 21,000 in Denmark and 35,000 in Spain, according to national statistics. There is a wide variety across European countries in the estimated total employment per MW installed. However, the average in Europe is around 12 individuals per MW installed. Employment projections in the wind power sector for 25 member countries of the EU (EU-25) for the year 2020 indicate: 153,400 direct and indirect employees for manufacturing, 27,400 for installation and 16,100 for maintenance. This gives a total estimated employment for 2020 in Europe of around 200,000. Governments that install more wind power capacity indirectly create jobs, economic activity and expert opportunities.

A BOOST TO THE ECONOMY

As such figures show, the use of wind power is more than compatible with a strong economy. Denmark, for example, gets over 20 per cent of its electricity from wind energy, and according to the World Economic Forum, Denmark is the fourth most competitive economy in the world. Over the coming two decades we will witness the largest turnover in electricity-generating technology the world has ever seen, as new plants are built and older ones cease activity; as new energy markets emerge and other ones evolve. We must use this as an opportunity to change our energy supply structure towards much larger shares of indigenous, renewable resources so we can develop our economies on the basis of known and predictable electricity costs.

PUBLIC OPINION

A Eurobarometer opinion survey in early 2007 confirmed overwhelming support for wind power from the European public. Seventy one per cent of EU citizens are "very positive" about the use of wind power in their country. This makes wind the second most popular energy source after solar across the 27 member states. A government or a business that chooses to support wind power will not be making an unpopular choice.

> **"Governments that install more wind power capacity indirectly create jobs, economic activity and expert opportunities."**

A PRESENT AND FUTURE NECESSITY

President Bush declared the United States' oil dependency an "addiction". It is an addiction his country shares with most other nations of the world, and it is not limited to oil. The battle for energy in this century will not be won by following the strategy that proved to be the winning one in the 20th century, ie of either producing fuel or of controlling fuel supplies. It will be won by those regions of the world that have the foresight to act now to cure their addiction to conventional power sources in order to protect their economies and the global climate. It will be won by the regions and governments that excel in developing, deploying and exporting renewable energy technologies, such as wind power, to a world that cannot afford to do without them. It is therefore not merely to the advantage of businesses and governments to embrace and support wind energy, but imperative that they do so. In this way, they can be at the head of the game in making the necessary shift from a reliance on fuel exporters to an indigenous power supply that is constantly renewable and everlasting.

Author

Christian Kjaer was appointed Chief Executive Officer of EWEA in March 2006. He previously held the post of Policy Director for EWEA having worked on a wide range of policy issues for the association. Christian steers the work of the organisation, drafting EWEA's direction, vision and long term strategy in collaboration with the President and the Executive Committee. He is responsible for ensuring EWEA's political impact and reputation as a trustworthy partner in the policy-making process is sustained. He represents the association in external forums, engages actively with international institutions, key stakeholders and NGOs, EWEA members and the media.

Organisation

The European Wind Energy Association (EWEA) is the voice of the wind industry – actively promoting the utilisation of wind power in Europe and worldwide. Its members are from 40 countries and include over 300 companies, associations and research institutions. The EWEA Secretariat is located in Brussels where it coordinates European policy, communications, research, and analysis. EWEA is a founding member of the European Renewable Energy Council which groups the eight key renewable industry and research associations under one roof, and of the Global Wind Energy Council.

Enquiries

EWEA – The European Wind Energy Association
Rue d'Arlon 63-65, B-1040 Brussels, Belgium
Tel: +32 (0) 2 400 10 55
Fax: +32 (0) 2 546 19 44
E-mail: communication@ewea.org
Website: www.ewea.org

Enel's hydro power plant, Veneto, Italy.

Enel's green revolution

International power company, Enel, shares the vision that in the future the world will be able to produce the necessary amount of energy at reasonable costs and zero emissions. With its proactive role in addressing environmental issues and a strong track record of environmental responsibility, Enel wants to be a part of this future. As a leading producer and vendor of electricity and natural gas, Enel understands that climate change is an enormous challenge that cannot be ignored, and explains here how technological and environmental leadership is one of its strategic targets.

Climate change is a global problem, which is already affecting our society, the way we conduct business and live our lives. It is an enormous challenge that we cannot ignore, and which is becoming increasingly important. Enel is keen to take a proactive role in addressing environmental issues and has a strong track record of environmental responsibility. Technological and environmental leadership is one of Enel's strategic targets. In our vision, we will be able to produce the necessary amount of energy at reasonable costs and zero emissions.

A COMMITMENT FOR THE PLANET

Enel has already carried out a programme to convert power plants from oil to combined-cycle natural gas-firing, a technology where we stand as national leader, with about 5,000 MW. This programme, together with investments in renewables, has enabled us to cut our CO_2 emissions by more than 15 million tonnes over the past six years, while Italian total emissions have increased.

> **"The industry should experiment with new solutions for energy production through renewable energy sources, nuclear and fossil fuels."**

Enel's wind farm, Sardinia, Italy.

Our Environmental Report epitomises our care for the environment. For 11 years, it has been a key vehicle for monitoring and communicating our environmental performance, as well as for reviewing and fine tuning our environmental strategy and policy. Enel has now become a multinational in the energy sector: with the growth of our international presence, the Environmental and Sustainability Reports have also been instrumental in extending the application of our environmental policies to all of our activities outside Italy.

Enel's inclusion in the Dow Jones Sustainability Index, as well as the Climate Leadership Index in 2005, clearly demonstrates the results achieved so far and our commitments towards sustainable development. In particular, the Climate Leadership Index acknowledges companies like ours, who distinguish themselves internationally in the reduction of greenhouse gas emissions, within the Carbon Disclosure Project. In October 2006 the Accountability Global 50+ list, published by Fortune magazine, ranked Enel 6th (3rd among world utilities) for its attention to sustainable development and respect for the environment.

Our effort has been thorough and systematic, and will continue to be so in the future. In particular, we firmly believe in the development of new technologies. The industry should experiment with new solutions for energy production through renewable energy sources, nuclear and fossil fuels. The choice cannot be limited only to the alternatives available today. No stone should be left unturned.

Successes

We have lowered our specific CO_2 emissions to 496 g/kWh, further improving on our 2005 value of 501 g/kWh (gram per kilowatt hour). This value had already allowed us, one year in advance, to better the target of 510 g/kWh, a voluntary commitment that we had taken with the Environment and Industry Ministers as early as 2000.

Enel is also committed to researching the best system to capture and store CO_2 and is a member of the European Platform for the generation of Zero Emission Fossil Fuel Power Plants. We hold positions in the Advisory Council as well as Working Groups, and have contributed to defining the Platform's vision: "To enable fossil-fuelled power plants to have zero CO_2 emissions by 2020".

Piero Gnudi
President of Enel

Through our global presence, we are well placed to make a viable contribution to the protection of the planet and new generations. We rank among the world leaders in renewables, with over 30 TWh (terawatt hour) generated every year (over 24 TWh in Italy and over six TWh abroad) – roughly 23 per cent (data as of December 2006) of our overall output. In 2006, our generation from renewables and nuclear energy (a sector in which we are once again present through the acquisition of Slovenské Elektrárne) avoided approximately 30 million tonnes of CO_2 emissions.

SCIENTIFIC RESEARCH AND INNOVATION

Our international presence will make it possible to transfer the best technologies and innovative solutions for power generation and distribution. Enel's commitment is not limited to reducing CO_2 emissions. The programmes of environmental enhancement that we initiated in the 1990s and that we have been supporting with substantial investments, have yielded significant results in terms of reduction of pollutant emissions into the atmosphere. From 2000 to 2006, we curbed specific emissions of sulphur oxides, nitrogen oxides and particulates by 63 per cent, 36 per cent and 70 per cent respectively.

We use leading edge technologies and the most efficient production processes, thereby optimising the use of primary energy sources, materials and precious resources, such as water. Over the years, our recovery of waste has attained levels of excellence. In Italy, we recover about 90 per cent of our waste.

The emphasis that we place on the environment is evident from the widespread use of environmental management systems. With regard to Italian and non-Italian operations, 97 per cent of our power grids and 75 per cent of our generating capacity are ISO 14001-certified.

Recently, we have vigorously renewed our commitment to the environment by launching a large investment plan called Environment & Innovation Project. Over €4 billion in five years for research, renewables, development and

Enel's man at work on a windmill, Sicily, Italy.

application of the most advanced technologies. The plan, which is unprecedented at Enel and almost unequalled in global terms, will add focus to our day to day attention regarding efficiency and continuous improvement.
Our final goal is zero emission power generation. The project doubles our financial efforts for developing renewables generation from €1.6 billion in the 2001-2005 period to €3.3 billion in 2007-2011.
We will invest another €800 million in the study of innovative renewable energy concepts, in research projects on CO_2 capture and storage, on the hydrogen frontier and in encouraging and promoting energy savings, ie the foremost and simplest way of protecting our environment. In this particular field, we plan to further improve on the excellent results that we have achieved so far.
We think that the most effective way of reducing consumption of primary energy, and thus of safeguarding the environment, is through more efficient energy, and, above all, the informed use of energy by consumers.

SMART, SUSTAINABLE, EFFICIENT

We are indeed convinced that in the immediate future, energy efficiency may be considered the most important 'renewable energy source'. Enel's industrial strategy, launched several years ago, and further strengthened recently, is in line with the EU's position, and is focused on promoting any possible operation that has high environmental compatibility.
In particular, over the past two years, Enel has developed a comprehensive information campaign on the intelligent use of electricity, with hundreds of advertisements in the media, millions of copies of the guide "Quando usi l'energia, usa la testa" ("When you use energy, use your head"), distributed to hundreds of thousands of people.
Environmental sustainability continues to be a key focus of our activities. As a global energy player, we have to work to the best of our abilities to deliver a better world to future generations and disseminate environmental awareness, so as to respond to the challenge of not changing the world; that's the real revolution.

Organisation

Today Enel is an international Group producing and distributing electricity and gas in 21 countries on 4 continents. Among the utilities listed in Europe, Enel is the second largest in terms of installed capacity and, with about 1.9 million of them, the largest in terms of number of shareholders. It also ranks among the largest utilities in the world in terms of market capitalization and is Listed on the Milan and New York stock exchanges since 1999.

Enquiries

Website: www.enel.com

Power to empower

AKANKSHA CHAUREY
ASSOCIATE DIRECTOR, DISTRIBUTED GENERATION
ENERGY ENVIRONMENT TECHNOLOGY DIVISION,
TERI

Women managing a solar workshop.

Access to reliable energy services is one of the biggest impediments in the socioeconomic development of communities across the world. While local governments are committing to provide basic services such as water, health, sanitation, electrification to all its population, lack of energy and electricity in particular, is proving to be a serious hindrance in this process. Distributed generation and supply of electricity based on renewable energy technologies is fast emerging as a viable option. This article describes distributed generation-based initiatives undertaken by TERI that have benefited rural communities and provided a platform to link public and private commitments towards a common goal.

INTRODUCTION

Rural electrification is one of the key drivers in achieving Millennium Development Goals as it facilitates economic and sociocultural development of the target population. Distributed generation of electricity, particularly in decentralised mode is one option to implement rural electrification schemes. The synergy of distributed generation and rural electrification is therefore critical to achieve the targets of universal service obligations, particularly by enhancing last mile access, which is one of the least addressed challenges in this sector. The viability of the technological solution, strength of institutional innovations, robustness of designed financing models and effectiveness of enabling mechanisms for communities to derive direct and indirect benefits from electricity services, are some of the facets of this challenge.

Through government and non-government efforts in a country such as India, renewable energy technologies are used in a variety of applications. However, there are very few dissemination models that are entrepreneur driven and incorporate innovations in delivery of services. The viability of such models relies on:

- Their ability to provide income opportunities on both the supply and the demand side.
- Their flexibility to bundle a variety of energy-related and energy-driven services, such as mobile telephone charging, water purification, sale and servicing of energy efficient appliances.

These models would also provide a platform for public-private-people partnerships. A number of such initiatives are described below.

> "By providing initial training and capacity building on several techno-commercial aspects, the project has helped six women become solar entrepreneurs who are now engaged in a variety of businesses."

WOMEN AS SOLAR POWER ENTREPRENEURS – PILOT PROJECT IN THE SUNDERBANS

The Sunderbans is a part of the world's largest delta formed by the rivers Ganges, Brahmaputra and Meghna. The area is interwoven by a network of small rivers, creeks and waterways and 70 per cent is under saline/brackish water. To protect the natural habitat, environment and the biodiversity of the region, the Sunderbans Biosphere Reserve was established in 1989. It covers 9,630 square km including about 4,444.33 square km of human habitat along with the entire forest area. Waterways are the major means of communication in this region.

Women managing a solar lantern charging station.

As most of the southern Sunderbans is separated from the mainland and divided by wide rivers and creeks, it is extremely difficult to extend high tension transmission lines and a distribution network for electricity. The remote villages thus suffer from chronic energy shortage, adversely affecting the socioeconomic development of the region. Against this background, energy interventions based on alternate technologies, such as solar, are relevant and viable. A large population in the unelectrified islands is currently using solar PV electricity.

TERI has been working in this region for the past 10 years in partnership with the state government as well as with NGO sector to promote the use of solar PV. The project outlined here has created viable enterprises on the supply side targeting women entrepreneurs. These enterprises have been created not only to provide solar PV based services in remote and interior villages, but to provide repair and maintenance services to already existing products and systems.

This pilot project, sponsored by the National Renewable Energy laboratory, USA, is implemented by TERI in partnership with the Ramakrishna Mission (RKM), a reputed NGO. The project has also created an organisational set up: MFEDO (Market Facilitating and Enterprise Development Organisation) within the cluster organisation Kalpartaru of the RKM to incubate these enterprises. Kalpartaru now acts like an anchor to these enterprises by helping them procure raw material, facilitate market linkages and formulate self-help groups.

> "Over 78 million households, roughly 390 million lives, in India lack access to electricity, causing life to come to a standstill after dusk."

By providing initial training and capacity building on several techno-commercial aspects, the project has helped six women become solar entrepreneurs who are now engaged in a variety of businesses, such as charging and renting of solar lanterns, designing and assembling small electronic items and repairing solar home systems. Apart from financial independence, a sense of pride arising from the newly acquired skill and confidence are the biggest benefits of the project.

Based on this pilot initiative, TERI has now launched a larger project called Lighting a Million Lives.

LIGHTING A MILLION LIVES

Over 78 million households, roughly 390 million lives, in India lack access to electricity, causing life to come to a standstill after dusk. Inadequate lighting is not only an impediment to progress and development opportunities, but also has a direct impact on the health, environment, and safety of millions of villagers as they are forced to light their homes with kerosene lamps, dung cakes, firewood and crop residues. Recognising the need to change the existing scenario in rural India, TERI initiated the Lighting a Million Lives' (LaML) campaign.

The sustainability of the LaML campaign stands on two important pillars:

- Providing solar lanterns to unelectrified and poorly electrified rural households and facilitating livelihoods as well as direct or indirect income-generating opportunities through provision of enhanced lighting.
- Creating local entrepreneurial driven delivery channels to ensure distribution and charging of solar lanterns, repair, maintenance and other aftersales services.

The Campaign is based on an entrepreneurial model of energy service delivery, developed by TERI and successfully tested in pilot projects across Rajasthan, West Bengal, and Uttarakhand, India. Through this model, institutional mechanisms are being created for energy services delivery in remote villages where kerosene is the predominant fuel for lighting. This is the case for households, small enterprises such as shops, local bazaars, tuition and coaching centres and various cottage industries.

The model is designed to benefit both user and supplier. For instance, rural entrepreneurs can be trained to manage and run a central solar lantern charging/ distribution centre where lanterns are charged during the day and rented in the evening. While this creates financial opportunities for entrepreneurs, the user of the lantern does not have to bear the burden of purchasing and maintaining it.

Apart from providing reliable and ensured lighting to households at an affordable rate, this model also facilitates entrepreneurial development among rural communities. The solar charging platform can, for example, also provide other services such as telecommunication services, purification of potable water and mobile telephone charging.

Grassroots-level partners install the hardware, identify, train and provide initial support to prospective rural entrepreneurs, create awareness among communities about the use and upkeep of lanterns and develop livelihood activities through access to lighting. The entrepreneur, in addition to providing the centre facility,

© TERI

Solar powered ICT based Rural Knowledge Centre.

> **" Over the past decade, India has become one of the leading innovators in information and communication technologies (ICT) serving the rural user. "**

also contributes equity in cash and pays for the training associated with the solar power platform and solar lanterns.

SOLAR POWERED ICT-BASED RURAL KNOWLEDGE CENTRES

Energy needs in India, however, go beyond lighting, and a key focus is now on livelihood creation in villages. Over the past decade, India has become one of the leading innovators in information and communication technologies (ICT) serving the rural user.

ICT in rural areas has been related to rural kiosks/ knowledge centres, which are technologically endowed access points to provide a variety of services catering to the needs of the rural populations in different ways. Services provided include those that are 'e' in nature, as well as computer education, spoken English courses and financial services, making these centres hubs for capacity building and training. Most ICT centres include, but are not limited to, a PC, connectivity solution, photo printer and a digital camera.

Such a rural ICT programme faces two specific challenges for its roll out:

- Unreliable supply of electricity.
- Inadequate entrepreneurial capacities.

Providing reliable electricity supply has been one of the most striking impediments. As the availability of quality power in rural areas is not assured, battery is used as a backup. However, the battery often does not get fully charged due to the unreliable supply of grid for the required charging period. The rural kiosk will then lose business which ultimately affects its viability.

The sustainability of the kiosk also depends on the multiplicity, quality and saleability of services that it offers. The onus of running the kiosk viably lies on its operator, who, apart from having business acumen and good communication skills, is also required to know basic business management. In the absence of customised entrepreneurship development programmes, imparting such skills to the kiosk operator is yet another challenge that the ICT programme faces.

The 'Solar powered ICT based rural knowledge centre' initiative has been set up to address these challenges by providing:

- Solar PV based decentralised electricity supply to ICT centres.
- Comprehensive capacity building and support of kiosk operators in managing and operating the centres.

The business model is based on three guiding principles:

- Use of energy efficient hardware and appliances, including locally available ecofriendly building material wherever possible.
- Hybrid power supply solution using solar PV and gird as two main sources.
- Bundling of services dealing with decentralised energy, water, health, education.

The use of ecofriendly, energy efficient design, materials, hardware and appliances help reduce electricity consumption, a prerequisite for providing solar PV based power solutions. Further, in order to give quality power supply for up to 4-6 hours, a hybrid of solar PV and grid is provided for. This hybrid system is based on an integrated solar PV based power supply system that has been developed and customised by TERI specifically for the rural ICT kiosks. The knowledge centre provides a platform for integrating and bundling services related

Biomass gasifier for village electrification.

© TERI

to employment and livelihoods schemes, decentralised energy, health, water and overall sustainable development. It also demonstrates the efficient use of energy through various energy efficient systems, such as LED and compact fluorescent-based lighting, solar fan and lanterns and solar powered UPS for the computer.

BIOMASS GASIFIER FOR RURAL ELECTRIFICATION

Since the 1990s, TERI has been developing small capacity (10-20kWe) biomass gasifiers that run on 100 per cent producer gas engines. These systems are ideally suited for electrification of remote locations, as they provide enough electricity for domestic lighting, street lights and some productive applications, such as rice milling and water pumping in a small village of about 80-100 households. In one such initiative, TERI along with NTPC Ltd, a public sector power producing company, provided a 10 kWe biomass gasifier system run on fuel wood collected from the forest by the villagers. The project has institutionalised a Village Energy Committee with women representatives to oversee the daily operation and management of the project. Two trained operators operate and troubleshoot the system. The system provides electricity to two light points in every household, lights up street lights and runs a flour grinder. The initiative is being replicated in more than a dozen villages across the country.

WAY FORWARD

These initiatives provide a platform for companies, corporations and business houses to channel their social responsibility commitments in a manner that catalyses socioeconomic growth of rural communities. They also provide evidence of public-private-people partnership effectiveness, where each entity owns the initiative and contributes in making it a success. For instance, governments (or public sector) and private sector would provide initial funding support considering the high upfront cost of solar PV technology; private companies and entrepreneurs while leveraging the government support of seed funding, would also add their knowledge and services to enhance the viability of the initiative; NGOs and communities would work towards sustaining these initiatives at grassroots level. The synergies of such commitments ensure that power is provided to empower people in the fight against poverty.

A household enjoying the benefits of biomass gasifier.

This article duly acknowledges the efforts of several colleagues at TERI, specifically those who have worked on the projects included in this article.

Author

Ms Akanksha Chaurey is Associate Director of the Distributed Generation Energy Environment Technology Division of TERI and has worked in the field of renewable energy for sustainable development for over 20 years. Her area of expertise is offgrid rural electrification and distributed generation. She began her career with TERI in 1987 and was involved in projects related to technology and market assessment of renewables, specifically solar photovoltaics. After five years, Ms Chaurey became Director of Sairam Solar, a company, designing and installing solar and wind energy systems for decentralised applications. Following two years as a renewable energy consultant she joined up with TERI again in 1999 and now works on projects dealing with distributed generation and supply of electricity for offgrid applications.

Organisation

A dynamic and flexible organisation with a global vision and a local focus, TERI is deeply committed to every aspect of sustainable development. From providing environment friendly solutions to rural energy problems, to helping shape the development of the Indian oil and gas sector; from tackling global climate change issues to enhancing forest conservation efforts among local communities; from advancing solutions to growing urban transport and air pollution problems to promoting energy efficiency in the Indian industry, the emphasis has always been on finding innovative solutions to make the world a better place to live in. All activities in TERI move from formulating local and national level strategies to global solutions to critical energy and environment related issues.

Enquiries

TERI, Darbari Seth Block
Habitat Place, Lodhi Road
New Delhi 110003, India
Tel: 91 11 24682100
Fax: 91 11 24682144
E-mail: akanksha@teri.res.in

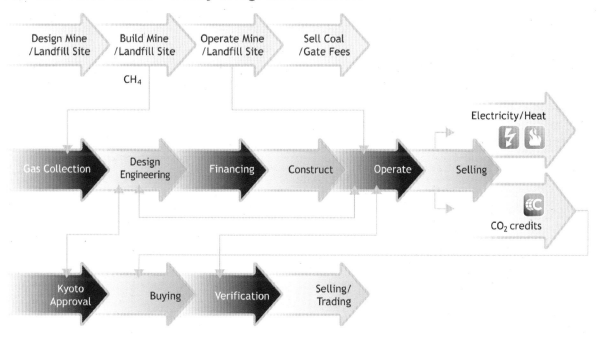

World Energy Outlook 2007: China and India insights

IMPLICATIONS FOR GLOBAL ENERGY AND EMISSIONS TRENDS

The continued and predicted increase use of coal primarily in China and India, will continue to drive up global energy related CO_2 emissions.

DR FATIH BIROL, CHIEF ECONOMIST, HEAD OF ECONOMIC ANALYSIS DIVISION, INTERNATIONAL ENERGY AGENCY

China and India's economies continue to grow at a staggering pace, pushing up energy needs sharply but drastically improving quality of life; a need that must be supported by the rest of the world. However, worldwide global energy demand is increasing alarmingly, especially when set out against existing and future climate changes. Here the outlook for world energy and, in particular, China and India contributions to this are considered. With a possible scenario of global energy needs being well over 50 per cent higher in 2030 than today, collective world action is the only long term solution to the threat of climate change.

CHINA AND INDIA – EMERGING GIANTS

Energy developments in China and India are transforming the global energy system by dint of their sheer size and growing weight in international fossil fuel trade. The staggering pace of Chinese and Indian economic growth in the past few years, outstripping that of all other major countries, has pushed up their energy needs sharply, a growing share of which has to be imported. These developments are contributing to a big improvement in their quality of life, a legitimate aspiration that needs to be accommodated and supported by the rest of the world.

The consequences for China, India, the OECD and the rest of the world of unfettered growth in global energy demand are, however, alarming. If governments around the world stick with current policies, (the underlying premise of our Reference Scenario) the world's energy needs would be well over 50 per cent higher in 2030 than today. China and India together account for 45 per cent of this increase in demand. Globally, fossil fuels continue to dominate the fuel mix leading to continued growth in energy related emissions of carbon dioxide (CO_2) and to increased reliance on imports of oil and gas, much of them from the Middle East and Russia.

THE WORLD FACES A FOSSIL FUEL FUTURE TO 2030

The world's primary energy needs in the Reference Scenario of the *World Energy Outlook 2007* (in which only those policies already enacted as of mid 2007 are considered) are projected to grow by 55 per cent between 2005 and 2030. Developing countries, whose economies and populations are growing fastest, contribute 74 per cent of this increase. Fossil fuels remain the dominant source of primary energy, accounting for 84 per cent of the overall global increase in demand between 2005 and 2030. Oil remains the single largest fuel, reaching 116 million barrels per day (mb/d) in 2030, 32 mb/d up on 2006. World oil resources are judged to be sufficient to meet the projected growth in demand to 2030; with output becoming more concentrated in OPEC countries.

The Reference Scenario projections are based on the assumption that the average IEA crude oil import price falls back from recent highs to around US$60 (in year-2006 dollars) by 2015 and then recovers slowly, reaching US$62 (or US$108 in nominal terms) by 2030.

In line with the spectacular growth of the past few years, coal sees the biggest increase in demand in absolute terms, jumping by 73 per cent between 2005 and 2030. Most coal increase use arises in China and India. Higher oil and gas prices are making coal more competitive as a fuel for baseload generation. China and India, which already account for 45 per cent of world coal use, drive over four-fifths of the increase to 2030 in the Reference

Table: Top five countries for energy related CO_2 emissions in the Reference Scenario.

	2005		2015		2030	
	Gt	rank	Gt	rank	Gt	rank
US	5.8	1	6.4	2↓	6.9	2=
China	5.1	2	8.6	1↑	11.4	1=
Russia	1.5	3	1.8	4↓	2.0	4=
Japan	1.2	4	1.3	5↓	1.2	5=
India	1.1	5	1.8	3↑	3.3	3=

Scenario. In all regions, the outlook for coal use depends largely on relative fuel prices, government policies on fuel diversification, climate change and air pollution, and developments in clean coal technology in power generation.

CHINA'S SHARE OF WORLD ENERGY DEMAND EXPANDS

In the Reference Scenario, China's primary energy demand is projected to more than double, growing on average by 3.2 per cent per year. China, with four times as many people, overtakes the US to become the world's largest energy consumer soon after 2010. In 2005, US demand was more than one-third larger. In the period to 2015, China's demand grows by 5.1 per cent per year, driven mainly by a continuing boom in heavy industry. In the longer term, demand slows, as the economy matures, the structure of output shifts towards less energy intensive activities and more energy efficient technologies are introduced.

Oil demand for transport almost quadruples between 2005 and 2030, contributing more than two-thirds of the overall increase in Chinese oil demand. The vehicle fleet expands seven fold, reaching almost 270 million. New vehicle sales in China exceed those of the US by around 2015. Fuel economy regulations, adopted in 2006, nonetheless temper oil demand growth. Rising incomes in China underpin strong growth in housing, the use of electric appliances and space heating and cooling. Increased fossil fuel use pushes up emissions of CO_2 and local air pollutants, especially in the early years of the projection period.

China's energy resources, especially coal, are extensive, but will not meet all the growth in energy needs. More than 90 per cent of Chinese coal resources are located in inland provinces, but the biggest increase in demand is expected to occur in the coastal region. This adds to the pressure on internal coal transport and makes imports into coastal provinces more competitive. China became a net coal importer in the first half of 2007. In the Reference Scenario, net imports reach three per cent of its demand and seven per cent of global coal trade in 2030. Conventional oil production in China is set to peak at 3.9 mb/d early in the next decade and then gradually decline. China's net oil imports jump from 3.5 mb/d in 2006 to 13.1 mb/d in 2030, while the share of imports in demand rises from 50 to 80 per cent. Natural gas imports also increase quickly, as production growth lags demand over the projection period. China needs to add more than 1,300 GW to its electricity generating capacity; more than the total current installed capacity in the United States. Coal remains the dominant fuel in power generation.

INDIA'S ENERGY USE POISED FOR RAPID GROWTH

In the Reference Scenario, primary energy demand in India more than doubles by 2030, growing on average by 3.6 per cent per year. Coal remains India's most important fuel, its use nearly tripling between 2005 and 2030. Power generation accounts for much of the increase in primary energy demand, given surging electricity demand in industry and residential and commercial buildings. Among end use sectors, transport energy demand sees the fastest rate of growth as the vehicle stock expands rapidly with rising economic activity and household incomes. Residential demand grows much more slowly, largely as a result of switching from traditional biomass, which is used very inefficiently, to modern fuels. The number of people in India relying on biomass for cooking and heating drops from 668 million in 2005 to around 470 million in 2030, while the share of the population with access to electricity rises from 62 to 96 per cent.

Much of India's incremental energy needs to 2030 will have to be imported. India will continue to rely on imported coal for reasons of quality in the steel sector and for economic reasons at power plants located a long way from mines but close to ports. In the Reference Scenario, hard coal imports are projected to rise almost seven fold. Net oil imports also grow steadily, to six mb/d in 2030, as proven reserves of indigenous oil are small. Before 2025, India overtakes Japan to become the world's third-largest net importer of oil, after the US and China. Yet India's importance as a major exporter of refined oil products will also grow, assuming necessary investments are forthcoming. Power generation capacity, most of it coal-fired, more than triples between 2005 and 2030.

AN ALTERNATIVE ENERGY FUTURE

In the Alternative Policy Scenario, in which policies under consideration are assumed to be fully implemented over the projection period, global primary energy demand grows by 1.3 per cent per year over 2005-2030 , 0.5 percentage points less than in the Reference Scenario. Global oil demand is 14 mb/d lower in 2030, equal to the entire current output of the US, Canada and Mexico combined. Coal use falls most in absolute and percentage terms. Energy related CO_2 emissions stabilise in the 2020s and, in 2030, are 19 per cent lower than in the Reference Scenario.

China

China is already making major efforts to address the causes and consequences of burgeoning energy use, but even stronger measures will be needed. In the Alternative Policy Scenario, a set of policies the government is currently considering would cut China's primary energy use in 2030 by about 15 per cent relative to the Reference Scenario. Energy efficiency improvements along the entire energy chain and fuel switching account for 60 per cent of the energy saved, eg policies that lead

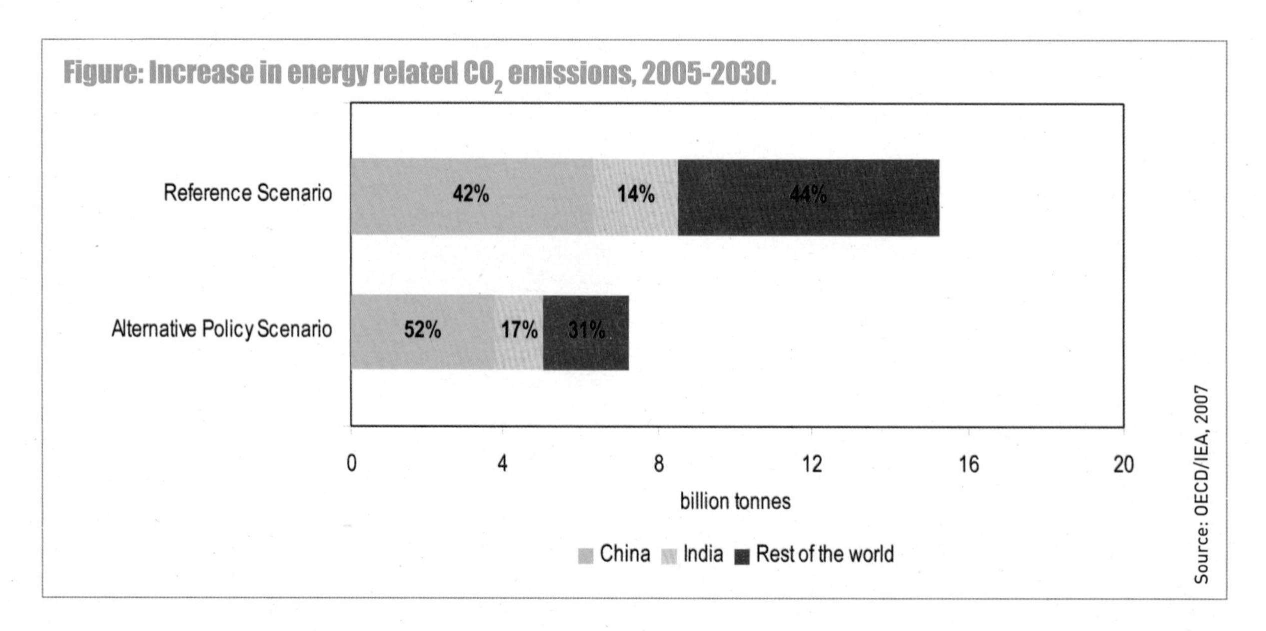

to more fuel efficient vehicles produce big savings in consumption of oil based fuels. Demand for coal and oil is reduced substantially but demand for other fuels, eg natural gas, nuclear and renewables, increases.

India

Stronger policies that the Indian government is now considering could also yield large energy savings. In the Alternative Policy Scenario, India's primary energy demand is 17 per cent lower than in the Reference Scenario in 2030. Coal savings, mainly in power generation, are the greatest in both absolute and percentage terms, thanks to lower electricity demand growth, higher power generation efficiency and fuel switching in the power sector and in industry. As a result, coal imports in 2030 are little more than half their Reference Scenario level. Oil imports are 1.1 mb/d lower in 2030 than in the Reference Scenario, but oil import dependence remains high at 90 per cent

THE WORLD'S ENERGY SECURITY

Rising global energy demand poses a real and growing threat to the world's energy security. In the Reference Scenario, China's and India's combined oil imports surge, from 5.4 mb/d in 2006 to 19.1 mb/d in 2030, representing more than the combined imports of Japan and the US today. Inter-regional oil and gas trade grows rapidly over the projection period, with a widening of the gap between indigenous output and demand in every consuming region. The Middle East, the transition economies, Africa and Latin America export more oil while all other regions have to import more oil.

Increasing import dependence in any country does not necessarily mean less secure energy supplies, any more than self sufficiency guarantees uninterrupted supply. However, it could carry a risk of short term energy insecurity for all consuming countries, as geographic supply diversity is reduced and reliance grows on vulnerable supply routes.

Longer term risks to energy security are also set to grow. The increasing concentration of the world's remaining oil reserves in a small group of countries, notably Middle Eastern members of OPEC and Russia, will increase their market dominance and may put at risk the required rate of investment in production capacity. The greater the increase in the call on oil and gas from these regions, the more likely it will be that they will seek to extract a higher rent from their exports and to impose higher prices in the longer term by deferring investment and constraining production. Higher prices would be especially burdensome for developing countries still seeking to protect their consumers through subsidies.

UNCHECKED GROWTH IN FOSSIL FUEL USE AND CLIMATE CHANGE

Rising CO_2 and other greenhouse gas concentrations in the atmosphere, resulting largely from fossil energy combustion, are contributing to higher global temperatures and to changes in climate. Growing fossil fuel use will continue to drive up global energy related CO_2 emissions over the projection period. In the Reference Scenario, CO_2 emissions jump by 57 per cent between 2005 and 2030. The US, China, Russia and India contribute two-thirds of this increase. China is by far the biggest contributor to incremental emissions, overtaking the US as the world's biggest emitter in 2007. India becomes the third-largest emitter by around 2015. However, China's *per capita* CO_2 emissions in 2030 are only 40 per cent of those of the US and about two-thirds those of the OECD as a whole in the Reference Scenario. In India, they remain far lower than those of the OECD, even though they grow faster than in almost any other region.

More efficient use of energy has positive environmental benefits. In 2030, SO_2 emissions in China are 20 per cent lower in the Alternative Policy Scenario, compared with the Reference Scenario. An associated benefit is the dramatic reduction in CO_2 emissions, by an impressive 2.6 gigatonnes (Gt). In the Alternative Policy Scenario, CO_2 emissions stabilise soon after 2020. Lower fossil fuel use results in a 27 per cent reduction in CO_2 emissions in India in 2030, most of which stems from energy efficiency improvements on the demand and supply sides.

Urgent action is needed if GHG concentrations are to be stabilised at a level that would prevent dangerous interference with the climate system. The Alternative Policy Scenario shows that if measures currently being considered by governments around the world were enacted, they could lead to a stabilisation of global emissions in the mid 2020s and cut this level in 2030 by 19 per cent relative to the Reference Scenario. Yet global emissions would still be 27 per cent higher than in 2005. Assuming continued emissions reductions after 2030, the Alternative Policy Scenario projections are consistent with stabilisation of long term CO_2 equivalent concentration in the atmosphere at about 550 parts per million. According to the best estimates of the Intergovernmental Panel on Climate Change, this concentration would correspond to an increase in average temperature of around 3°C above pre-industrial levels.

© Anjo Eijeriks/Fotolia

> **"The primary scarcity facing the planet is not of natural resources nor money, but time."**

In order to limit the average increase in global temperatures to a maximum of 2.4°C, the smallest increase in any of the IPCC scenarios, the concentration of GHGs in the atmosphere would need to be stabilised at around 450 ppm. To achieve this, CO_2 emissions would need to peak by 2015 at the latest and fall between 50 and 85 per cent below 2000 levels by 2050. This would require energy related CO_2 emissions to be cut to around 23 Gt in 2030, 19 Gt less than in the Reference Scenario and 11 Gt less than in the Alternative Policy Scenario. In a '450 Stabilisation Case', which describes a notional pathway to achieving this outcome, global emissions peak in 2012 at around 30 Gt.

Emissions savings come from improved efficiency in fossil fuel use in industry, buildings and transport, switching to nuclear power and renewables, and the widespread deployment of CO_2 capture and storage (CCS) in power generation and industry. Exceptionally quick and vigorous policy action by all countries, and unprecedented technological advances, entailing substantial costs, would be needed to make this case a reality.

ADDRESSING GLOBAL ENERGY CHALLENGES REQUIRES COLLECTIVE ACTION

The emergence of China and India as major players in global energy markets makes it all the more important that all countries take decisive and urgent action to curb runaway energy demand. The primary scarcity facing the planet is not of natural resources nor money, but time. Investment being made now in energy supply infrastructure will lock in technology for decades, especially in power generation. The next 10 years will be crucial, as the pace of expansion in energy supply infrastructure is expected to be particularly rapid. China's and India's energy challenges are the world's energy challenges, which call for collective responses. There can be no effective long term solution to the threat of climate change unless all major energy consumers contribute. The adoption and full implementation of policies by industrialised countries to address their energy security and climate change concerns are essential, but far from sufficient.

Author
Dr Fatih Birol is Chief Economist and Head of the Economic Analysis Division of the Paris based International Energy Agency. He is organiser and director of the *World Energy Outlook* series, the IEA's flagship publication. Dr Birol worked for six years in the Secretariat of the Organization of Petroleum Exporting Countries (OPEC) in Vienna, before joining the IEA in 1995. He has received several distinguished awards including the International Association of Energy Economics' Outstanding Contributions to the Profession Award, a Chevalier dans l'ordre des Palmes Academique from the French Government and the Golden Honour Medal of the Austrian Republic in recognition of his outstanding contribution to the understanding of global energy issues.

Organisation
The International Energy Agency (IEA) acts as energy policy advisor to 26 member countries in their effort to ensure reliable, affordable and clean energy for their citizens.

The IEA publishes the *World Energy Outlook*, on which this article is based. The series is widely recognised as the most authoritative source for forward looking energy market analysis. More than 20 IEA analysts contribute to the publication, which also benefits from the input of distinguished energy and climate change experts from around the world.

Enquiries
Website: www.worldenergyoutlook.org

Coal's role in a global technology revolution

Today, 1.6 billion people do not have access to electricity. They and their fellow citizens around the globe depend on energy to enhance their lives. Coal and all the world's energy resources are being called upon to meet these expectations, and the demands are growing. Coal stands apart because it is the world's most abundant, affordable and geographically dispersed fossil fuel. As a result, the National Mining Association (NMA) and coal producers around the world have looked for ways to make increased coal use compatible with a shared desire to address emissions of greenhouse gases.

Through the NMA's participation in research and development efforts in the United States, domestic and international public-private partnerships, and information sharing forums, we believe the groundwork can be laid for the deployment of large scale projects around the world aimed at demonstrating the effectiveness of various technology approaches. This brief overview outlines the technology roadmap upon which we are embarking.

Coal is the world's most abundant and geographically dispersed fossil fuel. Nations such as the United States, China, India and South Africa, each with huge domestic reserves, rely on coal to meet their electricity needs because coal provides secure, reliable and affordable power. Coal is the primary provider of baseload power for nations around the globe, see Figures 1 and 2.

GROWTH IN ENERGY DEMAND

The International Energy Agency (IEA) estimates that worldwide demand for all energy resources, including coal, will continue to grow. The most rapid growth will occur in developing economies as they become increasingly industrialised and their citizens benefit from improving standards of living and demand products that are taken for granted in developed economies. As a result, IEA estimates that developing countries will account for 55 per cent of world primary energy demand in 2050, compared with 37 per cent in 2003 and 22 per cent in 1971, see Figure 3.

These are compelling forces that are unlikely to be reversed. Indeed, achievement of the objectives of the United Nations Universal Declaration of Human Rights depends on economic development that is fueled by energy. Increasingly, there will be greater competition for the world's energy resources. All of these factors underline coal's continued and growing importance as a worldwide energy provider. For many nations, having an abundant and secure source of energy is reason enough for continued use of coal. Those benefits are further augmented by an inescapable fact: projected energy demand simply cannot be met over the long term without significant coal based generation.

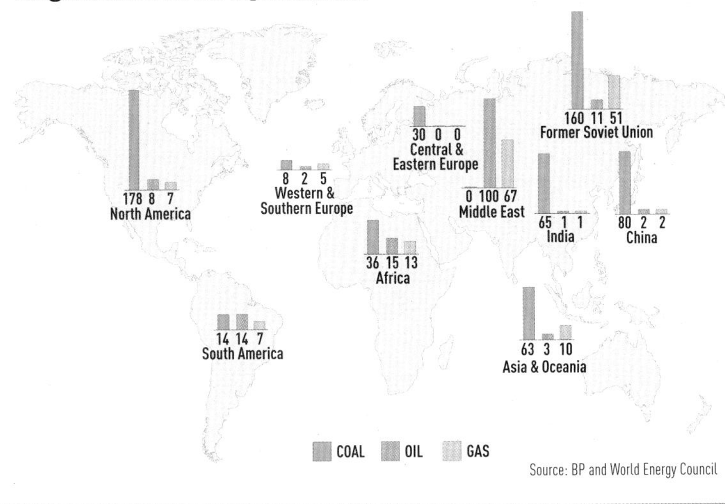

Figure 1: Location of the world's main fossil fuel reserves (Gigatonnes of oil equivalent).

Figure 2: Coal use in electricity generation, 2006 (e).

(e) estimated * 2005 data, source: World Coal Institute.

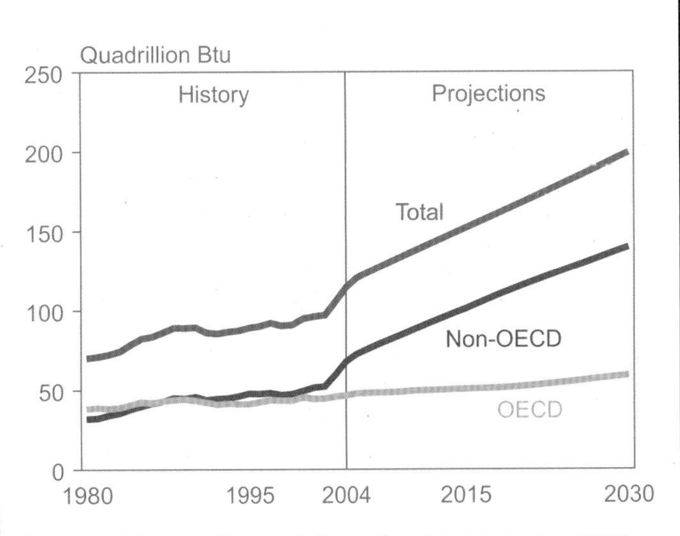

Figure 3: World coal consumption by region, 1980-2030.

Sources: **History:** Energy Information Administration (EIA), *International Energy Annual 2004* (May-July 2006), web site www.eia.doe.gov/iea. **Projections:** EIA, System for the Analy sis of Global Energy Markets (2007).

TECHNOLOGY INNOVATION AND TRANSFER

How can continued and growing coal use be reconciled with policies to reduce greenhouse gas (GHG) emissions? The NMA believes they cannot be, until there is widespread recognition that all frameworks for emissions reductions must rely on innovations in technology. Timetables, targets and market mechanism must be aligned with, rather than established apart from, the ability of technology to deliver solutions that are not financially punitive, that produce tangible results and that can be embraced globally.

The recently released study by the World Energy Council (WEC) in its June 2007 report, *Energy and Climate Change 2007*, makes the point that, using a broad definition of technology to include research, development, demonstration, technology choice and deployment and technology transfer, "technology is central to nearly all governments' approaches in both developed and developing worlds, though of course there are considerable differences of emphasis." With total investments in new energy technologies over the coming decades expected to be in the trillions of dollars, both in developed and developing countries, delivery mechanisms are receiving widespread attention.

WEC, along with others, has identified bureaucratic and political obstacles that often stand in the way of technology transfer. Among them are issues over maintenance and reliability and treatment of intellectual property rights. A number of international groups, including the Expert Group on Technology Transfer within the UN framework, are looking at these issues.

In concert with coal based stakeholders internationally, US coal producers are supporting a suite of approaches and technologies.

IMPROVEMENTS IN EFFICIENCIES AT EXISTING POWER PLANTS

Efficiency improvements result in lower greenhouse gas emissions. Governments should be encouraged to pursue policies that support, rather than inhibit, these improvements, which can be instrumental in flattening the growth in greenhouse gas emissions. This is a course of action we believe can produce substantial results in the US. The WEC also points out that "bringing a coal-fired plant in China and India up to the level of efficiency of a new German plant would deliver emissions savings commensurate with those expected from the whole Kyoto process, while improving the efficiency of resource use and enabling increased access to modern energy."

CONSTRUCTION OF FUTUREGEN

FutureGen is a public-private partnership to design, build, and operate the world's first coal-fueled, near zero emissions power plant, at an estimated net project cost of US$1.5 billion. The commercial scale plant will prove the technical and economic feasibility of producing low cost electricity and hydrogen from coal while nearly eliminating emissions. It also will support testing and commercialisation of technologies focused on generating clean power, capturing and permanently storing carbon dioxide, and producing hydrogen. Importantly, the FutureGen Industrial Alliance is made up of US and international coal producers and utilities, including China's largest coal based utility. This provides a built-in mechanism for technology sharing and transfer.

CARBON CAPTURE AND STORAGE TECHNOLOGY (CCS) AND CLEAN COAL TECHNOLOGIES

The IEA *Energy Technology Perspectives 2006* report concludes that, "Clean coal technologies with CCS offer a particularly important opportunity to constrain emissions in rapidly growing economies with large coal reserves, such as China and India. CCS is indispensable for the role that coal can play in providing low cost electricity in a CO_2 constrained world." IEA further estimates that coal equipped with CCS could contribute 12 per cent of the total reduction in CO_2 emissions in 2050- more than that achieved by hydroelectric power, biomass and other renewables, including solar and wind, combined.

Commercialisation of CCS is complicated by the large capital investments required, the long life of the assets and the need to ensure CCS technologies can be successfully integrated at least cost to consumers at coal based power plants that rely on varying combustion platforms, eg gasification technologies (IGCC) and clean combustion technologies (Subcritical, Supercritical and UltraSupercritical). Public policies are needed to help overcome these barriers and provide aggressive and sustained government support for technology demonstration and deployment through mechanisms such as capital grants, operational subsidies, tax incentives, accelerated depreciation and public investment in needed infrastructure.

The US is currently conducting CCS demonstration projects through partnerships managed by the US Department of Energy, and the National Mining Association and the industry's coal based customers are analysing funding mechanisms and policies aimed at providing the long term commitments needed to bring these technologies from the demonstration stage to widespread commercial deployment.

Coal is America's dominant energy resource

Coal is America's most abundant energy resource, making up 94 per cent of US fossil energy reserves (on a British thermal unit basis). At current rates of use, the US has more than 240 years of remaining coal reserves. Because of its abundance, reliability and affordability, coal generates half of US electricity, a share that is expected to grow to 57 per cent of the expanding electricity market by 2030.

Recent analyses by the US Environmental Protection Agency, academic institutions and energy consultants have all concluded that, even in a carbon constrained economy, the United States will continue to need coal to generate at least half of its electricity over the long term. Coal's prominent role in US energy supply suggests that arbitrarily restricting coal's use for a significant portion of electricity generation will invariably lead to an increasing reliance on imported energy, greater price volatility and an inability to meet energy needs.

Coal's role in meeting future US energy demand extends beyond electricity generation. Coal will be used to produce clean diesel transportation fuels ideally suited to power commercial and military aviation, long haul road transportation, railroads, ocean transport and other industrial sectors.

Amid increasing competition for scarce global petroleum resources, coal-to-liquid (CTL) transportation fuels are a viable option for reducing global petroleum use and limiting the economic risks associated with global oil price fluctuations, while achieving critical environmental improvement goals.

Coal's role in climate solutions

The NMA supports accelerated funding for research and development into technologies capable in the near term of reducing emissions through greater fuel efficiency and through carbon capture and storage technology that will provide a long term solution. NMA believes a national commitment to a technology programme for arresting and reducing CO_2 emissions is the strategy most compatible with economic growth and environmental improvement.

The widespread adoption of clean coal technologies since 1980 has already enabled US power plants to reduce major emissions by 40 per cent, at the same time the amount of coal used to generate electricity increased by 75 per cent and the economy grew by 93 per cent. Additional reductions of sulphur dioxide, nitrogen oxide, particulates and, for the first time, mercury, are expected as a result of new requirements.

The National Mining Association believes climate policies that incorporate a technology based approach will foster the development of alternative fuel technologies, promote economic growth and the efficient use of energy while addressing environmental concerns and promoting social advancement.

The US is also working with its partners in the Asia-Pacific Partnership (APP) on Clean Development and Climate Change to determine how these technologies can be deployed internationally. APP's founding partners (Australia, China, India, Japan, Republic of Korea and the US) and Canada, now an official partner, represent about half of the world's economy, population and energy use. The partnership is focusing on expanding investment and trade in cleaner energy technologies, goods and services in key market sections.

CONCLUSION

Other countries and governments may determine that alternative technologies and approaches are more suited to their individual needs and capabilities. The common need shared by all, are policies that give long term support for technology development and deployment, a clear understanding of the timelines and costs involved, and a commitment to so align emissions reduction strategies.

While coal producing interests around the world have focused on reducing emissions at power plants, greenhouse gas emissions arise either directly or indirectly from almost all human activity. Significantly reducing greenhouse gas emissions is a worldwide challenge requiring attention across all economic sectors.

Ultimately, solutions must balance energy affordability, energy security and availability and the preservation and enhancement of the quality of human life with our desire to achievement measurable results in reducing greenhouse gas emissions. Failure to do so will invite public resistance and loss of commitment.

Organisation

The National Mining Association (NMA) represents the interests of the US coal mining industry before government on legislative, regulatory and legal matters. NMA provides a forum for industry leaders to analyse issues of the day and to develop positions on public policy matters, including energy policy and climate.

NMA and its members are committed to the development and implementation of policies that address climate change, including reductions in greenhouse gas emissions from all sources. We also recognise the US's strong reliance on its coal reserves (the world's largest) to continue to provide American consumers and industries with abundant and affordable fuel to generate electricity, to serve as a feedstock for the future development of transportation fuels, and as a vital input to steelmaking.

Enquiries

Website: www.nma.org

Figure 1: Capture installation at a small gas-fired power plant in Malaysia.

Carbon dioxide capture and storage: a future for coal?

BERT METZ, SENIOR SCIENTIST, THE NETHERLANDS ENVIRONMENTAL ASSESSMENT AGENCY AND **HELEEN DE CONINCK**, SCIENTIFIC RESEARCHER, ENERGY RESEARCH CENTRE OF THE NETHERLANDS (ECN) AND INSTITUTE FOR ENVIRONMENTAL STUDIES, VU UNIVERSITY OF AMSTERDAM (IVM)

Worldwide coal use is rising dramatically and will remain an important source of energy for decades to come. It provides many developing countries with an abundant and affordable energy source to fuel their energy-hungry economies, and helps raise incomes for millions of the world's poor. But the increased use of coal significantly increases greenhouse gas emissions, far beyond sustainable levels, and is a great threat to controlling climate change.

Carbon dioxide capture and storage (CCS) is a new technology which offers the only way to significantly reduce greenhouse gas emissions without denying developing countries the use of coal. Assuming sufficient policy efforts and if risks and other public concerns can be addressed, CCS could be structurally deployed on new power plants worldwide in about 10 to 15 years.

INTRODUCTION

The economies of China, India and a number of other developing countries are booming. This rapid growth improves the life of millions of people and helps to reduce poverty but energy is needed for this development. In many of the fast growing developing economies, coal is the most important energy source. In 2004, 78 per cent of electricity produced in China was generated by coal. In India this was 69 per cent, and for all of developing Asia 65 per cent. In South Africa it was 94 per cent. China, India and South Africa are responsible for about half of all world coal consumption.

With growing concern about secure energy supply and abundant coal reserves, coal use is projected to increase worldwide, resulting in increasing CO_2 emissions. However, to control climate change, a drastic reduction of CO_2 emissions is needed with coal having to make a major contribution, thereby seriously limiting its use. A new technology, CO_2 capture and storage (CCS), may come to the rescue, making it possible, in principle, to continue using coal, while drastically reducing its emissions. But the technology still needs to be applied at large scale, is not cheap and has risks that need to be controlled. Can CCS indeed provide a future for coal in a carbon constrained world?

WHAT IS CARBON DIOXIDE CAPTURE AND STORAGE?

Carbon dioxide capture and storage comprises three components: capture of CO_2 from a large point source, transport to a storage location and storage in a geological formation.

Capture of CO_2

Point sources need to be large to make it economically attractive to capture CO_2. Examples of such sources include power plants and natural gas production wells, see Table 1. Capture can occur by separating CO_2 from flue gases or natural gas, eg through chemical absorption or membrane technology. CO_2 can also be captured before combustion by first gasifying and treating coal. A capture installation for a large CO_2 source is the size of a small chemical factory, see Figure 1, and the energy use of capture and compression is substantial, leading to lower power plant efficiencies if CCS is applied.

The capture step is by far the most costly component of CCS and research is underway to find more efficient capture processes. Although there are no full scale power plants with CO_2 capture facilities yet, several are planned and no major technological challenges are expected. Some CO_2 sources, such as existing hydrogen production facilities (in refineries and fertiliser plants) and gas recovery operations already provide a pure CO_2 stream, considerably reducing costs of CO_2 capture.

> **"Assuming sufficient policy efforts, CCS could be structurally deployed on new power plants worldwide in about 10 to 15 years."**

Table 1: An overview of global stationary CO_2 point sources larger than 0.1 million tonnes (Mt) of CO_2/yr.

(IPCC Special Report on CCS, based on The International Energy Agency's Greenhouse Gas R&D Programme)

Process	Number of sources	Emissions ($MtCO_2$/yr)
Fossil fuels		
Power	4,942	10,539
Cement production	1,175	932
Refineries	638	798
Iron and steel industry	269	646
Petrochemical industry	470	379
Oil and gas processing	N/A	50
Other sources	90	33
Biomass		
Bioethanol and bioenergy	303	91
Total	**7,887**	**13,466**

Transportation and storage of CO_2

Transport of CO_2 to the storage location is expected to occur through pipelines. Carbon dioxide transport in pipelines is already employed on a large scale in the US and Canada to transport CO_2 used in enhanced oil recovery operations. For long distance (more than 1,000 km) and overseas transportation, shipping is an option.

For underground storage, CO_2 could be injected in oil or gas reservoirs, deep saline formations or coal beds, Figure 2. Depleted oil or gas reservoirs have the advantage that the reservoir has contained oil and gas for a very long time, providing some guarantee for

Figure 2: Overview of geological storage options.

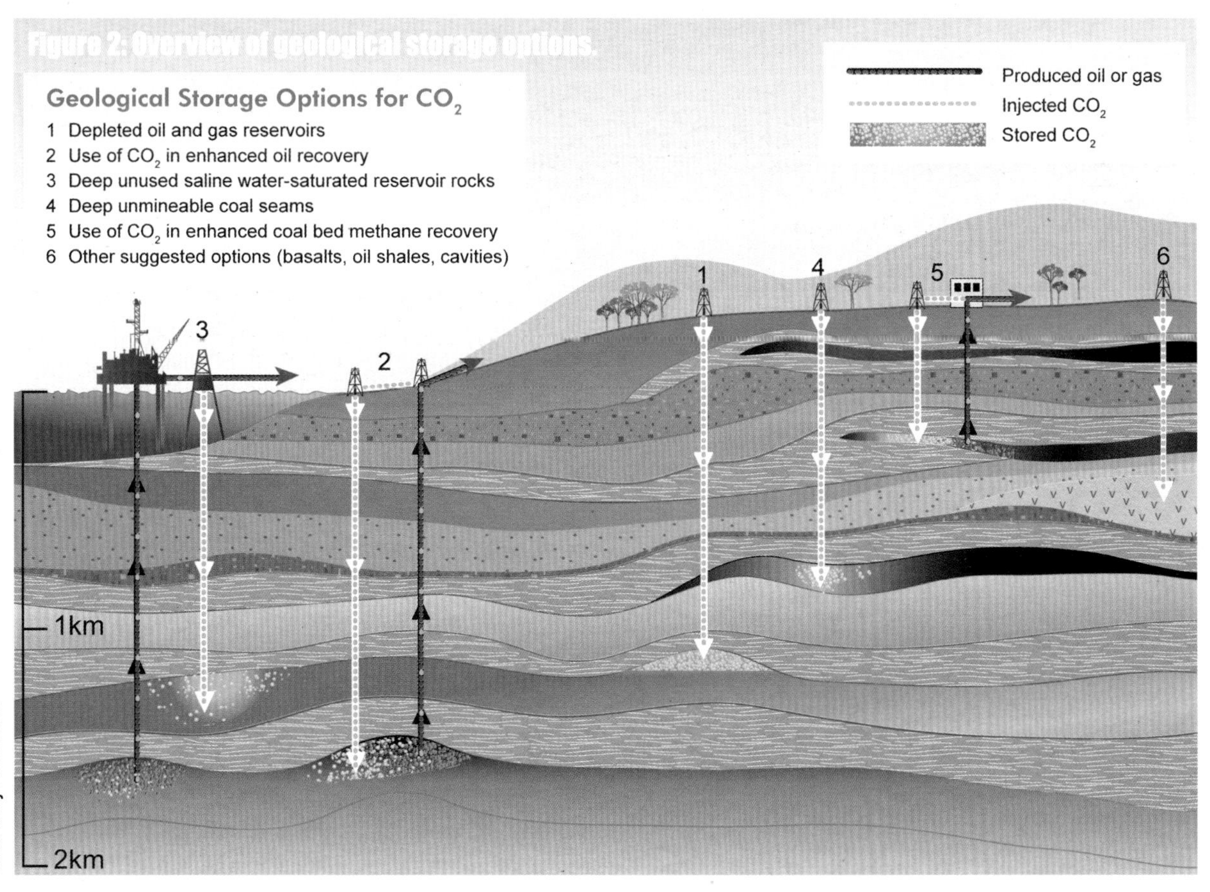

Courtesy: CO2CRC

Figure 3: Worldwide CO_2 point sources (top) and prospective storage locations (bottom).

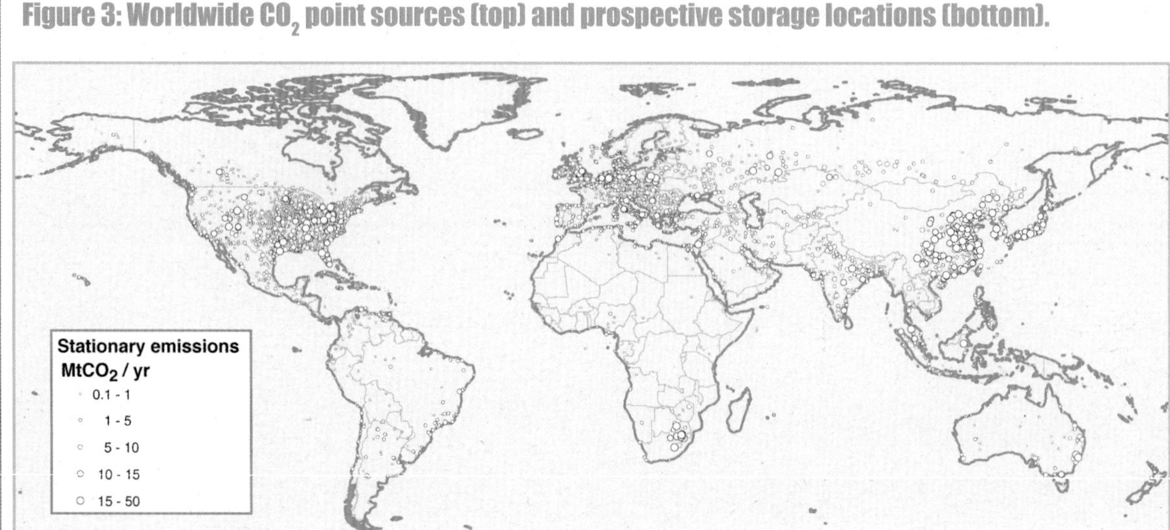

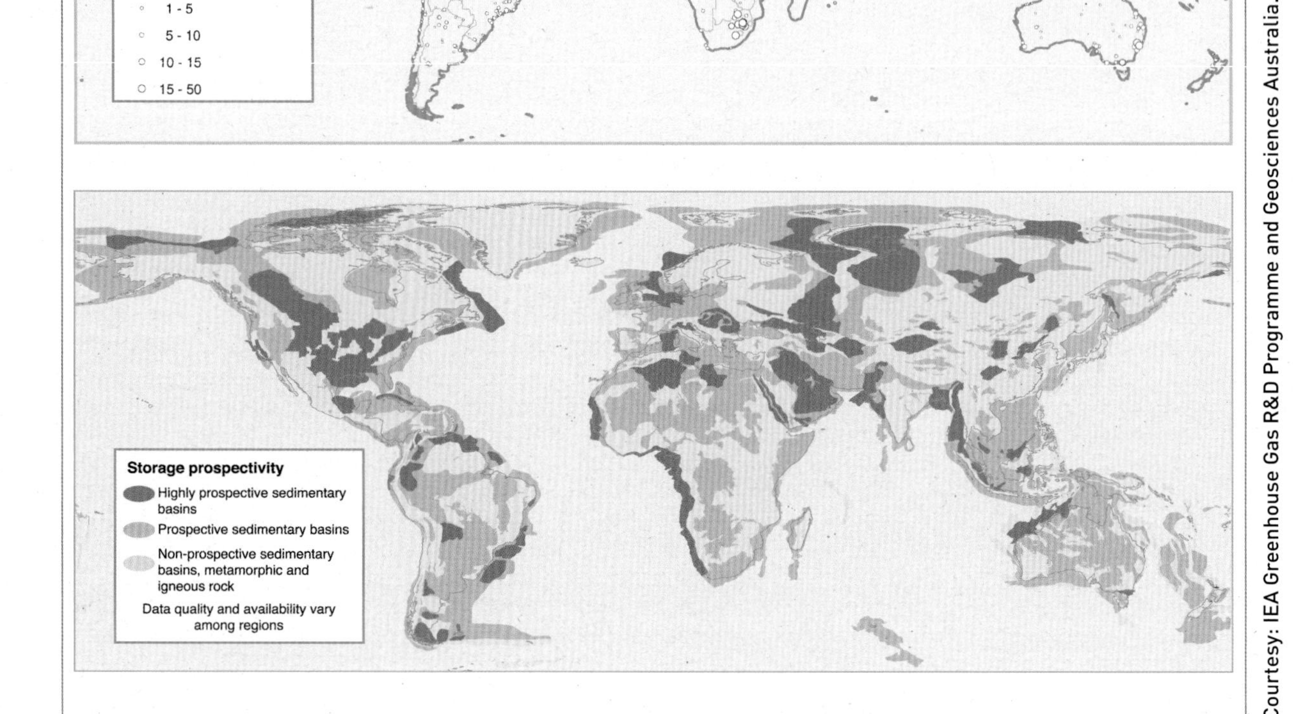

Courtesy: IEA Greenhouse Gas R&D Programme and Geosciences Australia.

storage permanence if old abandoned wells can be controlled. The injection of CO_2, a technique already commonly used by the oil and gas industry, can also enhance oil and gas recovery in almost-depleted reservoirs, reducing costs.

Probably the largest potential for CO_2 storage is in deep, saline water-bearing formations, with several large CO_2 injection projects currently being implemented, such as the Sleipner project in Norway. The feasibility of CO_2 storage in coal beds, with or without the recovery of coal bed methane, depends strongly on the permeability of the reservoir. Although small trials are happening, the potential of this storage option is likely to be limited.

Possibilities of capture and storage of CO_2 can be found all over the world and most point sources of CO_2 are within reasonable distance from promising storage locations, see Figure 3. Current knowledge also reveals that the overall capacity of geological storage is sufficient to store the CO_2 that is likely available to be captured. For some countries or regions, however, the situation may be less advantageous because of the lack of suitable underground formations.

> **“Possibilities of capture and storage can be found all over the world. Current knowledge reveals that the overall capacity of geological storage is sufficient to store the CO_2 available for capture.”**

ECONOMIC ASPECTS OF CCS

Adding CCS to a coal-fired power plant is not cheap. According to the Intergovernmental Panel on Climate Change (IPCC) Special Report it adds US$1-5 cents

Figure 4: Comparison of costs of CCS for early opportunities and in the power sector, with price levels for CDM and ETS based carbon credits.

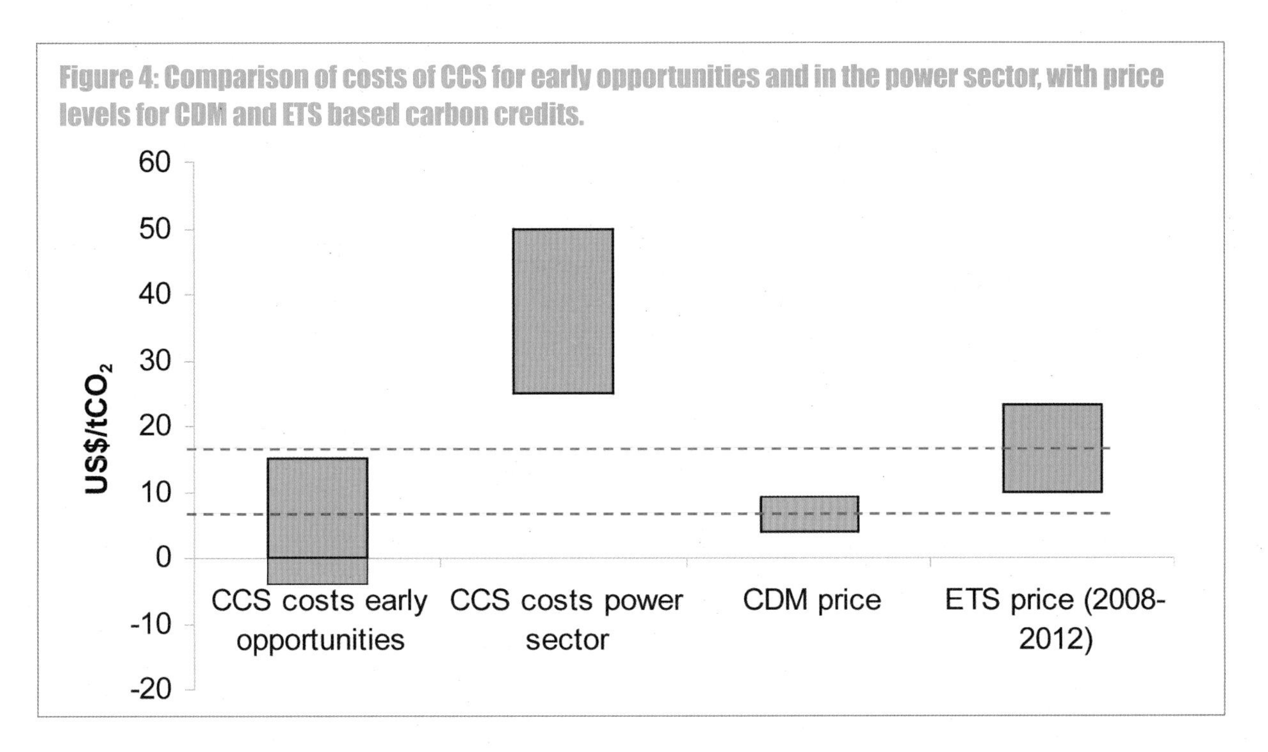

per kWh to the costs of electricity, depending on the technology applied and the local circumstances. For industrial customers who pay prices close to production costs this could mean a 25-100 per cent increase. Even for households normally paying a higher price, this is a substantial rise. Only a substantial carbon price increase through policy intervention could change the picture, enabling CCS to become economically attractive. According to the IPCC, carbon prices would have to be US$25-30per tonne CO_2 to see CCS deployed at a significant level.

> "Public perception studies on CCS show that the general public is reluctant rather than enthusiastic about CCS, and that 'not in my back yard' feelings play a role."

Several industrialised countries committed to emission reductions under the Kyoto Protocol are establishing a price on carbon emissions. The most extensive scheme is the European Union Emissions Trading Scheme (ETS). Carbon prices in the ETS have fluctuated heavily over the past two years, showing extremely low prices of US$5 and peaking occasionally around US$40 per tonne of CO_2. The current 2008 – 2012 forward price is about US$25 per tonne. In addition to the ETS, the Kyoto Protocol's Clean Development Mechanism (CDM) has established a carbon price in developing countries. These CO_2 prices are substantially lower, currently between US$5 and US$15 per tonne. CCS is currently waiting approval as a project activity under the CDM, which is an issue of considerable controversy and diverging political views in the climate negotiations.

Given the cost and limited incentives, it is not surprising that CCS has yet to be applied commercially at coal-fired power plants, see Figure 4, either in industrialised countries, or in developing countries with emerging economies. Although the economics of CCS look better for some other industrial sectors, particularly sources with pure CO_2 emissions, an incentive is still required to make it happen.

PUBLIC AND PRIVATE VIEWS ON CCS

Even if the costs are acceptable to society, risks of CO_2 leakage play a significant role in public perception of CCS. According to the IPCC, risks of CO_2 storage can be comparable to similar operations if appropriate governmental regulation regarding site selection and characterisation, monitoring and remediation is in place. Two major international marine treaties, the OSPAR and London Conventions, have recently reached agreement on such regulation, providing a useful precedent for other national and international policymaking. Regulation is also currently under development in the EU and Australia.

Despite this progress, public perception studies on CCS show that the general public is reluctant rather than enthusiastic about CCS, and that 'not in my back yard' feelings play a role. Education, transparent procedures and rule making, and public consultation seem essential to allow for public acceptance of CCS. Environmental organisations can play a role if they recognise the need for CCS, but they also see permanence of CO_2 storage as an uncertainty, and are mindful that CCS might become a replacement for, rather than a supplement to, energy efficiency and renewable energy.

Industry has generally welcomed the possibility of CCS rather than moving away from fossil fuels altogether. As early as 1996, the Norwegian oil company Statoil pioneered CCS in the Sleipner project, responding to a high CO_2 tax on offshore operations. BP operates a CCS project in In Salah, Algeria, and has announced several others in the US and Australia. Coal companies such as AngloCoal, formerly rejecting any climate policy, are lobbying for CCS. Coal, gas and even biomass-using electricity companies in Europe are announcing demonstration after demonstration. The private sector all over the world sees CCS as a major way to achieve extensive emissions reductions.

THE WAY FORWARD

What are the prospects for CCS application in the future? This depends strongly on the question of when a carbon price of US$25-30 per tonne CO_2 will be reached. Uncertainty about future carbon prices and heavy price fluctuation most likely play a role in the hesitating position of industry. Most CCS projects depend on government subsidies and will take at least another 10 to 15 years before CCS could become standard for coal-fired power plants in Europe.

What does that mean for the rapidly-rising CO_2 emissions form coal-fired power plants in emerging economies? Given their pressing economic development needs it is unlikely they can commit to stringent climate change policies any time soon. And unless CDM carbon prices skyrocket and CCS is approved under the CDM, it will not be sufficient to cover the costs of CCS in the power sector.

> **“The private sector all over the world sees CCS as a major way to achieve extensive emissions reductions.”**

There are, however, prospects for CCS when considered in a broader context. China, for example, is rapidly becoming one of the main suppliers of modern 'supercritical' coal-fired power plants. It is already building such plants in large numbers and is moving to the most advanced technology: integrated coal gasification systems. This technology has much lower air pollutant emissions and is the cheapest when it comes to adding CCS, possibly even through retrofitting. China could, with appropriate additional investment made available, eg through cooperation such as the zero carbon power consortium of China, the EU and the UK, become a leader in coal gasification-based CCS power plants, in conjunction with major international corporations.

CONCLUSION

Carbon dioxide capture and storage provides possibilities to make fossil fuels, and coal in particular, part of the solution to climate change. However, this will not happen automatically. Governments in industrialised countries and emerging economies alike should develop regulations to ensure that CO_2 storage occurs safely and permanently, with adequate financial incentives for CCS deployment at coal-fired power plants put in place.

Authors

Dr Bert Metz is currently a senior scientist with the Netherlands Environmental Assessment Agency (MNP) having been associated with the Agency since 1998. During the Netherlands presidency of the European Union in 1997 he chaired the Working Group that prepared the European position in the climate negotiations and the draft decisions on EU climate policy. Since 1996, he has been co-chairman of Working Group III on Climate Change Mitigation of the Intergovernmental Panel on Climate Change for the preparation of the Third and the Fourth Assessment Reports.

Heleen de Coninck is an atmospheric chemist and environmental scientist and works as a researcher at the Energy Research Centre of the Netherlands (ECN). Since joining ECN in 2001, she has worked on several issues, including rural electrification, the Clean Development Mechanism, CCS, and emissions trading. She was the coordinator of the IPCC Special Report on CCS and is currently pursuing a PhD with the VU University of Amsterdam on post-2012 climate policy.

Organisation

The Netherlands Environmental Assessment Agency is providing scientifically based information to national and international governmental bodies. It covers all aspects of environmentally sustainable development, environment and nature protection. The Energy Research Centre of the Netherlands carries out independent research on the application of energy technologies in all aspects of the energy system and provides energy policy advice to national and international governments.
private sector.

Enquiries

Bert Metz
Netherlands Environmental Assessment Agency
Division of Global Sustainability and Climate
PO Box 303
3720 AH Bilthoven
The Netherlands
Tel: +31 30 2743990
Fax: +31 30 2744464
Website: www.mnp.nl *and*
www.mnp.nl/ipcc for IPCC WG III related information.

Renewable energy in India:

STATUS AND FUTURE PROSPECTS

V SUBRAMANIAN, SECRETARY,
MINISTRY OF NEW AND RENEWABLE ENERGY,
GOVERNMENT OF INDIA

India's need to increase energy provision for its population and fast-growing economy coincides with increased concerns regarding climate change globally. This poses a formidable challenge but is perceived to be a great opportunity for the country to increase the share of renewables in the overall energy mix. India's approach to the global problem is to meet its energy needs in a responsible, sustainable and eco-friendly manner. A brief outline of government policies and issues related to renewable energy financing, large scale dissemination, research and development are given here, along with the role of renewable energy in contributing to national energy security.

INTRODUCTION

India is a developing and fast-growing large economy and faces a great challenge to meet its energy needs in a responsible and sustainable manner. India's task is to provide energy to over 600,000 human settlements, spread over 300,000 square km of territory, with a population of over one billion which is still growing, but expected to stabilise at around 1.6 billion during the next 40 years. The total primary energy supply in India has grown at a compound rate of around 3.4 per cent since independence to reach 537.7Mtoe (million tonnes of oil equivalent) in the year 2005 (IEA 2007). While commercial primary energy grew at 5.3 per cent over the period, non-commercial energy grew at only 1.6 per cent, which is a reflection of industrialisation. As a result, the share of commercial energy grew from

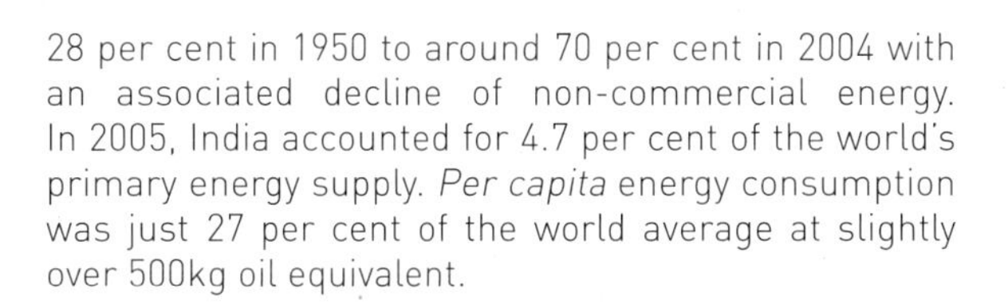

28 per cent in 1950 to around 70 per cent in 2004 with an associated decline of non-commercial energy. In 2005, India accounted for 4.7 per cent of the world's primary energy supply. *Per capita* energy consumption was just 27 per cent of the world average at slightly over 500kg oil equivalent.

ELECTRIC POWER

India accounted for 3.1 per cent of the world's electricity consumption in 2005 with an installed capacity of 135,780 MW as of September 2007. Of this, 87,200 MW is accounted for by thermal power plants, 34,200 MW by large hydro plants, 4,100 MW by nuclear, and the balance from renewable sources. The consumption of electricity in India rose from 4,157 GWh in 1950 to 38,6134 GWh in 2004/05. The *per capita* consumption was 612 kWh in 2004/05 as against 329 kWh in 1990 (CEA). Despite the significant growth in electricity generation, shortage of power continues to exist primarily due to the growth in power demand outstripping the growth in generation and generating capacity addition. In May 2007, the country experienced an estimated eight per cent energy shortage and 12.3 per cent shortage of peaking power. Even so, the 2001 census recorded 12 .5 per cent of urban households and 56.5 per cent of rural households as still unelectrified.

MODERN ENERGY PROVISION

One of India's major challenges is to provide a large proportion of the country's population with access to modern energy sources. Around 86 per cent of rural households and more than 20 per cent of urban households still rely primarily on traditional fuels, such as firewood, wood chips or dung cakes, to meet their cooking needs. The use of traditional fuels can cause health problems arising from indoor air pollution. Only five per cent and 2.7 per cent of rural households use LPG and kerosene respectively as a primary cooking fuel whereas 44 per cent and 22 per cent of urban households uses LPG and kerosene respectively. With low standards of living, ie below the per-person-a-day International Poverty Line of US$2, at Purchasing Power Parity (PPP) rates of about 75 per cent population, the task of providing modern energy services becomes severely compounded. This has resulted in low levels of *per capita* energy and electricity consumption on account of low levels of purchasing power.

> "India's task is to provide energy to over 600,000 human settlements, spread over 300,000 square km of territory, with a population of over one billion."

Projections made by the Integrated Energy Policy Committee of the Planning Commission have estimated that in order to meet the projected GDP growth of eight per cent per annum by 2031-2032, the demand for primary energy will increase to 1,836Mtoe representing almost a four fold increase since 2003-04. Commercial energy requirements would also be around 1,651Mtoe, which is an approximate five fold increase since the year 2003-04.

RENEWABLE ENERGY

India intends to provide a reliable energy supply through a diverse and sustainable fuel mix that addresses major national drivers. These include security concerns, commercial exploitation of renewable power potential, eradication of energy poverty, ensuring availability and affordability of energy supply and preparing the nation for imminent energy transition.

The country has an estimated renewable energy potential of around 85,000 MW from commercially exploitable sources: Wind, 45,000 MW; small hydro, 15,000 MW and biomass/bioenergy, 25,000 MW. In addition, India has the potential to generate 35 MW per square km using solar photovoltaic and solar thermal energy.

GRID-INTERACTIVE RENEWABLE POWER

By March 2007, renewable electricity, excluding hydro above 25 MW installed capacity, has contributed 10,243 MW representing 7.7 per cent of total electricity installed capacity. There has been phenomenal progress in wind power and, with an installed capacity of over 7,100 MW, India occupies the fourth position globally.

DECENTRALISED AND STAND ALONE RENEWABLE ELECTRICITY SYSTEMS

Over 3,000 remote and inaccessible villages and hamlets have been provided with basic electricity services through distributed renewable power systems. In addition, over 75 MW biomass based gasification systems in the capacity range of 10-100 kW are in use for small scale industrial applications and electrification purposes. Finally, over 1.3 million solar home lighting systems, including lanterns and street lights have been set up in different parts of the country.

HEAT ENERGY FOR COOKING PURPOSES

Since the 1970s, around 3.9 million family-type biogas plants have been set up to provide clean cooking energy options in rural areas. Biogas based cooking in rural areas has made cooking a pleasure with associated social and environmental benefits including zero indoor pollution.

PROCESS HEAT FOR DOMESTIC, INDUSTRIAL AND COMMERCIAL PURPOSES

Use of solar thermal systems has started gaining momentum, with a solar collector area of 1.9 million sq metres already installed to meet these needs.

> "Since the 1970s, around 3.9 million family type biogas plants have been set up to provide clean cooking energy options in rural areas."

LIQUID BIOFUELS FOR TRANSPORT APPLICATIONS

The large scale development of biofuels, including straight vegetable oil (SVO), biodiesel and bioethanol is still in its infancy. In 2004 around 0.1Mtoe ethanol was used for blending with petrol. Biodiesel use is still negligible. However, a policy framework for blending five per cent ethanol with petrol and the development of a biodiesel programme, based on non-edible oil, has been developed.

Source: http://www.remotepower.com/ramakrishna.html

RENEWABLE OUTLOOK

The Integrated Energy Policy Report of the Planning Commission of India has observed that the contribution of modern renewables to India's energy mix by 2031-32, excluding large hydro, would be around five-six per cent. However, our estimates indicate that by 2032, renewable power capacity, excluding large hydro could contribute up to 10 per cent of the total electricity generation in the country. About 25,000 remote villages could be provided with basic electricity services through renewable and seven per cent of the rural population would meet its cooking energy needs through biogas and other modern renewable energy systems. With focused biofuel programme, around seven to 10 per cent of oil needs could be met through biofuels. While this figure appears small, the distributed nature of renewables can provide many socioeconomic benefits. Further, its impact in abating greenhouse gas emissions would be significant. Widespread deployment of renewable systems would also create significant employment potential for unskilled and semi-skilled workers.

REGULATORY FRAMEWORK

India has been pursuing a three-fold strategy for the promotion of renewables:

- Providing budgetary support for research, development and demonstration of technologies.
- Facilitating institutional finance from various financial institutions.
- Promoting private investment through fiscal incentives, tax holidays, depreciation allowance and remunerative returns for power fed into the grid.

India's renewable energy programme is primarily private sector driven and offers significant investment and business opportunities. A large domestic manufacturing base has been established in the country for renewable energy systems and products. The annual turnover of the renewable energy industry, including the power generating technologies for wind and other sources, has reached a level of over US$10 billion. Companies investing in these technologies are eligible for fiscal incentives, tax holidays, depreciation allowance and remunerative returns for power fed into the grid. Further, the Government is encouraging foreign investors to set up renewable power projects on a 'build, own and operate' basis with 100 per cent foreign direct investment.

The most important legislative development which has stimulated the recent growth in renewable power is the Electricity Act of 2003. The Act recognises the role of renewable energy technologies for supplying power to the utility grid as well as in stand alone systems. The Act also has several provisions favourable for renewable power, including rural electrification. Its most important feature, however, is its empowerment of the State Electricity Regulatory Commissions (SERCs) to promote renewable energy and to specify a percentage of the total consumption of electricity in the area of a distribution licence that will be purchased from renewable energy sources. This is considered a major boost for renewable energy promotion in India.

> **"The annual turnover of the renewable energy industry, including the power generating technologies for wind and other sources, has reached a level of over US$10 billion."**

RENEWABLE ENERGY AND CLIMATE CHANGE

India's first National Communication (2004) reveals that the energy sector accounts for around 61 per cent of total national emissions. For fossil fuels, coal combustion had a dominant share of emissions, amounting to around 64 per cent of all energy emissions. With regard to India's emissions trajectory, the Integrated Energy Policy Report of the Planning Commission has observed that "Since GHG emissions are directly linked to economic activity, India's economic growth will necessarily involve increases in GHG emissions from the current extremely low levels. Any constraints on the emissions of GHG by India, whether direct, by way of emissions targets, or indirect, will reduce growth rates, and impair pollution abatement efforts."

Due to its vast market potential for renewable energy projects, and a relatively well developed industrial, financing and business infrastructure, India is perceived as an excellent country for developing Clean Development Mechanism (CDM) projects. As such, India has emerged as one of the most favoured destinations for CDM projects globally, with renewable energy projects having the major share. National renewable energy plans offer ample opportunity for CDM projects and technological innovations, such as biogas for transport application, offer new areas for project development.

TECHNOLOGY CONCERNS

The feasibility of a larger application of renewable energy, to that of the present assessments, would depend on how rapidly costs decline and efficiencies increase. As a result, research and technology development have been accorded high priority in the national renewable energy programme and mission mode research has been planned for developing solar, bioenergy and hydrogen technologies. India encourages international cooperation in renewable energy R&D, through well defined projects with proper division of labour and responsibilities for specific tasks with equitable financial burden and credit sharing arrangements. Bilateral, as well as multilateral, scientific and technological cooperation agreements could provide a framework for such R&D activities.

> "India has emerged as one of the most favoured destinations for CDM projects globally, with renewable energy projects having the major share."

Technology plays a central role in addressing climate change issues. In this context there is a need to treat renewable energy technologies as a 'global common' in the medium term. To begin with these technologies could be placed in the public domain and joint research and development projects could be taken up between the institutions of developed and developing countries. Technology transfer costs could be fixed at no-profit level and the expenditure to be incurred in these acquisitions could be made from a global funds under climate change mechanisms.

CONCLUSION

Indian efforts for promoting renewable energy are in harmony with global concerns. India's strategies focus on:

- Working towards lowering the relative price of new and renewable power technologies through a continuous and focused research and development effort.
- Improving access to reliable, affordable, economically viable, socially acceptable and environmentally sound energy services and resources.

The approach in India matches the global aim of ushering in a carbon free economy; an economy based on a fuel mix mainly provided by the green or renewable energy technologies.

For India, new and renewable energy development and deployment is of great importance from the point of view of long term energy supply security, decentralisation of energy supply particularly for the benefit of the rural population, environmental benefits and sustainability. In this context, the Indian renewable energy programme is a goal-oriented effort to meet the country's energy requirement in an environmentally sound way.

Author

Presently, Secretary to the Government of India since February 2006, Shri Subramanian was a commerce graduate of the University of Madras and a qualified banker who started his career as the Sub-Divisional Magistrate at Kalna and Barrackpore in the State of West Bengal. He moved to the Government of India in 1983 as Deputy Secretary, Department of Expenditure and was a Director in the Department of Economic Affairs during 1985-89. In 1990 he went on a Commonwealth assignment as Adviser on Loan and Grant Management to the Government of Mozambique. On his return to West Bengal, he was Power Secretary and Labour Secretary in the State Government.

Organisation

The Ministry of New and Renewable Energy (MNRE) is the nodal Ministry of the Government of India at the Federal level for all matters relating to new and renewable energy. The Ministry has been facilitating the implementation of broad spectrum programmes including harnessing renewable power, renewable energy to rural areas for lighting, cooking and motive power, use of renewable energy in urban, industrial and commercial applications and development of alternate fuels and applications. In addition, it supports research, design and development of new and renewable energy technologies, products and services.

Enquiries

Ministry of New and Renewable Energy
Block-14, CGO Complex
Lodhi Road, New Delhi-110 003
India
Website: www.mnes.nic.in

A roadside tuck shop.

©SELCO Solar Light-India

Renewable energy financing;

PV SOLAR HOME SYSTEMS FROM SOUTH INDIA

JYOTI PRASAD PAINULY, SENIOR ENERGY PLANNER, UNEP RISOE CENTRE, RISOE NATIONAL LABORATORY, DTU, DENMARK, AND **H V KUMAR**, DIRECTOR, CRESTAR CAPITAL INDIA PRIVATE LIMITED, MUMBAI, INDIA

The UNEP programme described here aimed to facilitate access to clean energy to households in South India using PV solar home systems. Accessing finance was found to be a critical barrier and to increase finance access, risk perception from partner banks had to be addressed. This was achieved through technical and market development support, including equipment specification development, vendor qualification, and development of a competitive market through a multivendor approach. More than 18,000 households have now received clean energy directly through the programme with a current vibrant market for SHS in Karnataka, India.

ENERGY OPTIONS IN INDIA

More than 60 per cent of households in South India are estimated to have no access to electricity and, even when connected to the electricity grid, rural and semi rural communities suffer power outages and fluctuating power quality. These households tend to rely on less efficient energy sources, such as kerosene for light and dung and wood for heat. This not only adversely impacts health but limits both economic and social development. Poor quality of electricity has led many rural households to consider generators, inverters and solar PV systems.

Solar energy using photovoltaics (PV) is a clean energy alternative to displace conventional polluting fuels, such as kerosene, for lighting and other applications. With significant PV sales and service infrastructure already in place, PV solar home systems (SHS) offers a viable alternative for lighting. But the lack of a credit market has constrained its growth. The households, who need it most, do not have access to credit needing subtle public interventions, just enough to drive the market forward in a sustainable way.

The general aim of the UNEP programme described here was to facilitate access to clean energy to households in South India, particularly poorer households in rural and semi urban areas with the most power shortages with no access to better, more expensive alternatives.

Is solar home system (SHS) a viable option?

Electricity produced using a PV system is not cheap and can only help households tide over the problem of access to reliable electricity, mainly for lighting purposes. Grid electricity is often subsidised making any comparison difficult. However, if subsidy was assumed to be eliminated and electricity prices assumed to increase consistently, an SHS, despite the high initial cost, can be an attractive long term option when compared to costly or unreliable power supplies, see Table overleaf.

Without financing, however, the high initial cost of PV will continue to constrain the growth of the Indian SHS market. But an increased access to credit from the banking sector could enable rural households to buy cleaner energy and pay for it with money that is currently being spent on less efficient, and often more polluting, forms of energy.

Table: Monthly costs in Rupees for a household with four lights.

Period	Existing grid customer	New grid customer	Kerosene	Inverter	SHS
First five years	115	297	212	465	325
10 years	148	298	272	465	200

Source: Economic and Financial Assessment of SHS for the State of Karnataka, Crestar Capital, March 2002.

INDIA'S BANKING SECTOR

India has a well developed rural banking infrastructure, but the links to the renewable energy sector are weak at best. For this UNEP project, the need was clear for some short term market intervention to assist banks in gaining confidence in renewable energy technologies and increasing their exposure to SHS. This would increase consumer access to credit and lower the cost of this credit. A market based intervention could remove key barriers and allow market forces to expand the SHS market without further external support.

Having identified access to finance as the main barrier, UNEP helped Indian banks develop lending portfolios of solar home systems. Potential customers for the loan facilities were households and small enterprises looking to use solar PV for either domestic or income generating activities. The programme was expected to help 18,000 customers directly and many more households indirectly, as a result of opening up the SHS credit market.

> **“Solar energy using photovoltaics (PV) is a clean energy alternative to displace conventional polluting fuels, such as kerosene, for lighting and other applications.”**

STAKEHOLDER CONSULTATION

During the first phase of the programme, a wide range of stakeholders in India were consulted, including SHS vendors, banks and other small scale financing institutions, governmental agencies, experts and NGOs. These consultations provided significant input on the state of the solar industry in South India, on the need for SHS financing and the best mechanisms for influencing growth of this financing.

FINDING THE RIGHT FINANCE PARTNERS

Based on consultations, the State of Karnataka was chosen as the target area for the initial phase of

A rural shop using solar light.

© SELCO Solar Light-India

A poultry farm using solar lights.

THE INTEREST RATE BUY-DOWN

The interest rate buy-down structure allows partner banks to offer consumer loans at a concessional interest rate of five per cent initially, seven per cent below the prime lending rate of 12 per cent.
While UNEP funded the interest rate buy-down, the banks provided the capital and carried 100 per cent of the credit risk. Banks are therefore motivated to maintain quality loan portfolios. Price distortion was minimal since the financing cost, not the capital cost was subsidised.

implementation mainly because of its strong rural banking system, established means of credit delivery and its vibrant solar industry with an extensive service infrastructure already in place.
After consulting with a number of banks, partnership agreements were signed with Canara Bank and Syndicate Bank, two of India's largest with branch networks in the target region.

Type of financing offered

The intervention model used was intended to encourage the market to grow and vendors to innovate in their product and service offerings while not substantially distorting the market. Importantly, a predetermined exit strategy was considered essential.
Feedback was solicited on a variety of possible finance support mechanisms, including capital subsidies, front-end and back-end interest rate subsidies, credit default guarantees, loan term extensions, beneficiary margin support and subsidised transaction costs.
After weighing the pro and cons of various alternatives, an interest rate subsidy was chosen for accelerating the market without overly distorting it in the long term. By providing loans with an interest rate buy-down (using the bank's capital but UNEP interest subsidies), both the high up front cost and high credit cost barriers could be addressed. The interest subsidy could be decreased progressively to make the exit easy.
This interest subsidy approach has in fact already been applied successfully in India, in the solar thermal sector by the Ministry of Renewable Energy (MNRE) and in the solar PV sector on a small scale by Selco, as a direct vendor initiative.

PROJECT'S APPROACH AND RESULTS

Based on discussions with partners, market development support was given by increasing awareness and providing information. Technical support was provided through vendor qualification and SHS specifications and training of bank staff (based on needs). Financial support was given through interest subsidy to end users (which also helped develop the market), and transaction cost support to the banks. A multivendor approach was taken to ensure quality products, competitive pricing and reliable after sales support. The programme was open to all vendors who fulfilled product quality and after sales service criteria. These were set through competitive forces, not programmatic regulations.
Since its launch, 18,000 customers have been served directly through the project over a period of four years by the partner banks via 2,000 branches. Many more were served outside the project as a result of the opening of the credit market for SHS in the region.
Interest subsidy was decreased in October 2004, and again in June 2005, bringing down the interest to nine per cent, which was close to the market rates. The interest subsidy reduction had no adverse impact on the programme, indicating its sustainability. The project won the prestigious Energy Globe Award in 2006 in recognition of its success.

ROLE OF THE BANKING SECTOR

Encouraged by the success of UNEP partner banks, several banks launched their own loan programmes independent of UNEP support and some broadly follow the practice established by the UNEP. Syndicate Bank continues to lend for SHS in Karnataka after the end of UNEP financial support and has extended it to all parts of India, although they have limited success elsewhere, mainly due to an under developed market. The Bank, however, registered good success with its 'Solar Grama' (Solar Village) initiative, where loans were provided mainly to lower income families for electrification of household clusters (through SHS).
Canara Bank successfully extended the loan programme to the neighbouring Kerala State. Although Canara continues to lend, the pace slowed down during 2005 and 2006 under competitive pressures from other banks. With the end of the formal partnership with Canara Bank in December 2006, the programme ended in South India.
New initiatives were taken up in the second half of 2006 and the programme was launched in early 2007 in Maharashtra in partnership with the Bank of Maharashtra, and in Gujarat in partnership with Sewa (a well known women's social organisation), and their financial arm, Sewa Bank. Technical support and support for awareness measures was provided to the partners under the programme. SHS sales at Sewa have picked up considerably, mainly on account of extensive product innovations by the vendor Selco to enhance livelihood opportunities using PV technologies. The Bank of Maharashtra is trying to put the programme on a sound footing. In addition, the project has supported international NGO, S3IDF, in providing solar lighting to the urban poor, eg hawkers and small shop owners.

Expanding into rural markets

Indian Banks have large rural banking networks and a wide range of products that cater to various customers.

© SELCO Solar Light-India

Light snack food stall stops using kerosene lamps after dark.

Adding to their range of asset products gives banks a competitive edge over the rest of the market, and solar home light loans address a critical need for their customers. Since most loans were sold in the rural markets, it helped banks fulfil specific business segment and geographical targets. The banks see the SHS business from a variety of perspectives, which includes extending their reach to rural markets, other business from the SHS customers, and satisfaction from meeting the needs of those who have no access to electricity for lighting, which they see as a part of their corporate social responsibility.

"Indian banks pursue stiff business targets for financial inclusion and solar light loans help them expand their reach in rural markets, especially among their lower income customers", said MBN Rao, Chairman and Managing Director of Canara Bank. "For small borrowers, the solar loan adds value to their livelihoods and increases cash flows in the form of higher productivity. Besides, our bank takes great pride in contributing to social change in association with UNEP. Putting solar home lighting systems and reliable energy supplies within the reach of the common citizen kindles hopes and aspirations of many for a better life and well being."

At grass roots level, bank managers see this as an opportunity to increase their business and derive a sense of satisfaction by meeting this crucial need for lighting. "This (SHS loan) helps the people to come to the bank and when they come ... they can get other financing facilities from the bank. We can do good business with them through dairy loans, education loans etc", says Vasudev Rao, Bank Manager of the Syndicate Bank, who was involved in the programme and advanced several SHS loans from his branch. "It is a very good programme, people are happy and as a bank, we are happy. We are also happy because we have done something good for the poor. It gives us satisfaction."

LESSONS LEARNT

Access to financing is a major barrier for rolling out renewable energy in India. The risk perception of financing institutions, and small or non-existent markets are two important issues that must be addressed for renewable energy to achieve its potential. Technical support and a code of best practice for vendors were used to address the issue of risk, and market development was supported through reduction of financial costs to consumers, and awareness programmes by the partners.

Stakeholders' involvement at every stage of the programme and in every decision was very important along with enough flexibility to incorporate changes as a result of consultations and changes in environment.

Renewable energy, especially PV electricity, is still expensive for the very poor. Therefore, it is important to integrate it with income generating activities, if the very poor are to be reached. Some were reached in the programme, through support for S3IDF's street hawkers lighting project, and the Sewa Bank's work in Gujarat.

Feedback mechanisms are other valuable tools that can make a programme successful. Annual audits of some branches of partner banks and customer satisfaction surveys were used in the programme as feedback mechanisms. Findings from these were used to prepare a good practices manual, and feedback was also used by the partners to improve the programme.

Although the programme is easily replicable, initial support may still be needed in other regions due to a need for market development through awareness programmes, and capacity development of financial institutions that may be new to this kind of programme.

Authors

Jyoti Prasad Painuly works as a Senior Energy Planner at UNEP Risoe Centre (http://uneprisoe.org), Denmark. He has wide-spread experience of working on a variety of issues in energy and environment area. This includes in energy and environment economics and policy, energy efficiency, renewables, technology assessment, climate change issues and bio-energy.

H V Kumar is Director of Crestar Capital India Private Limited and based at Mumbai, India. He advises on renewable energy finance and worked with the UNEP to manage the PV Solar Home Systems Project in South India.

Organisations

The UNEP Risoe Centre on Energy, Climate and Sustainable Development (URC) supports the United Nations Environment Programme (UNEP) in its aim to incorporate environmental aspects into energy planning and policy worldwide, with a special emphasis to assist developing countries.

URC is sponsored by UNEP, the Danish International Development Assistance (Danida) and Risoe National Laboratory. More information can be found at http://uneprisoe.org/about.htm.

Enquiries

Jyoti Prasad Painuly, Senior Energy Planner, UNEP Risoe Centre on Energy, Climate and Sustainable Development (URC), Risoe National Laboratory, DTU, Frederiksborgvej 399, PO Box 49, DK 4000 Roskilde, Denmark
Tel: +45 46 775167
Fax: +45 46 32 19 99
E-mail: j.p.painuly@risoe.dk

H V Kumar, Director, Crestar Capital India Private Limited, 4 Vishnu Mahal, D Road, Churchgate Mumbai 400 020, India
Tel: +91-22-22819944
E-mail: crestar@vsnl.net
Website: http://uneprisoe.org/IndiaSolar/index.htm; http://www.uneptie.org/energy/act/fin/india/

Because bioenergy can be developed in many areas with low input costs, it can benefit both small and large scale farming operations.

Sustainable bioenergy in Africa: issues and possibilities

MERSIE EJIGU, PRESIDENT & CEO
PARTNERSHIP FOR AFRICAN ENVIRONMENTAL SUSTAINABILITY (PAES) AND SENIOR FELLOW, FOUNDATION FOR ENVIRONMENTAL SECURITY AND SUSTAINABILITY (FESS)

Today, the majority of the African population finds itself trapped in a vicious cycle of low energy consumption, heavy dependence on traditional biomass energy, poverty and vulnerability to climatic variation. Modern bioenergy offers unique opportunities toward a strategy for unlocking this trap and developing smallholder business potential in Africa. With its tropical climate and huge land mass, Africa's bioenergy potential is almost unlimited. Harnessing it requires, at least, a two-pronged strategy: (i) promoting smallholder production and processing schemes; and (ii) encouraging socially and environmentally-sustainable large scale investment.

AFRICA AND TRADITIONAL BIOMASS ENERGY

Of the total energy Africa consumes, traditional biomass (solid wood, twigs and cow dung) accounts for 59 per cent providing fuel to about 320 million people and, according to forecasts by the International Energy Agency (IEA), this figure is set to increase. By 2015, the IEA predicts a further 54 million Africans will be dependent on traditional biomass. This surge in consumption comes primarily from two sources: (i) a rapid population growth estimated at 2.5 per cent (the world's highest), and (ii) slow industrial and service sector growth and subsequent failure to create off-farm employment opportunities.

BENEFITS OF MODERN BIOENERGY IN AN AFRICA SETTING

Modern bioenergy can be produced in liquid, gas, or solid form, with liquid biofuel the most common, and can be used for agro-industrial production, transportation, cooking, and lighting. Bioenergy technologies are expanding fast and have a number of key benefits for African production.

Broad-based development and greater multiplier effects

Widely available crops and plants are the basis for bioenergy production; crops that can be grown in many areas under rainfed conditions, with low input costs (fertiliser).

Because bioenergy can be developed in many areas with low input costs, it can benefit both small and large scale farming operations. It can open up new livelihood opportunities and domestic and export markets, enhancing rural economic transformation. This contrasts with the experience of countries with oil and mineral resources where wealth has typically been concentrated in the hands of the few, resulting in slow economic growth and continued deprivation of much of the population.

Suitable for smallholder agriculture business

More than 80 per cent of the population of many African countries is engaged in agriculture, which is primarily smallholder and largely subsistence. Modern bioenergy can be produced at this smallholder level, creating possibilities for generating cash income. It also encourages the formation of agricultural marketing and processing cooperatives, which, in turn, facilitate adoption of modern agricultural inputs and better farming practices.

Possibilities to export high value commodities

Modern bioenergy creates possibilities to produce a high value crop (biofuel/biodiesel) that has lucrative domestic and export market opportunities. Producing and processing biofuels at the smallholder level opens up new sources of income for a large segment of the population.

Low tech solutions to energy problems

Modern bioenergy technologies are simple and easily transferable since these are basically oil-pressing and alcohol distillation processes already known by local people. Most technologies can be easily operated and maintained with capacity that already exists at the village level by blacksmiths, and others developed with a little training.

Productive use of a vast area of uncultivated land

Africa's total cultivable land is estimated at 840 million hectares. Of this land, only 27 per cent is used (cultivated), compared to 87 per cent for Asia and 97 per cent for Southwest Asia. The development of bioenergy offers opportunities to convert the uncultivated area into energy wealth, given appropriate socially and environmentally responsible policy measures.

ENVIRONMENTAL BENEFITS OF MODERN BIOENERGY

With proper planning, bioenergy enhances the sustainable use of biological resources, ultimately supporting environmental recovery. Bioenergy also burns less carbon than traditional energy sources, helping to reduce greenhouse gas (GHG) emissions. Carbon dioxide is absorbed by new plants and recycled, rather than being released into the atmosphere. In contrast, carbon from fossil fuel combustion is fully released into the atmosphere.

CHALLENGES OF BIOENERGY IN AN AFRICAN SETTING

The promotion, production, marketing, and use of modern bioenergy face several constraints. However, ample opportunities exist to tackle these challenges.

High production costs

According to IEA, "grain-based ethanol costs on average around $0.30/litre ($0.45/litre of gasoline equivalent) in the US after production subsidies, which means competitiveness with gasoline at an average crude oil price of between $65 and $70 per barrel. In Europe, ethanol production costs, including all subsidies, are about $0.55/litre (0.80/litre of gasoline equivalent)". Brazil's ethanol industry, according to a presentation at the 2005 IIE biofuels conference, breaks even at $35 per barrel oil equivalent. However, Brazil produces almost all its sugarcane from rainfed agricultural practices and is believed to be the world's most efficient producer of sugarcane. In many African countries, which have lower agricultural productivity levels, production of both sugarcane and corn-based biofuel would entail much higher costs and therefore be far less price competitive with oil.

Food production and prices

Normally, when demand for corn, rapeseeds, soybean, and other biofuel crops rises, prices will also increase, thereby encouraging farmers to bring more land under cultivation. Under African conditions, however, the prevailing subsistence agricultural technology, poor market infrastructure, and low investment capacity may not permit expansion of both food and energy crops. The likely scenario would be for farmers to devote more of their land to cultivating biofuels feedstock and less to food and animal feed crops. This is likely to result in an across-the-board increase in crop prices and animal feed prices of animal feed.

Africa's energy consumption patterns

About 550 million people (75 per cent of the total population) in Sub-Saharan Africa have no access to electricity or any kind of modern energy services. Of the total energy Africa consumes, traditional biomass (solid wood, twigs, and cow dung) accounts for 59 per cent, electricity 8 per cent, petroleum 25 per cent, coal, 4 per cent and gas, 4 per cent.

Access to and efficiency of bioenergy technology

Much of the available knowledge on biofuels technology is based on large scale farming of two feedstocks: sugar cane and corn. Newer technologies that use a wide variety of feedstocks and operate at different capacities, particularly on a small and medium scale, are not yet widely available and easily accessible. For example, jatropha curcas has been a highly promising biodiesel feedstock in Zambia, India and other developing countries.

Land requirement

The land issue needs to be seen within the context of different production systems. For example, commercial farming runs the risk of crowding out small farmers to give space to large tracts of land needed by big business, and will clearly be constrained by land shortage. For smallholder agriculture, land availability would not be a serious constraint, particularly if appropriate land tenure and use policies are developed.

Policy and institutional weaknesses

The successful promotion, sustainable production, and marketing of bioenergy require strong policy and institutional support. However, policy and institutional weaknesses are often cited as primary constraints to the effective implementation of sustainable projects throughout Africa. What does make the bioenergy sector different is the magnitude and urgency of the energy problem, the supportive international climate, and the national, regional, and global processes that are promoting renewable energy development, in general, and bioenergy in particular.

Environmental impacts of monocultures

Supplying feedstocks to bioenergy plants requires growing the same crop year after year. Sugarcane, corn, sweet sorghum, and palm, currently the most known feedstocks, deplete soil nutrients. Continuous cultivation of these crops can turn arable land barren and eventually alter negatively the ecosystem and its services. Feedstocks grown on a small scale have greater possibilities for crop rotation, for example, switching between sweet sorghum and soybean, which replenish soil nutrients.

Threat to rainforest ecosystem

Sugarcane and palm have their best growth and productivity in high rainfall and humid areas, which are rainforest zones. Because investors usually hunt for good soils and rainfall, there will be a threat to Africa's small remaining rainforest. A good example is the highly publicised case of Mehta of Lugazi Sugar Works in Uganda that insisted on using a part of the Mabira Forest reserve, a pristine forest area richly endowed with biodiversity, for its sugar plantation. A government decision to grant the land was reversed following violent public protests that claimed some lives.

THE WAY FORWARD

The successful production, use, and marketing of modern bioenergy production require a long-term strategic vision of what feedstock to produce and how to produce it. A strong political commitment is needed also to translate the vision into concrete policy measures and implement them. Bioenergy investment and production need to be carefully planned and appropriate feedstocks chosen to ensure ventures are economically, socially, and environmentally sustainable.

Develop a national bioenergy strategy and policy

A national bioenergy policy is necessary to create an enabling environment for the production of sustainable bioenergy at both smallholder and large scale levels. Such a policy should take a holistic approach to the investment, production, processing and marketing of bioenergy. A well articulated bioenergy policy has huge multiplier effects and cross-sectoral impact that positively influences agricultural, industrial, and trade development. It is also an important means to mobilise external and internal resources.

> **"A national bioenergy policy is necessary to create an enabling environment for production at both smallholder and large scale levels based on a country's feedstock potential."**

Develop new and innovative funding mechanisms

In addition to accessing funding opportunities from traditional multilateral and bilateral sources, there is a need for bold new measures to generate funding, which may include: linking big investment and smallholder producers; micro-credit programmes; an infrastructure to reach widely-dispersed smallholder farms; public-private partnerships; concessionary loans; subsidies; cross-industry partnerships that tie the provision of one sector's services with funding to support bioenergy initiatives; and technical capacity to access global funds (e.g., Clean Development Mechanism and Global Environment Facility (GEF)).

Invest in research and development

Largely in response to high oil prices, extensive research is underway on first and second generation technologies. To fully benefit from this, African countries need to have the human resources and the institutional infrastructure to facilitate the transfer, adoption, application, and development of bioenergy technologies.

Increase the country's bioenergy potential through reforestation

Traditional biomass energy will continue to be the primary source of household energy in at least the next two to three decades. Tree planting is a low cost approach to increasing biomass density. Through improving vegetation cover, tree planting facilitates environmental recovery, reduces the hardship on women collecting firewood, mitigates climate change, generates income, and augments resources for modern bioenergy use.

CONCLUSION

Africa has huge bioenergy potential and technological advances make it highly promising. The current challenges of modern bioenergy, including the high cost of production of biofuels should not discourage our bioenergy outlook. The possibilities for developing non-food crops and perennial plants as biofuel feedstocks are broad and within reach, given the extensive research work underway at the global level. What is needed is not only political goodwill but also an enabling policy and institutional environment (on the part of key players) towards meeting Africa's sustainable energy and development objectives through promoting environmentally and socially sustainable large scale investments, while at the same time creating opportunities for smallholders to be energy producers.

Author

Mersie Ejigu is President and Chief Executive Officer of Partnership for African Environmental Sustainability (PAES) and Senior Fellow, Foundation for Environmental Security and Sustainability (FESS). Prior to that, he worked as Assistant Director General, Programmes and Policy of IUCN, The World Conservation Union based in Gland, Switzerland. He joined IUCN as the Regional Director of IUCN Eastern Africa. In his home country, Ethiopia, he was Minister of Planning and National Development of Ethiopia for nine years, during which time he guided, coordinated and managed the formulation of Ethiopia's annual, medium term, and long term development plans and the annual development budget.

Organisation

Partnership for African Environmental Sustainability (PAES) is a non-governmental organisation head-quartered in Kampala, Uganda, established to promote environmentally and socially sustainable development in Africa based on best practices.

Enquiries

Partnership for African Environmental Sustainability
5th Floor, Pan Africa House, PO Box 10273
Plot 3, Kimanthi Avenue, Kampala, Uganda
Tel: ++256 41 267068 | Fax: ++256 41 267041
E-mail: info@paes.org; mejigu@paes.org

PO Box 50914, Lusaka Zambia

PAES USA, 1120 19th Street,
N W Suite 600, Washington, D.C. 20036
Tel: +1 (202) 744 4357
Fax: +1 (202) 785 5904 or (703) 760 0797

Turning waste into energy

A NIGERIAN INITIATIVE TO REDUCE POLLUTION AND GENERATE CLEAN ENERGY

JULIA STEETS
PROJECT MANAGER,
GLOBAL PUBLIC POLICY INSTITUTE

There are many ways to improve the environmental performance of an organisation. One option is to work in partnership with other stakeholders. Often, this requires more time and greater initial investment than independent activities, but the solutions achieved are often more sustainable. This article highlights the 'Cows to Kilowatts' initiative in Nigeria, a partnership project to reduce the water pollution and greenhouse gas emissions of a slaughterhouse. With the help of innovative technology from Thailand, the project turns abattoir waste into household gas and organic fertiliser, providing local communities with a clean and cheap source of household fuel.

LOCAL PARTNERSHIPS AND THE SEED INITIATIVE

In the face of the challenges of climate change, environmental degradation and poverty, an increasing number of companies, governments and NGOs are opting to join forces. Large scale partnerships involving global players are the most visible examples of such cooperative approaches. Yet projects initiated at the local level are often more concrete and offer tangible outcomes.

The Seed Initiative (Supporting Entrepreneurs for Environment and Development) was founded by UNDP, UNEP and IUCN-The World Conservation Union to promote and support locally driven, entrepreneurial partnerships for sustainable development. It found that a huge number and variety of such local initiatives exist, often working to enhance environmental sustainability while at the same time alleviating poverty and hunger. In doing so, such initiatives often rely on the generation or transfer of relevant knowledge and technology.

The Nigerian project, 'Cows to Kilowatts', is an example of such a project, which aims to reduce water pollution and greenhouse gas (GHG) emissions and was one of five Seed Award winners in 2005. The Cows to Kilowatts initiative epitomises how innovative and cooperative approaches can have a real impact on the environment as well as local communities.

UNTREATED SLAUGHTERHOUSE WASTE

Slaughterhouses are a major source of water pollution and GHG emissions, especially in the developing world. Specific regulations for abattoirs often do not exist or are poorly monitored and enforced, with wastewater often remaining untreated and entering local rivers and water sources. This represents an immediate environmental problem, affecting the development of aquatic life. In addition, slaughterhouse waste often carries zoonotic diseases, animal diseases that can be transferred to humans, and the anaerobic degradation of wastewater generates methane and carbon dioxide, which contributes to climate change.

The biogas plant at Ibadan, due to open in 2008 is designed to produce 1,500 m³ of biogas per day and reduce daily GHG emissions by over 1.4 tonnes of CO_2.

© Cows to Kilowatts Initiative

In Ibadan, Nigeria, the NGO Global Network for Environment and Economic Development Research (GNEEDR) drew attention to this issue. Founder and Executive Chairman, Dr Joseph Adelegan, studied the effects of wastewater discharged from the Bodija Market Abattoir. He found that levels of organic pollution were highly elevated and had strongly negative impacts on nearby communities.

WASTEWATER TREATMENT

Dr Joseph Adelegan joined forces with two other Nigerian organisations to develop a solution. These two organisations were the Centre for Youth, Family, and the Law, a group of lawyers dedicated to environmental protection and community projects and the Sustainable Ibadan Project, part of UN-HABITAT's Sustainable Cities Programme.

> “The Cows to Kilowatts initiative epitomises how innovative and cooperative approaches can have a real impact on the environment as well as local communities.”

The first solution embraced by this group was to build an effluent treatment plant. Discussions with experts, however, revealed a need for improvement of this initial plan. Treating effluents with conventional methods effectively reduces water pollution, but leads to increased emissions of methane and carbon dioxide. The team therefore set out to find an alternative approach which would minimise the carbon footprint of the initiative.

COWS TO KILOWATTS

A solution involved capturing and transforming the gas emissions into a useful product. Relevant technology for achieving this had been created by a Thai research institution, the Center for Waste Utilization and Management at King Mongkut University of Technology Thonburi. Anaerobic fixed film reactors were developed to treat agroindustrial waste and produce biogas. With the help of an adapted form of this technology, slaughterhouse waste could be turned into clean household cooking gas, as well as organic fertiliser.

This approach offers at least three crucial advantages. Firstly, it minimises water pollution from slaughterhouse waste. Secondly, it significantly reduces the GHG emissions generated by the slaughterhouse and/or the treatment of its waste. Thirdly, it creates valuable biogas products. Through biogas sale, the project can not only become economically self sustainable, but turn into a profitable enterprise.

Implementation I: building the partnership

The Cows to Kilowatts initiative went through various phases, beginning in 2001 with the waste treatment and biogas production plant currently being built and expected to start operations in early 2008.

In a first crucial step, the project's initiator GNEEDR had to find relevant and competent partners. As head of an NGO, Joseph Adelegan needed others to contribute their expertise and resources to develop a sustainable solution to the problem of water pollution in Ibadan. Several organisations provide key inputs to the project:

- GNEEDR developed the initial project idea, conducted primary research on water pollution, represents the initiative and handles the construction of the plant.
- The Nigerian Center for Youth, Family, and the Law provides legal advice and helps engage local stakeholder groups, such as the local butchers' association and the Bodija market development association.
- The Sustainable Ibadan Project was crucial in securing the support of the government.
- The World Bank's Global Development Marketplace gave an important impetus to the initiative by suggesting the integration of a renewable energy component in its design.
- The Thai research institute provided the initial technology and adapted it for use with slaughterhouse waste.
- The Seed Initiative provided further input into the development of the project and brokered a crucial contact to UNDP Nigeria.

Implementation II: financing the initiative

Capital requirements for designing and constructing the waste treatment and biogas plant, as well as administering the project and consulting with local stakeholders amounted to around US$500,000. The project is designed to be commercially viable and envisages selling its household cooking gas at 25 per cent of current market prices: US$7.5 per 25 litres. Producing around 270 m³ of compressed biogas a month, the plant would generate returns on investment after two years. With an estimated lifespan of 15 years, the plant is expected to create substantial economic returns.

Despite these features, it proved difficult to get affordable commercial finance for a promising but untested project in Nigeria. The initiative gained international recognition through its selection as a finalist in the World Bank Global Development Marketplace and as a Seed Award winner, but no financial support. Finally, UNDP provided the necessary start-up capital through its Programme in Energy and Environment.

Implementation III: adapting the technology

The Thai research institute at King Mongkut University of Technology Thonburi had developed an innovative technology for treating agroindustrial waste and generating biogas. Through the use of anaerobic fixed film reactors, the institute managed to achieve much higher treatment efficiency, handling larger quantities of waste and generating high quality biogas at a faster rate than conventional technologies. It successfully applied this technology for treating waste from a rice starch factory and a fruit canning factory.
In an excellent example of South-South cooperation, the institute agreed to adapt its anaerobic fixed film reactor technology for use with slaughterhouse waste after being contacted by the Cows to Kilowatts initiative. Test results show that the adapted reactor can handle an organic load rate ranging from two to 10 kilograms of chemical oxygen demand (COD) per m^3 (COD is used as a measure of organic pollution in wastewater) and operates with a retention time of two to four days. The technology yields between 0.4 and 0.5 m^3 of biogas per kilogram of COD, containing 60 to 70 per cent methane. The tests also indicate that the technology is stable.

Implementation IV: plant construction

Even with secured finances, project implementation did not start immediately. UNDP's Programme in Energy and Environment is nationally executed, which means that funds are normally only disbursed to national governments. In the Cows to Kilowatts case, the Nigerian Federal Ministry of Environment agreed to receive and transfer the resources to the partnership. This, however, implied that the initiative had to go through complex formalities and deal with bureaucratic hurdles.
With the adaptation of the relevant technology completed and the design of the biogas and waste treatment plant finalised by the Thai research institution, construction began in 2007. The plant is scheduled to be completed by March 2008.

IMPACT OF THE INITIATIVE

Once the waste treatment and biogas production plant starts operating, it is expected to generate several positive environmental, economic and social impacts. The plant is designed to produce 1,500 m^3 of biogas per day and to capture 900 m^3 of pure methane per day. This is equivalent to a daily reduction of GHG emissions of the slaughterhouse of over 1.4 tonnes of CO_2. In addition, the plant will treat 3,500 m^3 of wastewater a day and produce 1,750 litres of organic fertiliser.
The captured methane will be upgraded and compressed for use as household cooking gas. It is expected to be sold locally, generating additional employment. The gas is expected to be distributed to around 5,400 households each month. The predominantly poor families will benefit from the gas because it constitutes a cleaner alternative to other commonly used fuels. At significantly lower cost than currently available sources of natural gas, it will reduce indoor air pollution and associated health hazards in poor communities.

© David Steets

ROLL-OUT AND REPLICATION

Many other cities in Nigeria and across the African continent are facing similar environmental and health challenges from untreated slaughterhouse waste. Once proven successful, the Cows to Kilowatts initiative carries great potential for further roll-out and replication.
Through its use of innovative technology, the Cows to Kilowatts initiative offers a solution to waste treatment which at the same time minimises the carbon footprint of slaughterhouse operations. Moreover, it is economically self sustainable and even profitable, generating a classic win-win situation. Finally, the pilot project in Ibadan is financed with the help of international donor money. Since the plant is expected to repay its start up capital within two years, the necessary financial resources will be available for replicating the project by 2010.

Author

Julia Steets is a Project Manager with the Global Public Policy Institute, Berlin. From 2004 – 2006, she was responsible for GPPi's work on the Seed Initiative Research & Learning Channel. She published the research results in *Partnerships for Sustainable Development: On the Road to Implementation* (2006), available at http://gppi.net/partnershipreport.

Organisation

The Global Public Policy Institute (GPPi) is an independent, nonprofit think tank based in Berlin and Geneva. Its research focuses on effective and accountable governance and its consulting practice offers services for public and private institutions. GPPi provides a platform for debate and fosters strategic communities bringing together the public sector, civil society and business. Until 2006, GPPi coordinated and led the Research & Learning Channel of the Seed Initiative, which supports locally driven partnerships for sustainable development.

Enquiries

Julia Steets
Global Public Policy Institute (GPPi)
Reinhardtstr. 15, 10117 Berlin
Germany
Tel: +49 178 298 5399
E-mail: jsteets@gppi.net
Website: www.gppi.net

Mitigating climate change: the Brazilian perspective

THELMA KRUG
SECRETARY OF CLIMATE CHANGE AND ENVIRONMENTAL QUALITY,
BRAZILIAN MINISTRY OF THE ENVIRONMENT

This article looks at the efforts made in Brazil to prevent climate change. These include originating the Clean Development Mechanism, the use of hydropower and the production of biofuels, namely bioethanol and biodiesel.

THE CLEAN DEVELOPMENT MECHANISM

The Clean Development Mechanism originated from a proposal by Brazil during the Kyoto Protocol negotiations in 1997 to ensure the participation of developing countries in helping Annex I Parties to achieve their emission reduction commitments under the Kyoto Protocol. Proposed originally as a Clean Development Fund that would be replenished every time an Annex I country did not fulfill its emission reduction commitment, the Clean Development Mechanism was finally agreed as part of the three mechanisms of the Protocol, along with emissions trading and Joint Implementation.

Brazil occupies third position relative to the total amount of greenhouse gas emission reduction achieved by all participating non-Annex I Parties in the Clean Development Mechanism in the first commitment period. Annual emission reductions of the order of 35 Gg of CO_2 equivalent (CO_2e) are expected during this period, corresponding to roughly 2.5 per cent of the total CO_2e emissions from Brazil in 1994.

By the end of October 2007, the total number of project activities in the Clean Development system totalled 2,618. Of these, 251 were from Brazil, third after China and India, which, taken together, accounted for approximately 61 per cent of the CDM projects activities (1,608).

The two most relevant gases in the Brazilian CDM project activities are carbon dioxide (65 per cent) and methane (35 per cent). Emission reductions are achieved mostly through electricity generation projects, corresponding to approximately 60 per cent of the total project activities in Brazil, distributed as follows:

- Biomass cogeneration, 54 per cent.
- Small hydro plants, 16 per cent.
- Wind energy, 13 per cent.
- Hydroelectric power plants, 12 per cent.
- Biogas, five per cent.

Two other sectors, in addition to the energy sector, are also relevant to the emission reduction efforts in Brazil: agriculture, through methane emission reduction from waste handling, and disposals in pig farming and landfill. The CDM is viewed as part of a broader effort that encompasses not only the traditional and long term investments in renewable energy, including hydro plants for electricity generation and the use of biomass to generate fuel for vehicles and small aircrafts but, most importantly, actions and policies to combat illegal deforestation, in particular in the Brazilian Amazon.

Unfortunately, the distribution of CDM project activities among non-Annex I Parties has, until now, been far from equitable, with some regions benefiting much more than others. Nonetheless, one has to bear in mind that the CDM was conceived as a market based stimulus for mitigation projects, regardless of the geographical balance. Eligibility constraints also influence the geographical distribution of the CDM, particularly in Africa.

China alone will account for than half of the expected emission reduction certificates in the first commitment period, followed by India. The leadership of China can partly be explained by the opportunity to develop specific projects that generate a large volume of carbon credits at a reasonably low cost, such as the reduction of hydrofluorcarbon 23 (HFC23) production.

HYDROELECTRIC GENERATION

Eighty-five per cent of the electricity generated in Brazil comes from hydroelectric sources, which is natural for a country with an estimated total water availability of 257,790 m^3/s. Out of the 403 TWh electricity generated in Brazil in 2005, nearly 84 per cent comes from hydroelectric plants, indicating the outstanding share of this source in electrical energy generation. The remaining is distributed among other sources, including renewables, natural gas, oil, nuclear and coal.

Actions and policies to combat illegal deforestation in the Brazilian Amazon are crucial to mitigate climate change.

THE NATIONAL ETHANOL PROGRAM – PROALCOOL

The National Ethanol Program was created in 1973, triggered by the first petroleum crisis. By then, the country had already more than 40 years of research in blending ethanol with gasoline for transportation purposes.

Substantial investments in R&D were provided by the government until in 1979 the first vehicle fuelled with 100 per cent hydrated ethanol was introduced in the market, completely replacing the use of gasoline. In order to guarantee the establishment of the ethanol industry, the government implemented a set of policies that ranged from provision of incentives to producers, through guarantee of payment and credit facilities to increase production, lowered price of the ethanol relative to gasoline, provision of tax incentives for ethanol fuelled cars, requiring ethanol pumps in all petrol stations and maintenance of safety stocks.

> "Out of the 403 TWh electricity generated in Brazil in 2005, nearly 84 per cent comes from hydroelectric plants."

Results from research indicate that on a well to wheel basis, the use of the non-hydrated ethanol from sugarcane produced in Brazil avoided the emission of nearly 2.7 kg CO_2 per litre of ethanol used to replace gasoline. This results from a number of measures that include:

- Investments in R&D to increase productivity.
- Intensive use of sugarcane residues to generate energy (co-generation).
- Use of the *vinhoto* (residue from the distillation process) as organic fertiliser in agriculture.
- Use of residues in the manufacture of biodegradable plastic and recycled paper production.
- Use of hydrated sugarcane bagasse as animal feed.

In addition to a number of co-benefits of replacing gasoline with hydrated ethanol in light vehicles, it is estimated that a total of 644 million tonnes of CO_2 were not emitted to the atmosphere during the period 1975 – 2005. This achievement has been acknowledged by the Intergovernmental Panel on Climate Change, which also refers to the introduction of flex-fuel vehicles in Brazil in 2003. This is expected to bring new impetus to ethanol use in transportation in Brazil due to their ability to run using any mixture of gasoline and ethanol as motor fuel. Sales of flex-fuel vehicles have increased since then to a level where they represented more than 75 per cent of the total number of cars sold in 2006.

Sugarcane production

Most Brazilian sugarcane production concentrates in Brazil's southeastern region (69 per cent), and its potential for expansion has raised several concerns regarding the potential impact on food security and pressure on forested land. Forests, including plantations, currently occupy around 464 million hectares while presently the sugarcane crop occupies approximately six million hectares, half of which is used for ethanol production. This represents less than one per cent of Brazil's national territory and less than 10 per cent of its total cultivated area. Brazil anticipates expansion of the sugarcane crop by an additional two million hectares, to ensure production of an additional eight billion litres of ethanol. Consistent investments in technological advances in cattle breeding are expected to reduce around 20 million hectares of pasture land.

THE NATIONAL PROGRAM FOR BIODIESEL PRODUCTION AND USE

The Biodiesel program was launched in 2004 to promote the introduction of biodiesel in the Brazilian energy matrix in a sustainable way: to ensure diversification of renewable sources of energy; to improve energy security; and reduce environmental impacts. The Program targets poor regions in the country, in particular the semi-arid and northeast regions, seeking to reduce regional inequalities. Specific policies, financing and technical assistance are provided to producers, as well as a flexible legislation which leave producers to choose the most appropriate raw materials to generate the biodiesel, eg palm oil, castor bean, soybeans, cotton, peanut etc. The Biodiesel Program is above all a social programme, with a strong focus on the sustainable development of the poorest regions in the country.

Since 2005, Brazilian legislation allows for the mix of two per cent biodiesel in mineral diesel in some types of engines. This mix will be mandatory by 2008, with an increased five per cent by 2013. The addition of two per cent biodiesel requires an annual supply of 800 million litres per year, whereas the amount needed for the five per cent mix is 2.1 billion litres per year. Presently, authorised production amounts to 1.6 billion litres per year, using 35 different plants. Around 5,000 petrol stations in Brazil sell biodiesel.

RENEWABLE CHARCOAL IN IRON AND STEEL PRODUCTION

Unlike most iron and steel production in the world which is heavily based on the use of coal coke as a reducing agent

Sugarcane is the primary crop used to produce ethanol in Brazil.

for the iron ore, in Brazil, most of the charcoal produced (85 per cent) is used in the iron and steel industry.
The production of charcoal from non-renewable biomass (native forests) was partially replaced in the 1940s by the introduction of coal coke. However, import restrictions in the 1960s, coupled with fiscal incentives to produce charcoal from planted forests, encouraged producers to invest heavily in energy plantations (normally eucalyptus and pinus) to meet the demand of the steel and iron industry. By the late 1980s, the fiscal incentives to plant were removed and producers were forced to reduce or abandon their charcoal production. The market opened again for imports and the use of coal coke was prioritised due to both availability and lower cost compared to that of charcoal from planted forests.
Presently, there is not enough charcoal from renewable biomass to meet the demand of the steel and iron industry. Hence, part of the charcoal used originates from non-renewable sources, such as wood from native trees in both forest and cerrado areas.
There is a genuine concern that, without additional investments, producers will not be able to meet the present and future demand of renewable charcoal for the steel and iron industry in the country. Investors from the private sector have sought the use of the Clean Development Mechanism as a means to provide the additional revenues needed to increase production, but were faced with a number of barriers. While many viewed the replacement of non-renewable biomass by renewable biomass as a process similar to the replacement of a non-renewable fuel source by a renewable one (such as replacing petrol by ethanol in transportation), many others understood that the emission reduction claimed would be associated with avoided deforestation, an activity which is not eligible under the CDM for the first commitment period.
Without pre-judging the results of negotiations for the post-2012 regime, it seems unlikely that avoided deforestation will be eligible under the CDM, since the concerns that led to its exclusion in the first commitment period are still valid today.

REDUCED EMISSIONS FROM DEFORESTATION IN THE BRAZILIAN AMAZONIA

The land-use change and forestry sector is the sector that contributes the most for the total CO_2 emissions in the country. In the period 1990 – 1994, the sector was responsible for approximately 75% of the total CO_2 emissions in the country, most of which was associated with the conversion of primary forests to other land uses, in particular cattle ranching and agriculture. It is unlikely that this contribution has changed drastically since.
As a response to the high rate of deforestation in 2004, the Brazilian government implemented an Action Plan for the Prevention and Control of Deforestation in the Amazon. The Plan introduced novel elements to tackle deforestation in the Amazonia region, engaging more than a dozen Ministries directly or indirectly involved in the issue, thus allowing deforestation to be tackled in an integrated manner.
The actions established in the Plan started to be implemented in 2004, and results have already been achieved in the following two years, with a significant reduction of the rate of deforestation and, hence, of the related CO_2 emissions. Assuming that the annual rate of deforestation would follow the trend of the three years prior to the implementation of the Plan, with a mean rate of gross deforestation of 2,463,000 hectares per year, the reduction observed in 2005 and 2006 corresponds to a reduction of emissions of more than 500 million tonnes of CO_2, assuming a mean carbon stock in aboveground biomass equal to 90 tonnes of carbon per hectare.

Author

Thelma Krug is presently the state secretary of the recently created Secretary on Climate Change and Environmental Quality of the Ministry of Science and Technology, Brazil. She co-chairs the Task Force on National Inventories of Greenhouse Gas of the Intergovernmental Panel on Climate Change (IPCC), and is a member of the IPCC Bureau. For many years she has been the general coordinator in the Earth Observation Coordination at the National Institute for Space Research, from the Ministry of Science and Technology. She has been deputy secretary in the Secretary of Policies and Programs in Science and Technology from the Ministry of Science and Technology, before joining the Inter-American Institute for Global Change Studies – IAI, as a visiting researcher. Her background is in mathematics, with a PhD on spatial statistics.

Organisation

The Brazilian Ministry of the Environment has competence in the following issues: the national environmental and water resources policy; preservation, conservation and sustainable use of the ecosystems, biodiversity and forests; development of strategies and economic and social tools to improve the quality of the environment and the sustainable use of the natural resources; policies for the integration of environment and production; environmental policies and programs for Amazonia; and ecological economic zoning. The Secretary has three Coordinations: one for Climate Change, another for Environmental Quality, and a third one on environmental licensing and environmental evaluation.

Enquiries

Thelma Krug
Secretária de Mudanças Climáticas e Qualidade Ambiental, Ministério do Meio Ambiente
Tel: +55 61.3317-1230Fax: +55 61.3317-1760
E-mail: thelma.krug@mma.gov.br

Innovative technologies to mitigate climate change

The most efficient way to combat climate change is to employ innovative technologies that increase the efficiency of power generation, transmission and utilisation in industry, buildings and transportation. With its broad portfolio, Siemens is uniquely positioned to address these challenges. The company invests more than €2 billion per year of its R&D budget in environmental technology, has 30,000 patents in this field, and the products it launched on the market eliminated in 2006 around 15 times the level of global CO_2 emissions that the company itself produces.

CARBON REDUCTIONS FROM EXISTING TECHNOLOGIES

The most effective way to reduce CO_2 emissions between now and 2050 is to employ innovative technologies that increase the efficiency of power generation, transmission, and utilisation. Some 26 billion tonnes (Gt) of CO_2 equivalent from an estimated 44 Gt of globally-emitted greenhouse gases (GHGs) caused by energy-related processes, can be addressed through technological innovations. Many solutions already exist. Estimates show, for example, that the global utilisation of the 10 most important environmental technologies that Siemens, and other companies, already have in their portfolios would reduce emissions by approximately 10 Gt of CO_2 per year by 2050. That figure corresponds to nearly 40 per cent of GHGs emitted due to energy-related activities today. Moreover, this estimate doesn't take into account other technological advances or growing markets.

Power plant retrofits

Power plant retrofits to achieve the highest levels of efficiency possible today would reduce annual CO_2 emissions by 2.5 Gt. Siemens is equipping a gas and steam combined cycle power plant in Germany with the world's largest gas turbine. This facility will achieve an efficiency rating of 60 per cent (by comparison, the best coal-fired plants reach 46 per cent). Building 20 such combined cycle power plants every year between now and 2050 would result in further savings of 1.6 Gt. All in all, retrofitting and new construction of state of the art fossil fuel facilities alone could reduce CO_2 emissions by 4.1 billion tonnes, which is the amount of CO_2 all of Europe currently emits.

Wind energy

Wind energy has the potential to reduce annual worldwide CO_2 emissions by 600 million tonnes by 2050. Establishing effective and safe processes for CO_2 capture and storage

© Siemens

(CCS) at fossil fuel plants could also eliminate a further 2 Gt of CO_2 per year from the climate balance, if 20 power plants per year were equipped with CCS technologies starting in 2020.

Buildings

An additional 2 Gt CO_2 reduction potential could be realised through better building insulation, modern heating and air conditioning systems, and more extensive building automation. Siemens has already modernised 6,500 buildings worldwide through performance contracting projects that offered guaranteed energy savings of more than €1 billion and reduced CO_2 emissions by 2.4 million tonnes. Investment in systems and equipment is financed via the energy savings achieved, creating a 'triple win' for customer, company, and environment.

Lighting systems

The savings potential harboured by lighting systems should not be underestimated as lighting accounts for nearly 19 per cent of global electricity consumption. The potential for savings is huge and easy to exploit because energy saving lamps and LEDs can reduce consumption by up to 80 per cent compared to conventional light bulbs. Added up over the entire product lifecycle, this amounts to several hundred euros per unit and an impressive 0.5 tonne reduction in CO_2 emissions for each energy saving lamp used. Siemens' Osram subsidiary estimates that annual CO_2 emissions could be reduced by 270 million tonnes if only 30 per cent of all lighting units, in houses, factories, and streets, were equipped with energy saving solutions.

NANO TECHNOLOGY FOR HIGH PERFORMANCE GAS TURBINES

Siemens is moving closer to the physical limits in its development of highly efficient power plant technologies. The materials in a gas turbine, for example, have to withstand very high temperatures of up to 1,500°C. The combustion chamber is also subjected to high pressure and a glowing hot tornado of gas that passes through at a speed of 100 metres per second. Finally, the turbine blades are exposed to centrifugal forces equivalent to 10,000 times the acceleration due to the Earth's gravity. Siemens researchers have developed extremely fine turbine coatings as well as new ceramics for the combustion chambers. They also designed new shapes for the turbine blades using 3D virtual reality simulations.

Due to these R&D efforts the world's biggest and most powerful gas turbine will have an output of 340 megawatts; as much power as 13 jumbo jet turbines. The combined cycle facility where it will be used will supply enough electricity for the population of a city the size of Hamburg and will achieve an efficiency rating of more than 60 per cent which will also be a world record. This power plant offers a good example of a groundbreaking innovation for environmental and climate protection and one that Siemens is developing to market maturity with its customer E.ON.

RENEWABLE ENERGY SOURCES

Siemens is committed to renewable energy sources, such as wind, water and geothermal energy. Offshore wind turbines, in particular, need to withstand extremely high stresses, which is why the company has developed a process that makes it capable of building a 52 metre long blade in one piece, without any adhesive joints. This enables the blades to withstand wind speeds of 10 metres per second and an air pressure of 100 tonnes. Siemens is not only building the world's biggest wind park in Lynn and Inner Dowsing (UK) but also many others. As the wind energy sector is growing rapidly in the US, Siemens recently opened a rotor blade factory in Fort Madison. More than 6,300 wind turbines built by Siemens are currently operating around the world. Their combined output of 5.5 gigawatts is helping reduce CO_2 emissions by 10 million tonnes per year, which is approximately the amount produced by one million Germans.

Siemens also has a long tradition in hydropower. In 1978, the company supplied the generators for the world's biggest hydroelectric power plant in Itaipú, which is located between Brazil and Paraguay. Today, one-third of all electricity produced with carbon neutral hydropower is generated with technology from Voith Siemens Hydro.

Geothermal energy sources are nearly inexhaustible reservoirs of heat in the earth. Siemens is using this technology to build what will be the world's most modern geothermal power plant near Munich, which will supply 6,000 households with electricity and 20,000 with heat.

FUTURE CARBON FREE POWER PLANTS

Achieving carbon free power generation with carbon capture and storage (CCS) poses the biggest challenge for power plant innovators. CCS can reduce CO_2 emissions by 80 per cent in coal-fired plants by capturing CO_2 and safely storing it, in old oil or gas fields, for example. Siemens researchers are working on the two most important technologies here: one for capturing CO_2 before it burns in integrated gasification combined cycle plants, and the other for extracting it from flue gas after combustion. Pilot projects for examining system feasibility and costs are now being planned.

EFFICIENT POWER TRANSMISSION

Siemens has developed high voltage direct current (HVDC) transmission systems able to transport electricity over great distances with low losses, ie in an environmentally friendly and economical manner. It's often possible to generate electricity using hydropower plants in remote regions, but the energy has to be brought to the consumers in the cities. Such is the case in China, where hydropower generated in Yunnan is transported over 1,400 Km to the Pearl River Delta. Without HVDC technology, fossil power plants that emit more than 30 million tonnes of CO_2 per year would have to be built near consumers.

SMART GRIDS – VIRTUAL RENEWABLE ENERGY NETWORKS

Small, distributed plants that use biomass or geothermal sources, combined heat and power plants, and plants with fluctuating outputs (wind, solar) need to be integrated into the existing network hierarchy without disturbances. This poses a major challenge, which is why Siemens developers are working on smart grid systems that bring together the most diverse electricity sources to form 'virtual power plants'. Smart grids can turn specific consumers on and off in order to balance out peak loads, thereby creating a dynamic balance in the power network rather than the static one that exists today. Siemens is working with RWE Energy on a pilot project that not only links cogeneration plants, biomass, and wind power facilities with a distributed energy management system but also monitors and controls system operation.

REDUCING ENERGY CONSUMPTION IN INDUSTRY AND TRANSPORTATION

Innovative technologies that increase efficiency can help reduce energy consumption in industry and households without negatively affecting economic growth, prosperity, or quality of life. Industrial motors alone account for 65 per cent of all the electricity consumed in industrial applications and also harbour the greatest potential for improvement. Siemens is now achieving energy savings of up to 60 per cent through the use of frequency converters and energy saving motors. If all motors suitable for retrofitting were to be re-equipped with such components, the resulting CO_2 savings would total several hundred million tonnes per year. The technology involved also substantially reduces energy costs, and the initial investment can be recouped in less than two years. Intelligent traffic guidance and management systems, such as those implemented by Siemens in London and the Ruhr District in Germany, reduce traffic jams and make a key contribution to climate protection. Achieving a climate friendly transport system also involves getting as many people as possible to use public transportation. Carbon dioxide emissions per seat and kilometre for passenger cars are three times higher than those for suburban and subway trains. Although trains are already one of the most environmentally friendly modes of transport, Siemens has identified new potential improvements. The Oslo subway system, for example, is now equipped with new technology that reduces the system's energy consumption by 30 per cent.

CLIMATE PROTECTION IS A PART OF CORPORATE RESPONSIBILITY

Siemens is very aware of its environmental responsibility, which is why it is committed to reducing its sales-related CO_2 emissions by 20 per cent between now and 2011. Due to its electricity and heat consumption, business travel, and use of industrial gases and corporate vehicle fleets, Siemens currently emits approximately 4.53 million tonnes of CO_2 equivalent per year. However, with its broad portfolio, Siemens is also uniquely positioned to address the challenges associated with climate change. The products Siemens launched on the market eliminated in 2006 around 15 times the level of global CO_2 emissions that the company itself produces.

Author

Dr Ulrich Eberl is head of the Technology Press and Innovation Communications department of Siemens. Eberl studied physics at the Technical University of Munich and wrote his dissertation about converting solar energy in the process of photosynthesis. In 1988, he became a science and technology correspondent for various newspapers and magazines. After working for Daimler's technology publications, he joined Siemens in 1996. Since 2001, Eberl has been editor-in-chief of the magazine *Pictures of the Future* (www.siemens.com/pof), which has been the recipient of several international awards.

Organisation

Siemens (Berlin and Munich) is a global powerhouse in electrical engineering and electronics. The company has around 475,000 employees and provides innovative technologies to benefit customers in over 190 countries. Founded 160 years ago, the company focuses on fields such as energy and environment, industry and infrastructures and healthcare. In fiscal 2006, Siemens had sales of €87.3 billion and a net income of €3.033 billion.

Enquiries

E-mail: Ulrich.eberl@siemens.com
Website: www.siemens.com/innovation

Transportation and greenhouse gas mitigation

NIC LUTSEY AND **DAN SPERLING**
INSTITUTE OF TRANSPORTATION STUDIES,
UNIVERSITY OF CALIFORNIA, DAVIS

Transportation presents a substantial and growing worldwide greenhouse gas (GHG) emission challenge. GHG mitigation strategies can be grouped into three categories: vehicle efficiency, low carbon fuels, and travel reduction. Potential GHG reductions are very large, with varying levels of cost effectiveness. Virtually all provide large co-benefits, including energy cost savings, oil security, and pollution reduction.

TRANSPORTATION SECTOR CHALLENGES

Transportation accounts for about one-fifth of total greenhouse gas (GHG) emissions worldwide, but close to 30 per cent in most industrialised countries. Worldwide, transport GHG emissions are growing faster than those from any other sector. Most are associated with motor vehicles, but air transport is an increasingly important source. Studies of cost effectiveness generally find transportation GHG reductions more expensive than reductions in most other sectors. The high cost is due to: low fuel price elasticity by passenger car owners (and light trucks); strong demand for personal travel, air travel, and goods transport; the difficulty of introducing new low carbon fuels and new fuel efficient propulsion technologies; deteriorating quality of public transport virtually everywhere; and the increasing share of goods carried by truck. In addition, petroleum fuel use is becoming more carbon intense. As easily accessed and high quality reserves are depleted, more carbon intense and remote sources of fossil energy are tapped and additional refining is required to upgrade fuel quality.

The analysis here focuses on the two largest components of the transportation sector: passenger automobiles and commercial freight trucks. Together, these make up about two-thirds of transportation GHG emissions.

Road pricing for city centres or highway congestion can moderate traffic and reduce GHG intensive travel.

ose/Fotolia

GREENHOUSE GAS STRATEGIES

Despite analyses that indicate transportation options tend to be less cost effective than others, there are many reasons why these are misleading. First, there are large, highly valued co-benefits. Most strategies to reduce transport GHG emissions also reduce petroleum use, thereby contributing to energy security. Most also reduce emissions of local conventional pollutants and those that involve reduced vehicle use also reduce traffic congestion. Second, many incremental, low cost technologies exist to reduce energy use. Innovations in engines, transmissions, aerodynamics and lightweight materials have continuously yielded greater efficiency. Many additional fuel saving innovations are being pursued. Third, the automotive industry has become highly competitive, with companies seeking ways to distinguish themselves. The halo created by the successful Prius hybrid has proved extraordinarily valuable to Toyota. It has shown that being first has great value. Toyota increased the value of its brand and the attractiveness of its other vehicles far more effectively than advertising. Companies are now increasing their investment in a wide variety of new low carbon fuels and efficient advanced propulsion technologies to achieve the same halo benefits. Fourth, there is substantial evidence that reductions in vehicle use are desirable and attainable for a wide variety of reasons. And fifth, there are many policies that could reduce fuel consumption and GHG emissions at zero net cost for the simple reasons that consumers do not highly value, or are unaware of, efficiency considerations in their vehicle purchase decision and ignore many simple practices to reduce fuel use.

VEHICLE EFFICIENCY

Available and emerging vehicle efficiency improvements can be categorised into three groups.

Incremental vehicle technologies

Incremental improvements include more efficient combustion, such as variable valve systems, gasoline direct injection, cylinder deactivation, more efficient transmissions such as 5- and 6-speed automatic, automated manual and continuously variable, and overall vehicle advances, such as aerodynamics and light-weighting. Greenhouse gas emissions rates can be reduced by 20-30 per cent with these technologies in new vehicles. Most studies show that fuel savings from these improvements more than outweigh the increased vehicle cost, often by a large amount. Similar technology packages yield substantial GHG reductions and net positive benefits for commercial freight trucks as well.

Advanced technologies

Much greater GHG reductions are possible with electric drive propulsion technologies. These include the increasingly popular hybrid gasoline-electric vehicles, plug-in hybrids which use both electricity and petroleum fuels, battery electric vehicles and hydrogen powered fuel cell vehicles. Such technologies can double vehicle fuel efficiency. The life cycle GHG emissions, considering the potential to use low carbon electricity and hydrogen, can be reduced by at least 80 per cent. However, these advanced technologies involve either larger initial costs, for electricity and hydrogen storage, and/or have high development costs and uncertain learned-out costs. Because vehicle turnover is slow and it takes a long time to deploy a new energy distribution system, it will take a long time to realise potential reductions.

On-road operational practices

On-road efficiency improvements involve a combination of consumer education, vehicle maintenance practices, and off-cycle vehicle technologies. Improvements to on-road vehicle efficiency can reduce GHG emissions by up to 20 per cent. Improved vehicle maintenance practices with regard to tires, wheels, oil and air filters ensure vehicles operate as efficiently as they were designed to do. Technologies in new vehicles that aid driver awareness of fuel use include dashboard instruments that present instantaneous fuel consumption, efficient engine rpm ranges, shift indicator lights, and tire inflation pressure. There are many possible policies to bring about these vehicle improvements. In addition to education and informational initiatives, incremental vehicle efficiency can be achieved with performance standards aimed at automakers, vehicle purchase and use taxes aimed at consumers and vehicle suppliers, and various actions aimed at assuring the supply of alternative fuels for the advanced vehicles. To make sure deeper GHG cuts in future years are achieved, government tax incentives to industry and consumers will be needed to overcome initial cost, institutional and infrastructure concerns and barriers.

LOW CARBON FUELS

Increased use of low carbon fuels, or fuels with lower life cycle GHGs, can greatly reduce overall transportation GHG emissions. Most alternative transportation fuels face a combination of infrastructural and economic barriers. The easiest action is to blend small proportions of biofuels into gasoline and diesel fuel. Biofuels are not necessarily less expensive, but the processes for converting abundant agricultural feedstocks, such as corn and sugarcane, into ethanol are well known and ethanol is easily blended into gasoline for use in conventional vehicles. The GHG benefits of sugarcane conversion are substantial, compared to gasoline, but only about 10-20 per cent for corn. Future biofuels, made from agricultural residue or cellulosic energy crops could have life cycle GHG benefits of 90-100 per cent. A similar array of biofuel feedstocks can be used to produce biodiesel, which can be mixed into conventional diesel fuel.

Table: Summary of transportation greenhouse gas mitigation options and policies.

Category	Today's measures (deployable 2007-2015)	Tomorrow's measures (deployable 2010-2030)	Supporting policies and practices
Vehicle efficiency	▸ Incremental efficiency improvements in conventional gasoline automobiles and diesel trucks. ▸ "On-road" improvements in maintenance practices, technology, driver education and awareness.	▸ Increased vehicle electrification (hybrid gas-electric, plug-in hybrid, battery electric). ▸ Fuel cell vehicles.	▸ Vehicle efficiency performance standards (fuel economy, CO_2 emission rate). ▸ Voluntary industry commitments. ▸ Vehicle purchasing incentives (rebates, feebates for low CO_2, high fuel economy). ▸ Government and company fleet efficient vehicle purchasing.
Low greenhouse gas fuels	▸ Mixing of biofuels in petroleum fuels. ▸ Use of lower GHG content fossil fuels (eg diesel, compressed natural gas).	▸ Electricity (in plug-in hybrids and battery electrics). ▸ Cellulosic ethanol. ▸ Hydrogen from renewable sources. ▸ Mobile air-conditioning (MAC) refrigerant replacement.	▸ Biofuel blending mandates. ▸ Low GHG fuel standards. ▸ Carbon tax on fuels. ▸ Government and company fleet incorporation of alternative fuels.
Vehicle demand reduction	▸ Intelligent transportation system (ITS) technologies to improve system efficiencies. ▸ Mobility management technologies. ▸ Inclusion of GHG impacts in land use and transport planning. ▸ Incentives and rules to reduce vehicle use.	▸ Greenhouse gas budgets for households and localities. ▸ Modal shifts (road to rail freight, public transit systems). ▸ ITS technologies to create more efficient transport modes.	▸ Road, parking, congestion pricing. ▸ Investment in public transit. ▸ Public awareness, outreach, education campaigns.

There are also other transport fuel options systems involving wholly different fuels and fuel distribution systems that can greatly impact GHG emissions. Marginally lower GHG fossil fuels, such as compressed natural gas and liquefied petroleum gas have continued to make small contributions to transportation, mostly in fleet vehicles. On the other hand, next generation fossil fuels produced from oil shale, coal, and tar sands would have much higher GHG emissions than conventional petroleum, unless the carbon from such fuels was captured and stored underground.

Large potential GHG benefits can be achieved by powering vehicles with hydrogen (and fuel cells) and electricity, with plug-in hybrids and battery electrics. Electric drive vehicles, powered by low carbon versions of these fuels made with biomass, wind, nuclear energy, or with fossil energy coupled with carbon capture and storage, could yield much greater GHG reductions than with vehicle efficiency improvements alone.

Lower carbon fuels have been subsidised and mandated by various governments, including biofuel mandates in Europe and ethanol subsidies in the US and Brazil. A new policy instrument gaining much attention worldwide is the low carbon fuel standard. In this case, government does not pick winners. It sets a greenhouse gas intensity target, eg 10 per cent reduction by 2020, and allows companies to meet the requirement however best suits them. Or companies can buy credits from those exceeding the GHG target.

TRAVEL REDUCTION

The same technologies and practices implemented by local governments to manage vehicle travel and traffic congestion can also be used to reduce GHG emissions. Strategies to reduce vehicle travel can be sorted into three broad groups.

Information and communication technologies

Information and communication technologies can be used to improve mobility and reduce transport GHG emissions. Incremental enhancements include: automating urban traffic signals to streamline traffic and reduce stop and go conditions; implementing integrated smart cards to facilitate multimodal travel and increase transit use; provide real time traffic data to traffic managers and vehicle users to improve efficiency. More substantial changes are possible by creating entirely new modes of travel, such as smart car sharing that allows convenient short term rentals, smart paratransit that provides door to door service without advanced reservations, and dynamic ride sharing that facilitates organised ride sharing.

Incentives and pricing schemes

Various incentive and pricing schemes can be designed to reduce GHG travel. Road pricing for city centres or highway congestion can moderate traffic and reduce GHG intensive travel. Parking policies, such as park and ride near transit facilities and parking cash-out programmes by employers, encourage higher occupancy travel modes. Incentives by workplaces to promote telecommuting and carpooling can also help mitigate peak time congestion travel. Vehicle pricing in conjunction with improved transit service programmes, such as bus rapid transit, attracts travellers to higher occupancy and thus lower GHG modes.

Denser land use

Densification of land use may be the most effective way to reduce the use of GHG intensive modes of travel. Research shows that residents in more densely populated areas and in areas with better mixes of land uses tend to emit far less GHG emissions from their travel. They tend to walk more, use public transportation more and drive less. Policies aimed at increasing density and influencing local governments to make land use development and zoning decisions based on likely impact on GHG emissions, could be highly effective at reducing emissions.

SUMMARY OF TRANSPORTATION GHG STRATEGIES

The Table categorises transportation sector GHG mitigation options into near and mid term options. For the near term, or 'Today's measures,' options highlighted are currently available and easily applied but would require policies or shifts in practice to achieve widespread adoption. For 'Tomorrow's measures,' the listed strategies offer deeper possible emission cuts,

Figure: Cost effectiveness supply curve of available GHG mitigation technologies in different sectors, with transportation options highlighted.

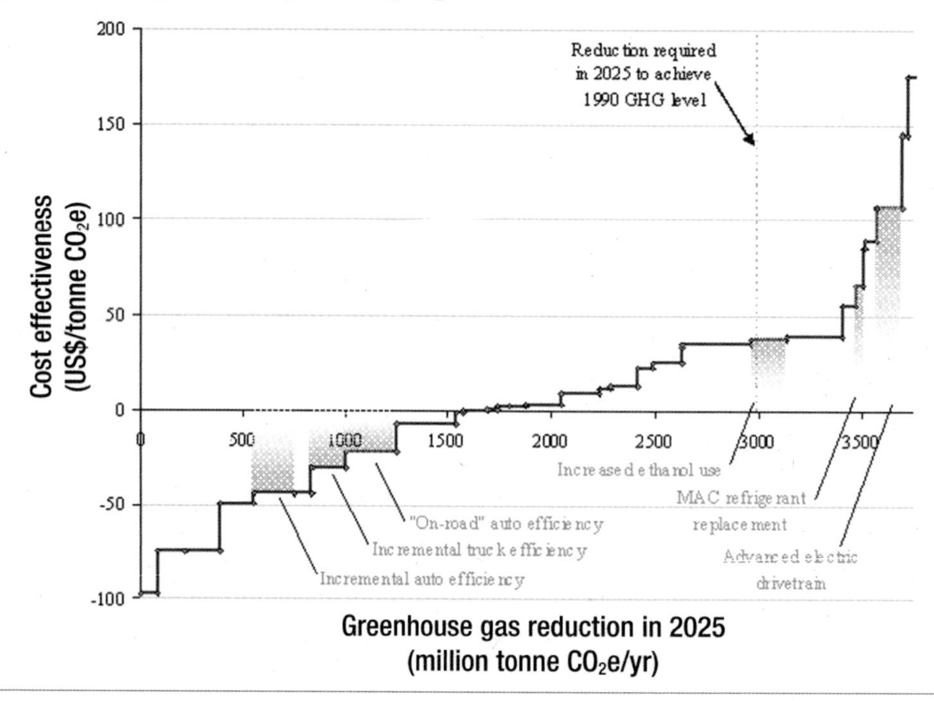

Lutsey, N., 2008. Prioritization of Technology Alternatives for Cost-Effective Climate Change Mitigation. Ph.D. Dissertation, Institute of Transportation Studies, University of California, Davis (forthcoming).

but there is greater uncertainty in how technology costs will drop over time and how industry will act.

GREENHOUSE GAS MITIGATION SUPPLY CURVES

Greenhouse gas mitigation strategies can be ranked using a supply curve framework. They are ranked according to their GHG reduction cost effectiveness, or cost per tonne of CO_2 equivalent emission reduction. Both the initial costs of the GHG technologies and the lifetime energy savings are included in the cost per tonne metric. The Figure is a supply curve of GHG mitigation actions for all sectors of the US economy, with transportation-specific measures highlighted (data from Lutsey, 2007). The non-transportation actions include electric power sector actions, eg coal to natural gas shift, carbon capture and sequestration, increased nuclear power, renewable electricity, more energy efficient buildings, including improvements in appliances, lighting, and air conditioning, and hydrofluorocarbon emission reduction technologies.

CONCLUSION

Many transportation strategies to reduce GHG emissions are highly cost-effective. Many generate cost savings over the life of an investment (in a particular energy-saving technology or product), when future energy savings are calculated using normal discount factors. When other co-benefits are included, such as improved energy security, many transport GHG mitigation options become attractive. These findings are counter to the conventional wisdom that often ignore co-benefits and emphasise near term resistance to the broad suite of technology and behavioural options.

Policies to bring low GHG technologies and practices to widespread deployment have already emerged and proven successful in limited venues. The available policy strategies are diverse, including local policies that integrate transport and land use decisions, fuel efficiency and GHG performance standards on vehicles, outreach and incentive campaigns to instil energy saving attributes into all consumer decisions and government and corporate research that promote technological breakthroughs and reduce infrastructural and implementation cost barriers.

Authors

Nic Lutsey is a PhD candidate in Transportation Technology and Policy at the University of California, Davis. He is the author of nine articles on transportation energy and greenhouse gas topics. His PhD evaluates GHG mitigation strategies for transportation.

Daniel Sperling is Professor of Engineering and Environmental Science and Policy and founding Director of the Institute of Transportation Studies at the University of California, Davis. Dr Sperling was recently honoured as a lifetime National Associate of the National Academies and is author or editor of 200 technical articles and 10 books. He was appointed by Governor Schwarzenegger to the California Air Resources Board in February 2007.

Organisation

The Institute of Transportation Studies at the University of California, Davis (ITS-Davis) is staffed by over 150 faculty, staff, and student researchers. It hosts an award winning interdisciplinary graduate program in Transportation Technology and Policy and is recognised as one of the premier university centers in the world for its studies of transportation energy, travel and vehicle purchase behaviour and advanced vehicle technology.

Enquiries

Daniel Sperling, Director, Institute of Transportation Studies, University of California, Davis, USA
Tel: +1 530 752-7434
E-mail: dsperling@ucdavis.edu
Nic Lutsey E-mail: nplutsey@ucdavis.edu
Website: www.its.ucdavis.edu

Reducing car carbon emissions: how low can we go?

DR PAUL NIEUWENHUIS
PROJECT MANAGER, SUSTAINABLE AUTOMOBILITY,
ESRC CENTRE FOR BUSINESS RELATIONSHIPS,
ACCOUNTABILITY, SUSTAINABILITY & SOCIETY,
CARDIFF UNIVERSITY

Chevrolet Volt Concept car 2007 – taking hybrid technology to the next logical step where the drive is purely electric with the internal combustion engine acting only as a generator.

The EU is the first major jurisdiction to propose regulation on car CO_2 directly, rather than fuel consumption. Meeting the proposed target of 130 g/km of CO_2 emissions from new cars is technically feasible with certain policy measures to bring this about. Many cars on sale in the EU already meet the proposed limit. It is also known from suppliers and manufacturers that more CO_2 reducing technologies are due to enter the market before the proposed EU deadline of 2012. But current technology can also deliver major improvements in car CO_2 emissions in countries such as China.

BACKGROUND

Cars are a significant source of carbon dioxide emissions and energy use. In the developed nations, emissions from other sources are reducing but those from transport are still rising. In order to tackle this problem a number of countries have attempted to reduce CO_2 emissions from cars, either through fuel taxes, restrictions on fuel consumption, or by reducing CO_2 emissions directly, such as in the EU.

Many regulators and politicians have unrealistically high expectations of alternative technologies such as fuel cells. Internal combustion (IC) will be the dominant technology for many years to come and even hybrids still use this technology. Any regulatory approach to carbon reduction should therefore consider IC technologies. Car manufacturers must also treat this issue with more urgency, even though the changes needed have the potential to change dramatically the nature of cars and of the industry that makes them.

THE EU CASE

In February 2007, EU Environment Commissioner Dimas proposed a mandatory reduction of average CO_2 emissions from new cars to 130 g/km by 2012. This reflected the industry's inability to meet the voluntary target set in the late 1990s of 140 g/km by 2008. Some manufacturers were on track, and the voluntary agreement had some impact. Figure 1 (overleaf) shows average new car emissions in different EU member states. Despite reductions in the average figure, we are still some way off the voluntary limit of 140 g/km agreed between the European Automobile Manufacturers Association (ACEA) and the European Commission.

> **"Internal combustion (IC) will be the dominant technology for many years to come and even hybrids still use this technology."**

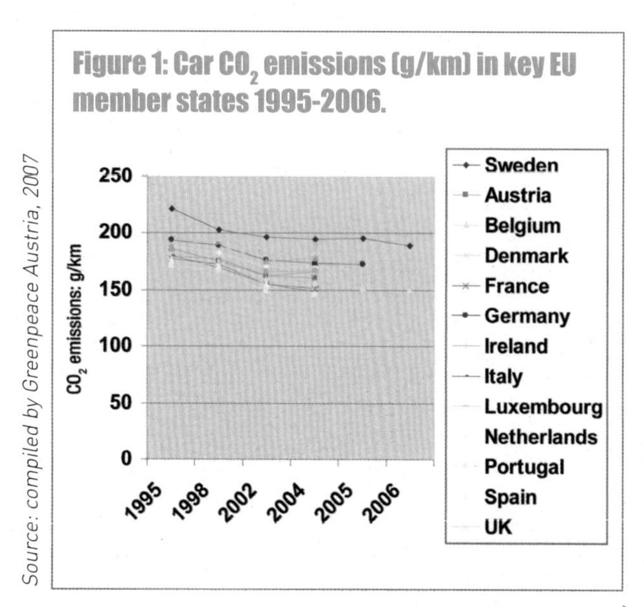

Figure 1: Car CO_2 emissions (g/km) in key EU member states 1995-2006.

Source: compiled by Greenpeace Austria, 2007

Much of this reduction in CO_2 output was achieved by greater reliance on diesel engines. The better efficiency of the diesel engine more than offsets the slightly higher carbon content of diesel fuel. Although there are health risks associated with diesel emissions, there is little doubt that any further reduction in CO_2 emissions will be achieved partly through a further increase in diesel, see Figure 2 below.

Cars have also become larger and heavier, so attempts to reduce CO_2 have in part been negated by weight gain. A typical example is VW's Golf, which has long been Europe's best-selling car. The Golf has added weight with each new generation as features were added to enhance its appeal in the market.

HOW LOW CAN YOU GO?

ACEA suggests that reducing CO_2 emissions to 130 g/km for new cars is a major challenge, yet the industry currently offers many vehicles that meet this standard. It is therefore far from impossible, see Table 1.

The large car market

Cars that currently emit up to about 170g/km can be brought down to 130g/km with reprogrammed engine management systems and low cost powertrain

Table 1: < 140g/km cars currently available in the EU.

Make	Models < 120g/km	Models 120-130g/km	Models 130-140g/km
BMW	MINI 1.4d	118d, 120d	118i, MINI 1.4, 1.6
Chevrolet		Matiz 0.8	Matiz 1.0, 0.8auto
Citroën	C1, C2 diesel, C3 diesel	C4 1.6d	C2 1.4 stopstart
Daihatsu	Charade, Sirion 1.0		Sirion 1.3
Fiat	Panda 1.3 Multijet; Grande Punto 1.3 Multijet 75	Grande Punto Multijet 90	Panda 1.1,1.2, Punto 1.2; Stilo 1.9 Multjet 90 3d
Ford	Fiesta 1.4 tdci, 1.6tdci	Focus 1.6 tdci; C-Max1.6tdci	Focus 1.8tdci
Honda	Civic 1.3 hybrid	Jazz 1.2dsi-s	Jazz 1.4dsi; Civic 1.4, 2.2cdti
Hyundai		Amica 1.1gsi; Getz 1.1, 1.5d	Amica 1.1cdx
Kia		Picanto 1.0, 1.1; Rio 1.5d; Cerato 1.5d	
Mazda		2 1.4d; 3 1.6d	
Mercedes-Benz		A160 cdi	A180 cdi
MINI			Cooper 1.6
Mitsubishi		Colt 1.1, 1.5d	
Nissan		Micra 1.5d	Note 1.5d
Perodua		Kelisa1.0	Myvi 1.3sxi; Kenari 1.0
Peugeot	107 1.0 urban; 206 1.4hdi	1007 1.4hdi; 207 1.4hdi, 1.6hdi; 206 1.6hdi; 206cc 1.6hdi; 307 1.6hdi 3d	307 1.6 hdi 5d
Proton			Savvy 1.2 street
Renault	Clio Campus 1.5dci; Clio 1.5 dci 86	Modus 1.5dci; Clio 1.5dci; Megane 1.5dci 86 & 106	Scenic 1.5dci 86 & 106; Grand Scenic 1.5dci 106 Privilege
SEAT		Ibiza 1.4tdi	Ibiza 1.9tdi 100; Leon 1.9tdi
Škoda		Fabia 1.4tdi pd	Fabia 1.9tdi; Roomster 1.4tdi pd
Smart	ForTwo Pure; all For Two diesels	ForTwo Pulse & Brabus; ForFour 1.0, 1.5cdi; Roadster, Roadster Brabus	
Suzuki		Swift 1.3d	
Toyota	Aygo; Prius	Yaris 1.0, 1.4d	Auris 1.4d
Vauxhall		Corsa 1.3 cdti; Tigra 1.3cdti	Agila 1.0; Corsa 1.0, 1.2 (some); Meriva 1.3 cdti; Astra 1.7cdti
Volkswagen		Polo 1.4tdi	Polo 1.9tdi
Volvo		C30 1.6d; S40 1.6d	C30

Source: Nieuwenhuis 2007 Car CO_2 reduction feasibility assessment, www.brass.cf.ac.uk; table does not show all models that comply

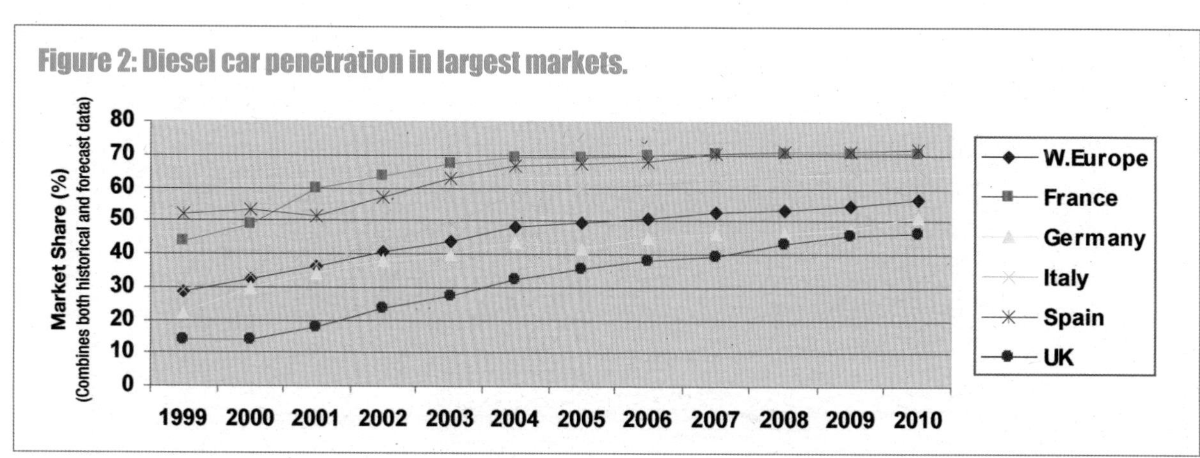

Figure 2: Diesel car penetration in largest markets.

Source: Power, J. D. (2005) Global Car & Truck Forecast, Fourth Quarter 2005, Oxford and Troy MI: J D Power and Associates Automotive Forecasting Services.

Table 2: Future powertrain developments.

Technology	Likely introduction	Likely CO_2 savings (source), %
Variable valve actuation	Now	10-15
Electronic valve actuation (no camshaft)	2010	15-20 (Valeo)
Direct injection petrol engines (GDI)	Now	15 (Bosch)
Cylinder switch off (available in US)	2010	15-20 (Chrysler)
Stop-start	2006	10-15 in urban driving (Citroen); 5 overall (Lotus); 20-25 in urban driving (Fiat)
Starter-generator	Now	
Variable compression	?	?
Turbocharging and supercharging combined with downsising	Now	Variable
Improved transmissions (CVT, DSG, AMT)	Now	Variable
Low rolling resistance tyres	Now	2-5 (Michelin)
Petrol electric hybrid (Connaught)	Now	18 (Honda); 22 (Lotus); 25
Diesel electric hybrid	2010-2012	35 (PSA)

Source: CAIR/BRASS

improvements, eg stop-start. Larger cars need more expensive technologies to bring them down to a CO_2 emission level which doesn't distort the industry average and this is where the challenge lies. The result may be a split in the market, whereby small to medium cars will continue to be available at price levels similar to today, while larger cars become more expensive. Although this may unduly affect certain manufacturers, they tend to operate higher margins and could pass on much of this extra cost. These manufacturers could introduce smaller cars which will still need to be premium priced in order for them to survive. BMW's Mini and Mercedes A and B Class are examples of how this might be done. The skill is in carrying traditional brand values into more compact cars, ie in marketing, not engineering.

There are other advantages. Large luxury cars lose value quickly because used car buyers are less able to afford high running costs. If luxury cars were smaller and lighter, their appeal to the used market would rise, boosting residual values. This impacts on lifecycle costs of luxury cars, making them generally competitive in economic lifecycle terms. This benefits consumers, but also manufacturers, as higher residual values enhance brand appeal.

NEWLY MOTORISED COUNTRIES

Countries such as China and India, with their large populations, are causing concern to environmentalists, not least in view of their growth in motorisation. The Chinese fleet will continue its rapid growth for the next 25 years, potentially reaching 50 million new passenger vehicle registrations per year. However, research carried out at ESRC Centre for Business Relationships, Accountability, Sustainability and Society (BRASS) suggests it is possible, with current technologies, to reduce the CO_2 emissions resulting from this growth at relatively little cost. Scenarios covered diesel, minicars and hybrids.

Although all three technology options provide some CO_2 reduction benefit, the widespread adoption of hybrid electric technology would reap the greatest benefits. If this could be installed in minicars and powered by a diesel engine as the internal combustion element, even greater benefits would accrue. Estimates suggest that in a mature market, such as the EU, achieving the CO_2 emissions target of 120 g/km desired by the Commission would translate into a cost to consumers of around €1,200-€2,000 on the price of an average car. With the expected longer term rise in oil prices, much of this is offset by fuel savings. In a lower cost economy, such as China, costs would be much reduced as Chinese consumers would either benefit from work carried out to meet EU targets, or carried out at lower cost in China itself.

NEW TECHNOLOGIES NEEDED

There are a number of technologies coming onto the market which will keep conventional IC engines more environmentally competitive, see Table 2. Petrol engines will become smaller, turbocharged and fitted with technologies for greater efficiency. This will make them competitive in fuel consumption, and, in CO_2 emissions terms, with diesel but with the advantage of cheaper

The new Toyota Yaris.

Mercedes F700 Concept Car 2007 – large and luxurious but powered by a small 1.8 litre engine, with CO_2 emissions of only 127 g/km.

emissions control than future generations of diesel engine. Improvements in the diesel combustion process are also being developed to avoid expensive and complex after treatment. As a result of widespread outsourcing of product development, much of this expertise now resides with suppliers, rather than car manufacturers, a fact not yet fully appreciated by regulators who still often interface with car manufacturers as the sole representatives of their industry.

Concept cars

At the 2007 Frankfurt motor show, Mercedes showed a concept car which gives an indication of what could be achieved with technologies currently under development. The F700 is a large luxury saloon, but powered by a small 1.8 litre engine. The engine uses a combination of diesel and Otto cycles to produce 258bhp, yet with CO_2 emissions of only 127 g/km. This is achieved by combining the IC engine with a hybrid powertrain, while the engine itself has two stage turbocharging and optimised IC technology.

The Chevrolet Volt concept car, shown at the 2007 Detroit show, takes hybrid technology the next logical step: a series-hybrid configuration, whereby the IC engine acts only as a generator. Drive is purely electric, with the option of additional recharging from an external power source, at home. A production version is expected around 2010-2012.

ROLE OF POLICYMAKERS AND CUSTOMERS

Technology is not the whole solution; enabling the take up of those technologies and changing cultures of automobility are equally important. It is possible to restrict car ownership and use as in Singapore, or access to urban areas as in Italy, Norway and London. European cities are trying to get citizens to return to cycling, yet several Chinese cities are trying to restrict cycling. With their new found automobility it may be challenging to engage public support for reversing these policies, though the negative consequences of this rise in car ownership may help build support for such measures, particularly in cities.

Dramatic reductions in CO_2 emissions are possible with current and new technologies. Countries leading the technologies – Japan, US and EU – should also lead in regulation to encourage the market. This provides a home market for low carbon technologies, and a competitive advantage to firms pioneering such technologies.

CONCLUSION

Traditionally, the car industry blames the customer for the products it makes. Increasingly it is recognised that the customer is not a car designer or automotive engineer. Ordinary citizens cannot track down lifecycle assessments for every product they buy or use. The customer can only choose from what is offered in the market place. Stuart Hart's influential article in the Harvard Business Review in February 1997 put the primary responsibility for greening products firmly on manufacturers:

"Like it or not, the responsibility for ensuring a sustainable world falls largely on the shoulders of the world's enterprises... corporations can and should lead the way, helping to shape public policy and driving change in consumers' behaviour". Hart encourages industry to help shape public policy not in its own short term interests but in the longer term social interests that are implied in the sustainability agenda.

The car of tomorrow will need to address resource depletion, waste generation, congestion and quality of life in the broadest sense. It is likely these cars may well be more likeable and more fun to drive than the often overspecified, overweight cars of today.

Author

Dr Paul Nieuwenhuis joined the Centre for Automotive Industry Research (CAIR) at Cardiff University in 1991 and became one of its two directors in 2006. He is a founder member of the ESRC Centre for Business Relationships, Accountability, Sustainability and Society (BRASS). He has written a number of books on environmental issues and the automotive industry including *The Green Car Guide* (1992), *The Death of Motoring?* (1997), and most recently, *The Business of Sustainable Mobility* (ed. 2006). He also contributed to the Beaulieu Encyclopaedia of the Automobile which won a Cugnot Award from the Society of Automotive Historians.

Organisation

CAIR is one of the leading academic centres in Europe that analyses economic and strategic issues affecting the world automotive sector. BRASS is the leading academic centre in the UK dedicated to analyse issues of CSR and sustainability as they relate to business. Both centres engage in academic research, policy advice and contract research for private and public sector clients.

Enquiries

Dr Paul Nieuwenhuis
Centre for Automotive Industry Research
Cardiff Business School, Cardiff University
Aberconway Building, Colum Drive
Cardiff CF10 3EU
Wales, UK
Tel: +44 (0)29 2087 5702
E-mail: nieuwenhuis@cardiff.ac.uk
Website: www.brass.cf.ac.uk

Future transportation systems in China

ALTERNATIVE FUELS AND ADVANCED VEHICLE TECHNOLOGIES OUTLOOK

© CPW/iStockphoto

DR LIXIN FU, DIRECTOR OF AIR POLLUTION RESEARCH AND PROFESSOR, AND CO-AUTHORED WITH **DR YE WU**, ASSOCIATE PROFESSOR, DEPARTMENT OF ENVIRONMENTAL SCIENCE AND ENGINEERING, TSINGHUA UNIVERSITY

With China's on-road transportation sector growing rapidly with projected carbon dioxide emissions in the billions of tonnes by 2050, China is looking at clean fuel alternatives in order to drive the sector and maintain growth.

INCREASING TRANSPORTATION DEMAND IN CHINA

Over the past three decades, China has experienced rapid growth in the population of motor vehicles with the annual growth rate over 10 per cent, a trend that is likely to continue. As a result, oil consumption and carbon dioxide (CO_2) emissions associated with on-road transportation are increasing fast. The Chinese government has been paying close attention to oil demand since the contribution of imported oil to Chinese total oil consumption keeps increasing; currently it is over 40 per cent. To relieve imported oil dependence, the Chinese government has been expanding considerable efforts to save oil use. In 2004, for example, China issued its first national fuel consumption standard for passenger cars (GB19578-2004) to improve vehicle fuel efficiency.

However, no matter how conservative the projection of future vehicle population is, current policy efforts still do not offset the increase in oil consumption. Carbon dioxide emissions from the on-road transportation sector will continue to increase. A recent study by Argonne National Laboratory and the Energy Foundation projected China's on-road vehicles consuming approximately 614-1,016 million tonnes of oil per year (12.4-20.6 million barrels per day) with associated emissions of 1.9-3.2 billion tonnes of CO_2 per year in 2050. This will put tremendous pressure on the balance of the Chinese and world oil supply and demand and could have significant implications on climate change.

FUTURE VEHICLE/FUEL SYSTEM OPTIONS

Recently, alternative transportation fuels and advanced vehicle technologies are being promoted to help solve urban air pollution problems, reducing greenhouse gas (GHG) emissions and relieving dependence on imported oil in the developed countries as well as the developing countries, such as China. These vehicle/fuel systems could be classified in two categories in terms of technology availability. The first includes those technologies which are readily available, such as compressed natural gas vehicle (CNGV), liquefied petroleum gas vehicle (LPGV), ethanol and biodiesel flexible fuel vehicle (E85 FFV) and grid-independent hybrid electric vehicle (GI HEV). The second category includes those technologies still under R&D and estimated to be ready in the next 10 to 20 years, such as hydrogen fuel cell vehicle, grid connected HEV (also called plug-in HEV), and pure electric vehicle (EV).

> "No matter how conservative the projection of future vehicle population is, current policy efforts still do not offset the increase in oil consumption."

© Gene Chutka/iStockphoto

Hydrogen fuel cell technology is considered to be one of the most promising because of its high energy efficiency, zero carbon emissions and tailpipe pollutant emissions.

CNG and LPG vehicles

Compared with conventional gasoline, CNG and LPG have around 20 per cent and 10 per cent lower carbon intensity, respectively, on the energy basis, eg per MJ of each fuel. As the upstream steps to produce a unit of LPG require less energy than gasoline, LPG can achieve more carbon reductions. If we consider from the fuel cycle point of view (also called well-to-wheels, including vehicle operation as well as upstream fuel production processes), both LPG and CNG vehicles could achieve 15 per cent -20 per cent CO_2 reduction versus conventional gasoline vehicle based on per km driven.

CNG and LPG vehicles in China have been increasing steadily over the past decade. For example, it was estimated that there were at least 200,000 CNG cars and buses in China by the end of 2006. In early 2007, more than 650 natural gas stations were built in China. In the near term, in those provinces with abundant natural gas resources, such as Sichuan Province, the penetration of CNG vehicles will continue to increase.

> "It was estimated that there were at least 200,000 CNG cars and buses in China by the end of 2006."

Diesel vehicle

Diesel engines are inherently more energy efficient compared with their gasoline counterpart. As a result, diesel vehicles emit around 15 per cent lower CO_2 than their gasoline counterparts based on per km driven. The penetration of diesel cars to China's vehicle stock is pretty slow, primarily due to the concern of relatively high particle/NO_x emissions. Because the leading criteria air pollutant in most Chinese cities is PM10, the promotion of diesel cars will be limited without stringent controls on tailpipe particle emissions. However, dieselisation has been largely observed over the past decade in the category of commercial vehicles all over China.

© CPW/iStockphoto

Ethanol and biodiesel flexible fuel vehicles

China is the third largest ethanol producer worldwide, producing around one million tonnes of ethanol per year. Most of this ethanol is produced from corn. Current production of biodiesel is pretty small in China at around 100,000 tonnes a year. The biodiesel feedstock is primarily from used cooking oil or other animal oils. Bioethanol and biodiesel are considered as renewable fuels and carbon free because the carbon in the fuel originally came from the atmosphere via photosynthesis. However, corn or other agricultural food based ethanol production will be limited due to the concern of food safety issue in China. Furthermore, as coal is the primary process fuel for ethanol production in Chinese mills, upstream CO_2 emissions may offset the carbon benefit from such biofuel production. In the long term, the Chinese government is pursuing cellulosic biomass based fuels with biomass as the process fuel, which could make such biofuels close to a real carbon free. This effort, however, is still in an early R&D stage.

> "China is the third largest ethanol producer worldwide, producing around one million tonnes of ethanol per year."

Grid-independent hybrid electric vehicle (GI HEV)

Compared with a conventional engine counterpart, GI HEV is usually 30-50 per cent more energy efficient. The significant improvement of fuel economy makes hybrid vehicles probably one of the most promising vehicle technologies in the near future. Domestic car manufacturers in China are also putting a major effort in developing hybrid technologies. The Chery Company, for example, is planning to sell at least 50 per cent of its new cars with a hybrid system after 2010. There are no major technical barriers in the gird-independent hybrid technology. The hot sale of hybrid vehicles, led by Toyota, in North America,

Japan and other developed countries proves the success of GI HEV. This will only help to cut the cost of HEV to compete with its conventional engine counterpart.

Hydrogen fuel cell vehicle

The hydrogen fuel cell vehicle is considered to be one of the most promising vehicle technologies in the long run because of its high energy efficiency, zero carbon emissions and tailpipe pollutant emissions at the vehicle operation stage. Furthermore, hydrogen is an energy carrier instead of energy source. It could potentially be produced by various feedstocks, which means every region in China has some indigenous fossil or renewable resource that can be used to make hydrogen. For example, in Sichuan Province, hydrogen could be produced from natural gas; and in Zhejiang Province, hydrogen could be produced from nuclear or cellulosic biomass.

The hydrogen fuel cell vehicle is still in the early R&D stage and there are several major concerns to be addressed before its mass penetration. For example:

- The investment to develop a hydrogen infrastructure is huge.
- Coal is the majority fossil energy source in China. Producing hydrogen from coal without carbon capture and storage (CCS) may result in significant CO_2 emissions.

> "The significant improvement of fuel economy makes hybrid vehicles probably one of the most promising vehicle technologies in the near future."

Plug-in hybrid electric vehicle and electric vehicle

Similar to hydrogen fuel cell vehicles, plug-in hybrid electric vehicles (HEV) and electric vehicles (EV) are promising vehicle technologies in the long term. These vehicle technologies use another energy carrier, electricity. The advantage of plug-in HEV/EV compared with hydrogen fuel cell vehicles is that the electricity infrastructure is currently available. The disadvantage is the fact that majority feedstock of power generation in China is coal. Without CCS, these technologies will have tremendous CO_2 emissions at the upstream electricity generation stage. The plug-in HEV and EV are still in the early R&D stage in China.

Authors

Dr Lixin Fu is the Director of Air Pollution Research and a Professor in the Department of Environmental Science and Engineering of Tsinghua University, Beijing, China. He has been leading national projects on mobile source pollution and control in China, and has supported a series of important national policies and regulations of vehicular emission reduction. Dr Fu has a bachelor's degree in fluid mechanics, a master's degree in Architectural Physics, and a PhD in environmental engineering, all from Tsinghua University.

Dr Ye Wu is an Associate Professor at the Department of Environmental Science & Engineering, Tsinghua University. He received his PhD in environmental engineering from Tsinghua University in 2002. He served as an assistant systems analyst in Argonne National Laboratory for five years. His research focuses on life cycle analysis of various vehicle/fuels systems and vehicle emission control.

Enquiries

Department of Environmental Science and Engineering, Tsinghua University
Beijing, 100084, P R China
Tel/Fax: +86 10 62771465
E-mail: fuchen@tsinghua.edu.cn

The greening of shipping

EFTHIMIOS E MITROPOULOS
SECRETARY-GENERAL, INTERNATIONAL MARITIME ORGANIZATION

Climate change and global warming pose a great threat to our environment. In a world dependent on motorised transport systems that emit large quantities of greenhouse gases, significant reductions in shipping emissions will help to reduce shipping's burden on global warming. A number of measures to reduce greenhouse gas emissions are now being considered by the International Maritime Organization within the context of the United Nations Framework Convention on Climate Change and the Kyoto Protocol. For measures to be effective, all those in the maritime industry must ensure they are implemented widely, working towards a global regulatory framework for shipping.

Of all the threats to our environment, the most significant concerns atmospheric pollution and the associated phenomena of climate change and global warming and, on this issue, the time for doubt has passed. Whether we like it or not, the modern world is utterly dependent on air, road and shipping transport systems that run largely on fossil fuels and emit greenhouse gases, principally CO_2 which is a significant contributory factor towards global warming and climate change.

SHIPPING'S GHG EMISSIONS

The shipping industry is currently a smaller contributor to the total volume of greenhouse gas (GHG) emissions than road vehicles, aviation and public utilities, such as power stations. Moreover, significant improvements in engine and propulsion system efficiency, improved hull design and the use of ships with larger cargo-carrying capacities have led to a reduction in ships' emissions per tonne-mile over the past decade and a noteworthy increase in fuel efficiency.

There are, however, predictions that the situation will deteriorate quickly unless action is taken now. A study in 2000 by the International Maritime Organization (IMO) looked at GHG emissions from ships and cautioned that if none of the GHG-reduction measures identified were applied, projected annual growth in fleet size could lead to a sizeable increase in fuel consumption between 2000 and 2020. This would lead to a consequential increase in CO_2 emissions.

In addition, as shipping activities grow apace with the global economy, the contribution of ship emissions to air quality problems in many parts of the world is growing and becoming more conspicuous. Some argue that for exposed coastal areas of Europe and the United States, shipping will, in 10-15 years, become the biggest single source of sulphur and particulate emissions if the trend continues unabated.

Although these are only forecasts, the shipping industry has developed a clear sense of responsibility with regard to its environmental credentials and continues to make efforts to make shipping greener and cleaner.

EMISSIONS REGULATIONS

In the 1960s, the most serious environmental problem was oil spillages into the seas, through accidents or poor operating practices. In 1973, IMO adopted the International Convention for the Prevention of Pollution from Ships, known universally as MARPOL. This deals with the prevention of pollution from ships by oil, noxious liquid substances carried in bulk, harmful substances carried by sea in packaged form, sewage and garbage.

In 1997 a new MARPOL Annex VI was adopted which covered the prevention of air pollution from ships. For the first time these regulations set limits on sulphur oxide (SOx) and nitrogen oxide (NOx) emissions from ship exhausts, prohibited deliberate emissions of ozone-depleting substances, and put a global cap on the sulphur content of fuel oil. Provisions allowed for special SOx Emission Control Areas to be established (as have already been agreed for the Baltic and North Sea areas), with more stringent controls on sulphur emissions. The Annex entered into force in 2005, and a comprehensive review of its provisions is currently underway.

Recording and indexing GHG emissions

Greenhouse gas emissions from ships are not covered in the 1997 Annex, but in December 2003 IMO adopted its Assembly resolution A.963(23) on *IMO Policies and Practices related to the Reduction of Greenhouse Gas Emissions from Ships*.

Work has focused on GHG reductions and developing a GHG Indexing Scheme for ships. *Interim Guidelines for Voluntary Ship CO_2 Emission Indexing for Use in Trials* were approved in July 2005, with the objective of establishing a common approach for trials on voluntary CO_2 emission indexing. This enables shipowners to evaluate their fleet performance with regard to such emissions. As the amount of CO_2 emitted from a ship is directly related to the consumption of bunker fuel oil, CO_2 indexing also provides information on a ship's performance with regard to fuel efficiency.

Results from hundreds of trials now exist with a huge volume of CO_2 data and in November 2007 a central database was established in IMO's Global Integrated Shipping Information System to make the information accessible for comparison by Member States and the shipping industry. There are interesting differences between results from identical ships in similar trades for example. These differences may result from different weather conditions or operational differences concerning the specific use of individual ships involved in the trials. Issues such as the length of time spent waiting in port areas, the length of ballast voyages, whether the ship is fully laden or not, can all make a difference.

Reducing GHG emissions

In looking at overall levels of GHG emissions from shipping, it should be borne in mind that the total amount of shipping activity is not governed by shipping itself, but by global demand for shipborne trade. Over the past four decades, estimates suggest that total seaborne trade has more than quintupled, from less than 6,000 billion tonne-miles in 1965 to over 30,000 billion tonne-miles in 2006. Today, world trade continues to grow and the international shipping industry has responded to the demand for its services.

But the glare of international publicity, fuelled by today's global communication infrastructure, ensures that environmental issues are played out on a worldwide stage. The broader concerns of society mean that pressure to be green and clean is mounting.

IMO continues to make progress on reducing GHG emissions. In 2007, work began to update the 2000 study on GHG emissions from ships, and a timetable adopted for future work on reducing GHG emissions. Such work must be seen in the context of the United Nations Framework Convention on Climate Change (UNFCCC) and the 1997 Kyoto Protocol and be part of a broad-based effort in which everyone has a responsibility and role to play; a maxim reflected in the well known environmental call to action 'think globally – act locally'.

While there is an impressive track record of continued environmental awareness, response and action by the maritime community over many years, the industry must not rest on its laurels. The IMO continues to pursue a long term strategy to ensure shipping maintains and improves its contribution to global sustainability. Such a strategy involves governments, the shipping industry, environmental organisations, engine manufacturers, oil producers, scientists and other relevant interests, so that all parameters can be taken into account when the key decisions are made and implemented.

CONCLUSION

Now that there is widespread recognition that the greenhouse effect represents a real, present, clear and serious threat to the environment and to Earth as a whole, public opinion, stimulated by information about the depletion of the ozone layer, GHG emissions and climate change, is impatient for action. The ball is clearly in the court of politicians, both in developed and developing countries, in established and emerging economies. By action in the maritime field now, they will also respond positively to the wishes of the 2000 Millennium Summit and the 2005 World Summit, and work toward environmental sustainability in which the world community has quite rightly placed so much hope.

The protection of the environment is something that can be, and must be, addressed by all. The work of the IMO at the global regulatory level, is only really effective if the measures we adopt are widely implemented, on a daily basis, by ordinary people in the industries that we serve. Everybody, no matter who they are, can, and must, do their bit to make a difference. When it comes to the environment, what we do, what you do, every day really does matter.

Author

Efthimios E Mitropoulos (Greece) is Secretary-General of the International Maritime Organization (IMO). Mr Mitropoulos was born in Piraeus into a maritime family and served on merchant ships before joining the Greek Coast Guard. He headed the Greek delegation to various meetings at IMO and joined the IMO Secretariat in January 1979. In May 1992 he was appointed Director, Maritime Safety Division and, in May 2000, Assistant Secretary-General. He was appointed Secretary-General of IMO for the period 2004 to 2008. In November 2006, the IMO Council decided to renew his mandate for another four years, concluding 31 December 2011.

Organisation

The IMO is the United Nations specialised agency with responsibility for safety and security of shipping and the prevention of marine pollution by ships. It is also involved in legal matters, including liability and compensation issues, and the facilitation of international maritime traffic. The IMO currently has 167 Member States and three Associate Members.

Enquiries

E-mail: info@imo.org | Website: www.imo.org

A single European sky would save over 12 million tonnes of CO_2 annually.

A sustainable vision for the aviation industry

PHILIPPE ROCHAT
DIRECTOR, AVIATION ENVIRONMENT,
INTERNATIONAL AIR TRANSPORT ASSOCIATION

Air transport provides many benefits around the world including job and wealth creation and unprecedented global mobility. The challenge for the industry is to keep such benefits while eliminating any negative climate change impacts. The International Air Transport Association's vision of the future is for carbon neutral growth in the medium term and of zero carbon emissions technology development within 50 years. To achieve this, however, development in the four areas of technology, operations, infrastructure and economic measures, will be vital.

AIR TRANSPORT: GOOD OR BAD?

Air transport is an important catalyst for globalisation and world economic growth. The industry contributes to sustainable development by facilitating tourism and trade, generating economic growth, creating jobs and increasing tax revenues. For example, a US$350 million investment in aviation infrastructure and services in Kenya boosted long term GDP by 0.4 per cent per annum. Air transport often serves as the only means of transportation to remote areas, thus promoting social inclusion. It facilitates the delivery of emergency and humanitarian aid relief and the swift delivery of medical supplies, as well as organs for transplantation.

In Africa, the role of air transport is especially important given the absence of effective ground transport networks and the continent enjoyed an 8.6 per cent growth in passenger traffic in 2006, against a global average of just five per cent. Air transport generates 470,000 jobs and US$11.3 billion in economic activity across Africa.

The benefits of air transport are indisputable but the industry is coming under increasing pressure over its environmental impact. The arguments are numerous: aviation is a major contributor to global warming; it is the fastest growing source of greenhouse gas (GHG) emissions; flying is destroying the planet. So what is the truth?

The IPCC, winners of the 2007 Nobel Peace Prize, estimates the level of global CO_2 emissions from aviation at two per cent, with a total climate change impact of three per cent. This includes not only commercial traffic, but also general aviation and military flights. These figures have remained largely unchanged over the past two decades, despite the growth in air traffic. Projecting forward to 2050, the IPCC has aviation at three per cent of global CO_2 emissions and five/six per cent of climate change impact. Aviation is but a small part of the big problem of climate change.

BALANCING ACT

Despite these figures, aviation's carbon footprint is growing and that is politically unacceptable for any industry. The challenge for aviation is to keep its many benefits, such as unprecedented global mobility supporting 32 million jobs and US$3.5 trillion worth of economic activity, while eliminating its negative impacts.

In September 2007, government leaders responsible for civil aviation gathered for the Triennial Assembly of the International Civil Aviation Organization (ICAO). Environment was the hot topic of the week with all 190 ICAO contracting states reaffirming ICAO's leadership on aviation and environment by endorsing a comprehensive strategy on climate change, and embracing the International Air Transport Association's (IATA) 25 per cent goal to improve fuel efficiency by 2020.

Unilateral European proposals to include aviation into its emissions trading scheme have put economic measures at the centre of the political debate, partially fuelled by the upcoming Kyoto deadlines. The solution is not to return to the days when flying was reserved for the well off by making it artificially expensive with even more taxes. Punitive economic measures, such as emissions trading will not have a big impact on aviation's environmental performance. With 28 per cent of costs coming directly from fuel, the airline industry is the most incentivised in the world to keep fuel consumption low. Positive measures, such as tax credits to encourage faster refleeting or grants to fund alternative fuel research, would deliver better results.

If emissions trading is to be imposed on airlines, it must be effective and ICAO has promised to work towards a global emissions trading scheme that states could implement on a mutual consent basis.

A VISION FOR AVIATION

The industry is pushing governments for much loftier goals. In June 2007, IATA proposed a vision for the industry to aim for carbon neutral growth in the medium term and to develop zero carbon emissions technology within 50 years. Aircraft manufacturers, engine makers, fuel suppliers and airlines will all be involved in making this goal a reality. A goal that is definitely possible from an industry that went from the Wright Brothers to the jet age in just five decades. The critical question is how to turn the vision into reality?

The first step will be for aviation to decouple carbon emissions from traffic growth, to stop its carbon emissions from growing in absolute terms. The second step must see aviation further reduce its remaining carbon emissions to eventually become a totally carbon emissions free transport mode. To achieve this, IATA has challenged the industry to build and operate a commercial airliner that produces no net carbon emissions within the next 50 years.

Although complete solutions do not exist today, some of the building blocks, ie materials and designs, alternative energy sources, advanced IT solutions, are already taking shape. IATA is confident that with a commitment from all parties, and with ICAO leadership, this vision can be made reality. The vision is built on four pillars.

1: TECHNOLOGY

Technology is a main driver of progress. It is the only way to zero emissions. Quite simply we need to build better planes and more efficient engines powered by non carbon sources. Some potential building blocks already exist: solar power, hydrogen cells and biofuel. Accelerated development of alternative fuels and more advanced technology for airframe, engine and air traffic management is absolutely essential.

IATA, manufacturers and fuel suppliers are jointly working on an action plan focusing on short, medium and long term measures. In the short term, the potential exists to realise emissions reductions by identifying and applying product enhancements and modifications for the current fleet. Airlines are investing billions of dollars in new, more fuel-efficient aircraft. In the past four decades, fuel efficiency increased 70 per cent and will improve a further 25 per cent by 2020.

For the medium term, possibilities must be explored to accelerate fleet renewal and to introduce the latest technologies as early and as widely as possible. Now that the giant A380 and the Dreamliner B787 have moved from design to production, the main challenge lies with the replacement of the single aisle A320 and B737. The decisions made by Airbus and Boeing, in coordination with engine manufacturers, will have a substantial impact on aviation's future carbon footprint. Development of cleaner, alternative aviation fuels must also take priority, while for the longer term, joint initiatives should be launched to identify and develop radically new technologies and aircraft designs.

2: OPERATIONS

According to the IPCC, improved aircraft operations can save fuel and CO_2 emissions by up to six per cent. IATA is compiling industry best practices, publishing guidance material, conducting airline visits and establishing training programmes to improve existing fuel conservation measures.

In 2007, IATA updated its fuel efficiency goal. It expects airlines to improve their fuel consumption per Revenue Tonne Kilometre (RTK) by at least 25 per cent by 2020, compared to 2005 levels. This will save around 345 million tonnes of CO_2 emissions.

IATA is determined to raise environmental standards across the industry and is aiming to extend its existing fuel conservation programme to promote environmental management systems across all airlines.

3: INFRASTRUCTURE

Infrastructure improvements present a major opportunity for fuel and CO_2 reductions in the near term. By addressing airspace and airport inefficiencies, governments and infrastructure providers can eliminate up to 12 per cent of CO_2 emissions from aviation, according to the IPCC.

Unfortunately, politics often gets in the way of good common sense. Uniting Europe's skies offers the biggest single opportunity to improve aviation's environmental performance. In Europe, there are 34 different air traffic control authorities. In the US, a similar land area, there is just one. If Europe had a single European sky, travel would be more efficient with planes spending less time in the air. This alone would cut CO_2 emissions by 12 million tonnes each year. But after 15 years of talks, a single European sky is still just an idea.

The flexible use of airspace must also become a reality, especially in Asia where traffic growth is particularly strong, and in most oceanic regions where navigation, communication and surveillance should greatly benefit from satellite signals. Thanks to new technologies and procedures, new routes are being opened, including polar ones that shorten long haul flights between Asia and North America or Europe. Moreover, pilots are increasingly allowed to select the most fuel efficient routes in real time, thus avoiding unfavourable wind or other meteorological conditions.

4: ECONOMIC MEASURES

Economic measures should be used to boost the research, development and deployment of new technologies rather than as a tool to suppress demand. The use of tax credits and direct funding must be explored as incentives to drive new technology programmes.

Punitive taxes do not improve environmental performance. Emissions trading would be a more cost effective solution as part of a global package of measures including technology, operations and infrastructure improvement. But the trading system must be properly designed and implemented on a global and voluntary basis. It must also be an open trading system, allowing permit trading with other industries. Economic measures can further be used to engage airline customers in climate change initiatives through carbon-offset mechanisms.

CONCLUSION

The world's airlines, IATA's 240 members, are committed to the vision outlined here. Manufacturers, airports and air navigation service providers are aligned. All are working hard. It is now time for governments to come on board. Through ICAO, governments must set challenging but realistic targets in two important areas. First, targets must be set to improve air traffic management and eliminate unnecessary fuel burn. Then technology targets should be included in a roadmap. This will provide regulatory certainty to back major investment decisions by manufacturers and airlines.

Airlines are investing billions of dollars in more fuel-efficient aircraft.

Author

Philippe Rochat is Director of IATA's Aviation Environment Department. As such, he supervises all activities related to emissions, sustainability and noise and is Secretary of the IATA Environment Committee. He is also Executive Director of ATAG (www.atag.org), a worldwide alliance of the many sectors of the industry. ATAG has recently launched a cross industry website www.enviro.aero with information about what the industry is doing to tackle climate change.

Philippe Rochat began his aviation career as assistant to the Director General of Civil Aviation of Switzerland, moving on to direct the commercial and financial departments of Geneva International Airport, with additional responsibility for environmental and facilitation issues. After serving as the Representative of Switzerland on the Council of the International Civil Aviation Organization (ICAO), he was elected Secretary General of ICAO and served from 1991 to 1997. Philippe Rochat has a Doctorate of Law from the University of Lausanne and teaches Air Law and Air Transport Economics at several universities.

Organisation

IATA is an international trade body, created some 60 years ago by a group of airlines. Today, IATA represents over 240 airlines comprising 94 per cent of scheduled international air traffic. The organisation also represents, leads and serves the airline industry in general.

Enquiries

Philippe Rochat, Director Aviation Environment
International Air Transport Association
Route de l'Aéroport 33
PO Box 416
1215 Geneva 15 Airport
Switzerland
Tel: +41 22 770 2670 | Fax: +41 22 770 2686
E-mail: rochatp@iata.org | Website: www.iata.org

InterCall

The alternative to business travel

Recognising the damage business travel is doing to our climate, business leaders are increasingly looking for cost effective alternatives, such as audio, web and video conferencing. For the vast majority of internal and external meetings, says Carolyn Campbell, Senior Director of Marketing Communications at InterCall, a web conference can be the best solution to save money, cut down on unproductive travelling time and reduce carbon emissions.

PLANES, TRAINS AND WEB CONFERENCING?

We all know how much damage travel, and its accompanying carbon dioxide emissions, cause to the environment. We know that a flight from London to New York emits carbon dioxide (an average of 1.2 tonnes) into to the upper atmosphere and that this is very bad for the environment. This is why a growing number of informed and conscientious consumers now think carefully about the environmental effects of their holiday destinations and modes of transport.

Yet, how many of those consumers think about the effect of their business travel? According to Wagonlit Travel the global travel market is now worth approximately US$870 billion, of which US$350 billion is business travel. So, around 40 per cent of our travel related emissions comes from business travel.

Table: Travel to business meeting – kg of CO_2 output per trip.

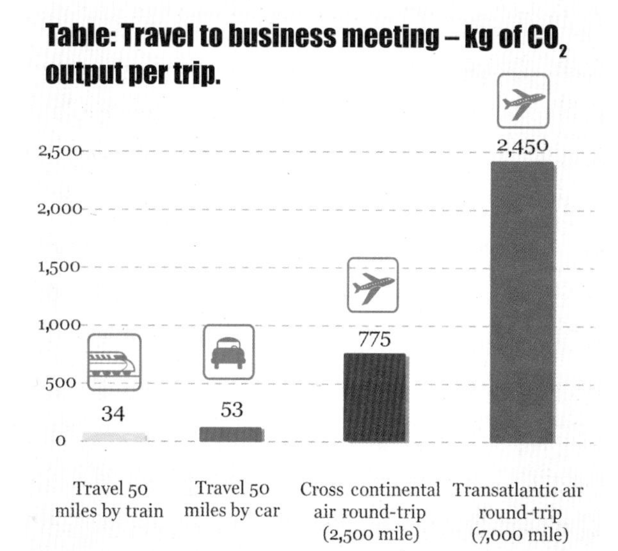

> **"Around 40 per cent of our travel related emissions comes from business travel."**

However, few of us can simply decide not to visit clients, attend overseas internal meetings, or only liaise with suppliers by telephone.

Furthermore, there appears little we can do about the environmental damage we cause travelling to and from work. Using an average commuting distance of 18 miles each way and an average mileage of 23.4 miles per gallon, the daily fuel consumption due to commuting is at least 1.5 gallons for a round trip. So, a five day commuting working week releases more than 5,154 lbs

INTERFACEFLOR

InterfaceFLOR is the largest manufacturer of modular carpet in the world. Based in LaGrange, Georgia, the company creates tiles for commercial use. It prides itself on its ecofriendly manufacturing process, which incorporates everything from using corn fibre as its raw material to using methane gas from a Georgia landfill and solar power as its energy source.

Every creative, manufacturing and business decision is guided by its 'Mission Zero' commitment: to have zero harm on the environment by the year 2020. As one of the first in its industry to make this commitment, InterfaceFLOR is not only manufacturing high quality carpet, it is also working to reduce its footprint on Earth.

Twelve years into Mission Zero, InterfaceFLOR integrated its two websites into one. This new site contains interactive features and allows customers to have a more sensory experience when choosing their products. The challenge the company faced was training its 120 salespeople, who are scattered across the US, without producing harmful carbon emissions by flying or driving.

With help from InterCall's Meeting Consultants, InterfaceFLOR set up Microsoft Live Meeting web conferencing. This allowed executives to give a live tutorial of the new web site to employees right across the country.

During the tutorial, instructors gave real time demonstrations of how the site works, and gave immediate answers to employees' questions. The web conference even allowed leaders to poll participants or monitor their engagement with the mood indicator button.

InterCall's web conferencing solution enabled InterfaceFLOR to hold four training sessions for 30 employees at once, altogether training more than 120 company employees. As a result, the company avoided more than 120 plane flights and prevented the release of thousands of pounds of harmful carbon emissions in the air.

of carbon dioxide into the atmosphere each year, per employee.

Increasingly, then, businesses are looking at these figures, recognising the damage they are doing to our climate through business travel, and looking for cost effective solutions that not only reduce emissions, but also maintain employee productivity. For a rapidly growing number of businesses, the answer is found in collaboration tools such as audio, web and video conferencing.

THE VIRTUAL MEETING

Most businesspeople are now familiar with the concept of virtual meetings. Many of us have attended them and we all tend to think we know what they are. However, according to Carolyn Campbell, Senior Director of Marketing Communications at InterCall, the world-leading conferencing company, there is confusion and misunderstanding about the different types of conferencing.

She explains: "Video conferencing has been around for a long time and has an image of being expensive and cumbersome. It uses specialised video equipment and is helpful in situations where high definition visuals are useful. However, people are increasingly using web conferencing. It is all done over the web with plug-and-play technology that allows people to talk to each other, see each other, and even collaborate on documents."

Over the past few years, as internet connections have grown faster and broader, the quality of web conferencing has vastly improved. This is encouraging a growing number of companies to spend more time in virtual meetings and less time on expensive and polluting journeys. Web conferencing is also enabling more and more people to work from home. According to research conducted by WorldatWork through The Dieringer Research Group, the number of Americans whose employers allow them to work from home at least one day per month increased from 9.9 million to 12.4 million in 2006.

"As internet connections have grown faster and broader, the quality of web conferencing has vastly improved."

Campbell comments: "Web conferencing can't replace every type of meeting. It's still helpful to meet in person to kick off a new team, for example, or to get together if the event will take the better part of a day or many days. But for the vast majority of internal and external meetings a web conference is now the best solution. You don't need anything more complex than a webcam to get started. It saves you money, cuts down on unproductive travelling time and, because it uses computers and telephones that would be switched on anyway, has almost no impact on the environment."

CHOOSING THE RIGHT PROVIDER

Like many of its competitors, InterCall began offering conferencing facilities in the early 1990s and saw steady growth during the economic globalisation of that decade. It was no surprise that demand soared in the wake of 9/11. What has surprised Campbell and her colleagues is how that demand has continued to grow even after people have once again become comfortable flying.

Between 2001 and 2003 business travel dipped, but since 2005 it has risen by 24 per cent. Yet, according to Wainhouse Research, more than 60 per cent of all meetings today are done virtually, and that number is expected to rise to 70 per cent in the next two years. The conferencing industry has grown to US$3.7 billion, a 32 per cent increase in revenue from 2000. This growth is apparent at InterCall as well. With its 35 sales offices worldwide and more than 500 Meeting Consultants, InterCall generated US$439 million in revenues in 2005 and exceeded US$600 million in 2006.

Today, Chicago based InterCall serves more than 400,000 individual conference leaders in more than 70,000 organisations. Its global operations reach Canada, the UK, France, Germany, Australia, New Zealand, Hong Kong, Singapore, Japan, China and India. The company handles more than three billion minutes annually and provides conferencing services to 80 per cent of Fortune 500 companies and 50 per cent of Fortune 1000 companies.

“More than 60 per cent of all meetings today are done virtually, and that number is expected to rise to 70 per cent in the next two years.”

A GROWING MARKET

To some extent InterCall's growth has occurred as a result of growth in the market as a whole. However, Campbell believes that there are three factors that have set it apart from the competition. She says: "Firstly, we're the world's largest conferencing provider because we're not tied to any one system. We work with all the major systems available in audio, web and video services. Also, we offer our customers a fully integrated global VoIP system that allows easy connections to meetings without the usual hassles of international dialing. Wherever you are in the world, you can be confident that InterCall will be able to meet your needs in the way that you need it to."

She continues: "Secondly, there's a perception that virtual meetings are cold and impersonal, but we've invested in an extensive global service network. This ensures you are taken care of and feel comfortable with your service technicians and sales teams at all levels of the provider relationship. Thirdly, we're solution evangelists. We have a passion for training people on how to use our products. We know all the tips and tricks to make the technology work as well as possible, and we're keen to share them with our customers."

LOOKING AHEAD

Campbell expects the market for web conferencing to continue to grow. "The use of conferencing has grown at over 15 per cent a year since 2000 and doesn't show any sign of slowing," she says. "It's growing at different rates in different regions. Most notably Europeans and Asians are now realising what North Americans have known for some time – that it's just not feasible, economically or environmentally, to get in your car or on a plane and see everybody. Especially not when there are such good alternatives out there."

She expects that, aside from this increased take-up in Europe and Asia, the main development in the web conferencing market will be the increasing focus on unified communications. Campbell explains this by saying: "At the moment, my system knows, from looking at my Outlook Calendar, that I'm in a meeting. So, it's ensuring that I'm not disturbed. In the future we'll be seeing more of this sort of integration between programmes, allowing us all to focus on the job rather than the technology."

So, there clearly are options for businesses that want to maintain their productivity while also minimising the environmental impact of their business travel. Forward looking businesses are rapidly adopting technology that in the space of a few years has become more effective and available. As the rest of the business world joins them, so we will all become as environmentally responsible in our working lives as we already are in our personal lives. That can only be good for everyone.

Organisation

InterCall provides the broadest choices of audio conference calls and web/video conferencing for business collaboration. Using InterCall, you can connect with your customers, teams and vendors without the significant carbon footprint (and expense) caused by commuting or air travel. Collaborate better, communicate faster and reduce environmental impact with InterCall.

Enquiries

InterCall
8420 West Bryn Mawr
Suite 400
Chicago
IL 60631
USA
Tel: US: 800 820 5855;
Canada: +1 877 333 2666
APAC: +612 8295 9197;
UK and EMEA: +44 1452 556200
Website: www.intercall.com

© Steve Johnson/iStockphoto

DENNIS PAMLIN
GLOBAL POLICY ADVISOR, WWF,
CO-AUTHORED WITH **KATALIN SZOMOLANYI**
MAGYAR TELEKOM (HUNGARY), ETNO

The high tech sector –

LEADERSHIP TOWARDS A LOW CARBON ECONOMY

The possibility to provide new and better services with the help of information and communication technology (ICT) is well known. Less known is the significance of these new services to reduce carbon emissions. This discrepancy in understanding has resulted in a situation where focus has been on ICT's internal emissions and the energy efficiency of equipment. While these two areas are important, their contributions to global emissions are not the most significant. Solutions provided by ICT can and must play an important role as we move towards a low carbon economy if we are to reach carbon reduction targets set by scientists and policy makers.

Midway through the first decade of the 21st century, the world is rapidly approaching a situation where, for the first time in human history, more people will live in cities than rural areas. 2008 will mark this historic moment and the future of this urban millennium very much depends on the decisions made today in preparation for such continued growth. By 2030, approximately two billion additional people will live in cities with a resulting increase in demand for urban solutions that can improve quality of life without consuming excessive natural resources. Parallel to this trend, global energy and natural resource use is increasing rapidly, with energy demand expected to increase by more than 50 per cent by 2030 if current trends continue.

ICT AND THE FUTURE OF THE PLANET

For too long the environmental agenda has been dominated by a reactive and negative approach. The focus has been to identify what's wrong and how to stop things. The language and approach to our global challenges has been developed in a context that has often failed to incorporate the perspective of innovation and global development. From a long term perspective, this makes little sense. It is time for a shift in focus from reactive to proactive and from an incremental to disruptive perspective on change.

Over the coming decades we will see the rise of China and India as two new global economic superpowers. By the middle of the century, it is likely that the Chinese economy, then the largest in the world, will be four and a half times bigger than the US economy is today. India will be the third largest, just after the US at around three and a half times larger than the current US economy.

Many are worried by the implications of such growth, and the difficulty of tackling the problems it will generate; problems which will have ramifications for the rest of the world. Increasingly, other poor countries are pinning their own economic development hopes on providing the natural resource base for these emerging economies: soy from Latin America, oil from Africa, timber from South East Asia. The need for ICT solutions that can provide welfare without increased resource consumption should be obvious.

GROUNDS FOR HOPE

The world should see current changes as an opportunity and a ground for hope. If we are to meet global targets, such as the Millennium Development Goals, the Kyoto Protocol, The Convention of Biodiversity, we must start to think outside the box. It is important to see how progressive companies who are willing to engage in the challenges ahead can be supported. How these companies can be included in low carbon policies should be high on any country's low carbon agenda.

BEYOND A SIMPLE POLARISED PERSPECTIVE

ICT has the potential to radically influence transport patterns, energy consumption, overall resource usage and, to an unknown degree, our culture and even the way we perceive the world, our relationship to it, and our actions. Although ICT will have an enormous effect on tomorrow's society, surprisingly little research has been conducted regarding its future environmental and social consequences, and even less concrete policies exist to support sustainable ICT use.

Most existing work in this area has reach one of two conclusions: either ICT will bring only good things, from solutions to world hunger and the elimination of all transportation problems to a revitalised democracy; or ICT will bring nothing but problems, accelerating resource consumption, introducing new toxic materials and resulting in greater inequity by introducing a digital divide that will worsen the already unequal distribution of wealth and influence. The first challenge, if we want to include ICT for the future, is to go beyond this polarised perspective.

NOT A PANACEA

However positive ICT use can be if aimed in a sustainable direction, it is important to acknowledge that it is not a panacea to the world's environmental and social problems. Our society will continue to have an environmental impact upon the world and social tensions will continue to exist. There is no such thing as a weightless or frictionless economy: we will always need food and shelter, will own material objects, and will need transport with their consequent environmental impacts. In the same way, ICT will result in new social situations where new actors emerge and new values are shaped. Even if the new society will be more just and equitable, people's views on how society should develop will still differ, as will their influence.

The information society will be layered atop the industrial society in the same way that the industrial society is layered over the agricultural society. Needless to say, the development of the new form of economy will not take place simultaneously, or look the same everywhere. It is therefore pointless to try and draw a distinct line between the so-called old and new economy, as this blurred line is drawn in time, includes all sectors and contains gradual changes mixed with sudden shifts.

THE USE OF ICT FOR A LOW CARBON ECONOMY

The numbers are meant to give an overview of the opportunities that exist for economic growth whilst combating climate change in the same time. The opportunity numbers are only related to EU-25, but can easily imagine their potential in e.g. China.

A new and more efficient meeting culture: Travel replacement

Possible target for EU-25: 24 million tonnes CO_2/year

Business travel replacement (video-conference)

If 20% of business travel in EU-25 Countries is replaced by a non-travel solution (e.g. video-conference), around 22 million tonnes of CO_2 can be saved per year.

Audio-conference

If 50% (96.512 million) of EU-25 countries' employees replaced one meeting with one audio-conference call per year, then 2 million tonnes of CO_2 can be saved per year.

Sustainable consumption: e-dematerialisation

Possible target for EU-25: 4 million tonnes CO_2/year

Online phone-bills

If all households with internet access, in EU-15 countries, and all mobile customers in EU-25 countries would get an online phone-bill, then 500 thousand tonnes of CO_2 can be saved per year.

Web-based tax return

If all employees in EU-25 countries (193 million) deliver their tax return via the Internet, then 200 thousand tonnes of CO_2 can be saved per year.

E-dematerialisation has much more potential than the examples above.

Sustainable community/city planning: combined measures

Possible target in EU-25: 22 million tonnes CO_2/year

Flexi-work

If 10% (19,3024 million) of EU-25 countries' employees became flexi-workers, then 22 million tonnes of CO_2 can be saved per year.

(based on ETNO-WWF roadmap: 'Saving the Climate @ the Speed of Light")

A UNIQUE OPPORTUNITY

As the infrastructure for this new society begins to come into place, there is a unique opportunity to shape the whole system in a way coherent to humanity's needs and visions for the future. The complexity of ICT makes it difficult to approach in a traditional manner, which focuses only on the direct impacts of extraction, manufacturing, use, and disposal. It is vital that all due caution be taken when responding to ICT based challenges. The impact of ICT must be viewed in a very broad sense, from cultural changes caused by the use of new technologies, to the appearance of new possibilities for shaping a new economy in which production and consumption patterns look fundamentally different.

The impact of eshopping, for example, cannot be reduced to the computers used for transactions, or even to the eliminated car trip to the store. Instead, the whole distribution chain must be examined to determine the long term impact this kind of consumption will have on the economy, the development of the infrastructure, and on habits.

FOUR STRATEGIC AREAS FOR ACTION

Corporations, organisations and political parties are all very enthusiastic about ICT and its possibilities to deliver resource efficiency, but judging from the actual results so far, these ideas and thoughts are seldom transformed into concrete action. The lack of results often seems to stem from a lack of focus. Backed only by vague ideas, the process leading up to actual implementation often winds to a halt somewhere before a concrete result has been achieved. In order to focus, a limited number of areas need to be prioritised initially.

ICT products

The importance of ICT products is two fold. First, even if ICT products themselves have only a marginal environmental impact, there is a great risk that the political system and mass media will judge the whole sector as environmentally unfriendly if the sector as a whole does not have a clear environmental strategy. Thus, ICTs broader credibility is threatened. Second, the rapid increase and penetration of ICT products can, if no action is taken, result in increased energy demand and bigger quantities of toxic substances.

Transport of goods

One area where ICT can contribute significantly in the short to medium term relates to the transport of goods. It is clear that ICT can be used to find more optimal transport routes and keep track of large fleets of vehicles. However, this could also result in further investment in a system that is inherently unsustainable. Due care must be taken to avoid a situation where investments only result in marginal short term reductions and create an infrastructure that makes it hard to reach necessary long term solutions.

When considering a sustainable transportation system, it is important to think in terms of service. Is it possible to deliver the same service without a physical product? Can the Internet be used to transport a service instead of moving a product, to move bytes instead of atoms?

MOVING ICT FORWARD

For governments

1. Appoint a responsible person in the government for ICT and climate change and allocate a budget.
2. Set targets for use of ICT in key areas, not only CO_2 reductions, but jobs created, number of patents etc.
3. Support companies that set targets for CO_2 reductions for the use of their products/services.
4. Ask companies for a product/sale catalogue for CO_2 saving services.
5. Set targets for export of ICT solutions that reduce CO_2 emissions.
6. Review rules and legislation from a dematerialisation perspective.
7. Explore a sustainable innovation zone.
8. Ensure economic policies support incremental improvements and disruptive new solutions that ensure more than marginal improvements of CO_2 reductions.

For business

Generic

1. Move focus from risk to profit – look at how sustainable services can increase profit.
2. Move focus from reactive to proactive – create new markets and support rules that help customers to use sustainable services.
3. Move focus from an internal and product perspective to an external and service perspective. Focus on how the a service can help customers to become sustainable companies/individuals.

Specific

1. Report on targets for CO_2 reductions for the use of ICT products/services.
2. Produce a product/sale catalogue for CO_2 saving services using ICT
3. Ask the government to set target for use of ICT in key areas
4. Ask the government to set targets for export of ICT solutions that can reduce CO_2 emissions.
5. Ask the government to review current legislation and organisation from a dematerialisation perspective.
6. Establish an internal (and later and external) sustainable innovation zone that allows staff, customers and developers to collaborate in developing new ICT solutions.

This process, along with the encouragement of local production, could set in place a less ecologically damaging economy.

Business travel

Not only would a reduction in business travel result in significant environmental gains; but done in the right way it could also contribute to a corporate culture where avoiding physical travel would be an option. The most important change needed in order to shift from a flying and car culture to a video conferencing and

telecommuting culture is a mental and institutional shift. High levels of travel reflect higher status in many sectors, and frequent flyers mileage is an extra bonus for many. In many countries, there are also tax incentives in place that encourage travel.
The significant shift towards these technologies will come when a majority recognise that high tech, high resolution video conferencing and telecommuting are better and more intelligent solutions than physical travel. Full wall high resolution projections for virtual meetings should be a standard in all major corporations, and all major cities should supply this kind of facility for groups and corporations that do not have sufficient inhouse resources. The technology exists; it is just a matter of coordinated investments.

Converging technologies

With rapid technological development, difficult ethical, environmental and economic questions beg to be discussed. Ethical questions regarding new technologies will be even more important in the years to come, with probable breakthroughs in areas, such as biotechnology, nanotechnology, robotics and quantum computing.
If technologies like these, and especially the combination of these technologies, are to be used in a responsible way, and if technological development is to contribute to solving at least some of the challenges faced today, there is a clear need to approach the issue in a strategic way. It is important to create incentives to direct development towards meeting the most basic and important needs of this planet.
In order to direct ICT and broader technological development in a sustainable direction, independent agencies should be created at both national and international level to evaluate emerging technologies. One way to address this issue would be to develop an International Convention on New Technologies, to assess the societal and political implications of emerging technologies before their commercial release. Due care must, however, be taken so as not to create a culture that believes all problems may be solved thanks to technology.

LET'S START NOW

There is no time to waste and the world desperately needs a new generation of business leaders who are not afraid of a global perspective.

© Terraxplorer/iStockphoto

Authors

Dennis Pamlin has a background in engineering/industrial economy/marketing. He has worked for WWF with global policy issues since 1999.
Examples of ICT projects:
2002: Edited the book "Sustainability at the speed of light".
2004: Initiated the project "Saving the climate @ the speed of light" with ETNO.
2007: Project supervisor for a WWF/HP project that will identify the first billion tonnes of CO_2 reductions with ICT solutions.

Katalin Szomolanyi is the Head of Corporate Sustainability Group, Magyar Telekom, Hungary. She has a background in environmental sciences – geography, nature conservation, and international business.
She has worked for Magyar Telekom since 1997.
Examples of climate projects:
2002: Launched and managed the ETNO project of ICT's effect on climate change
2004: Initiated ad managed on ETNO-side the joint initiative together with WWF on ICT & climate change
2006: Co-author of the "Saving the climate @ the speed of light" roadmap for reduced CO_2 emission in the EU and beyond

Organisation

For more than 45 years, WWF has been protecting the future of nature. The largest multinational conservation organisation in the world, WWF works in 100 countries and is supported by 1.2 million members in the US and close to five million globally. Using the best available scientific knowledge and advancing that knowledge where we can, we work to preserve the diversity and abundance of life on Earth and the health of ecological systems by: protecting natural areas and wild populations of plants and animals, including endangered species; promoting sustainable approaches to the use of renewable natural resources; and promoting more efficient use of resources and energy and the maximum reduction of pollution. ETNO – the European Telecommunications Network Operators' Association – is the industry's leading policy voice. It represents 41 major European telecommunications companies in 34 countries, inside and outside the EU. The Association launched the Environmental Charter in 1996 and the Sustainability Charter in 2004.

Enquiries

Websites: www.panda.org/ict
www.magyartelekom.hu/english/aboutmagyartelekom/sustainability/main.vm
www.etno.be

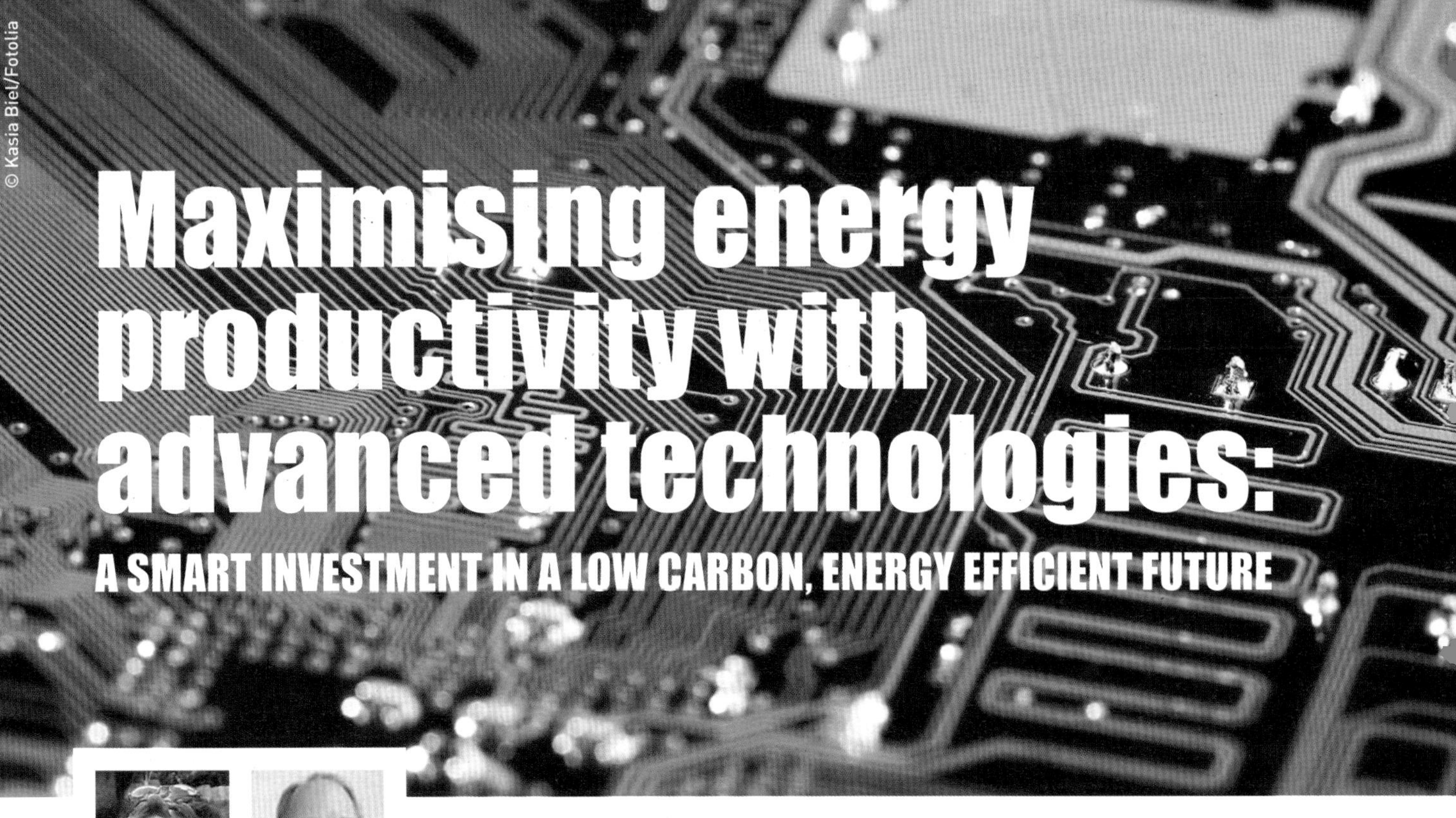

Maximising energy productivity with advanced technologies:

A SMART INVESTMENT IN A LOW CARBON, ENERGY EFFICIENT FUTURE

KAREN EHRHARDT-MARTINEZ AND
JOHN A 'SKIP' LAITNER
AMERICAN COUNCIL FOR AN ENERGY-EFFICIENT ECONOMY (ACEEE)

The decoupling of energy consumption from energy services, eg lighting, heating, hot water, mobility, through increased energy productivity provides a potential, albeit partial, solution to global climate change. Among the principal drivers of increased energy productivity during the past 15-20 years has been the widespread adoption of advanced technologies. Further enhancement and deployment of high tech electronics and information and communications technologies can provide the foundation for the many new innovations that can set a trajectory for a low carbon path to the future. Greater levels of investment can maximise the potential cost saving benefits of increased energy productivity.

Are technologically advanced countries of the world simply the primary source of global climate change, or might they also be part of the solution? There is no doubt that greenhouse gas emissions from the more advanced countries have contributed disproportionately to global climate change. However, without past technological innovations and supportive government policies, emissions levels today would be vastly greater. More importantly, the thoughtful and widespread application of many of the advanced technologies that are currently available could lead to even more dramatic reductions in future greenhouse gas emissions.

DECOUPLING

Compared to less developed countries, *per capita* energy consumption in the US and other wealthier nations is disproportionately large. Current US energy consumption is in the order of 105.6 exajoules annually (roughly 21 per cent of global consumption) – despite the fact that the US population represents only five per cent of the global population. Conversely, however, US energy consumption per dollar of economic output has declined by 50 per cent over the past 35 years from 18.98 megajoules in 1970 to 9.61 megajoules in 2005. In other words, as shown in Figure 1, current US energy consumption is only half of what it would have been if levels of energy productivity had remained unchanged. A similar decoupling has been experienced in Japan and the European Union.

> **"Current US energy consumption is only half of what it would have been if levels of energy productivity had remained unchanged."**

Understanding this phenomenon necessitates a clear distinction between energy commodities, such as fuels and electricity, and energy services, such as lighting, heating, hot water, mobility and an acknowledgement that growth in energy services need not be directly linked to growth in energy consumption. This decoupling of energy consumption from energy services illustrates the potential contribution of energy efficiency in reducing energy consumption while maintaining a high standard of living associated with our reliance on a wide variety of energy services. But how has this decoupling been achieved?

Figure 1. Historical gains in US energy efficiency 1970-2006.

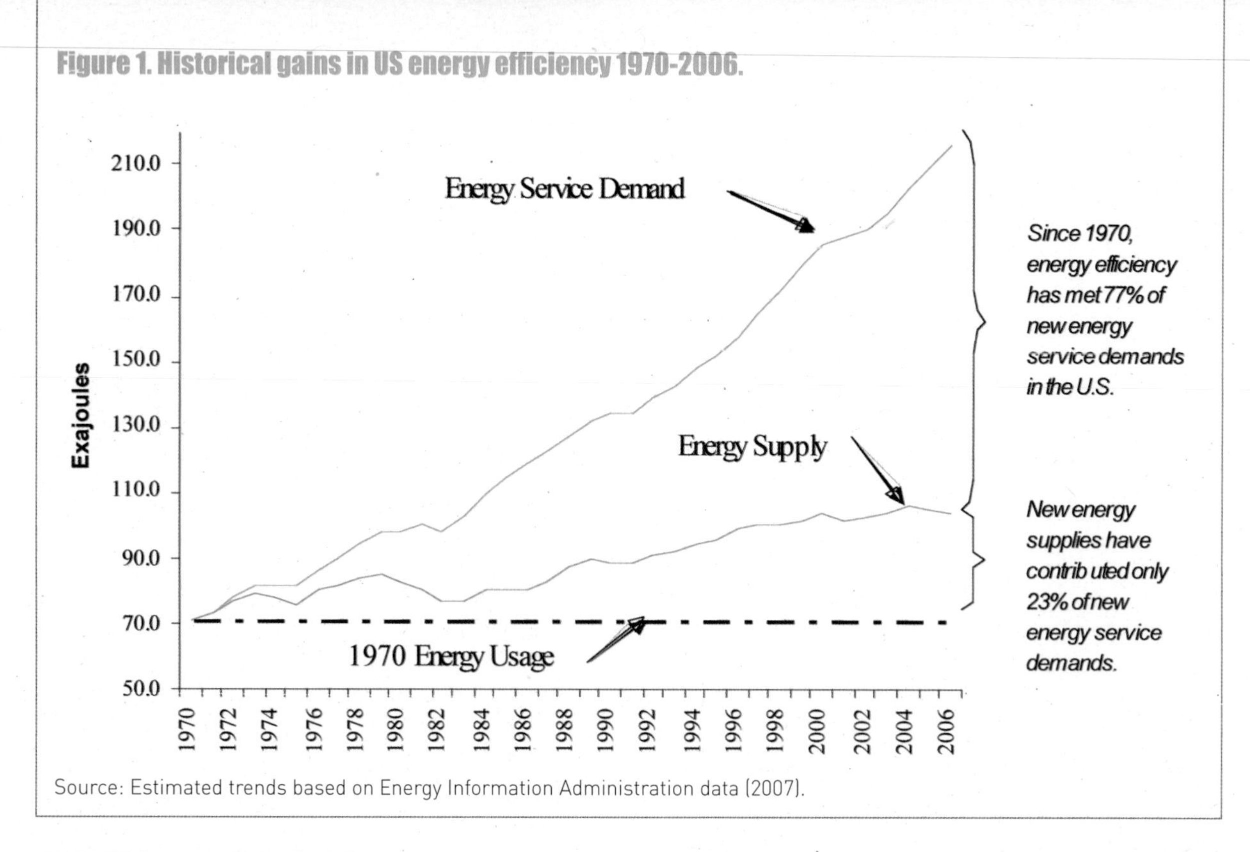

Source: Estimated trends based on Energy Information Administration data (2007).

ADVANCED TECHNOLOGIES AND ENERGY EFFICIENCY

Among the principal drivers of increased energy productivity during the past 15-20 years is the emergence and widespread adoption of advanced technologies, including high tech electronics and an array of information and communications technologies (ICT). These technological drivers of energy efficiency range from stand-alone products, such as computers and cell phones, to numerous types of sensors, microprocessors and other technologies embedded in everyday products, such as cars, lighting systems and appliances.

While it is easy to imagine that the proliferation of advanced ICT technologies would lead to an *increase* in power demand in all sectors, calculating their *net* effect on energy usage requires a broader understanding of the ways in which new technologies have continued to displace and improve older processes and systems. On the one hand, evidence suggests that energy consumption, and especially the use of electricity, has increased as the development of new applications drives the expansion of new devices and appliances in our homes and businesses. On the other hand, the larger, economy-wide productivity gains and efficiency improvements that have been realised through the use of these technologies have more than offset the energy used to power them.

As such, the further enhancement and deployment of ICT and other advanced technologies can provide the foundation for the many new innovations that can set a trajectory for a low carbon path to the future. Such a path would provide strong momentum toward achieving the goals of the Kyoto Protocol and other efforts to reduce total carbon emissions throughout the world. Several studies by the American Council for an Energy-Efficient Economy (ACEEE) and others confirm that advanced electronic and ICT technologies, if given the appropriate mix of policy signals, could play an even greater role in reducing climate changing emissions, and in limiting the potentially negative economic consequences of efforts to simply curtail energy consumption.

ICT ENABLED EFFICIENCY TECHNOLOGIES

If people are asked to think about advanced electronics or information technologies, most are likely to imagine a desktop computer or a laptop with access to the Internet. Others might think about mobile phones, blackberries and iPods. But in fact, advanced technologies and ICT enabled devices and equipment are much more ubiquitous than a casual observer might first imagine. Microprocessors and other semiconductor devices are embedded in the most mundane consumer products, from toys and refrigerators to stoves, vacuum cleaners, and coffee makers. In fact, microprocessors and other semiconductors have already successfully enabled numerous and significant energy efficiency improvements throughout the global economy.

Even a cursory review of recent news articles underscores just how much advanced electronics and ICT capabilities are already an embedded feature in numerous devices, software tools and networked systems. Digital control devices and sensors have been rapidly replacing analogue electronic devices and numeric controls in numerous industrial processes whenever economically and technologically feasible. Sophisticated analogue systems employing the latest advances in materials technology and mechanical systems have merged with digitally based and/or Internet addressable control devices.

Sensors have become increasingly solid state, with the capability of linking to analogue-to-digital converters within digital control systems. These microprocessors, using application-specific integrated circuits (ASICs) and other solid state components, can guide the actions or movements of ubiquitous mechanical systems such as motors, pumps, fans, compressed air systems and even more sophisticated industrial equipment such as machine tools, mixers, conveyors and robots. As such, sensors, automatic controls, and smart software programs all now have widespread use in every sector of the economy, including buildings, transportation, electric power generation, consumer appliances and entertainment.

Two noteworthy examples will highlight the many larger opportunities for energy efficiency improvements. The first deals with telecommuting and teleworking. The

Telework offers a high tech means of substituting telecommunications for work related travel.

second deals with the optimisation of building energy management operations.

Telecommunications

Cutting edge telecommunications technologies offer an alternative solution to growing levels of congestion, energy consumption and carbon dioxide (CO_2) emissions within the transportation and buildings sectors. Transportation related energy consumption alone was recently estimated to be responsible for nearly one-third of the EU-27's total energy consumption and 28 per cent of energy consumed in the United States. Throughout the world, more cars are driven more miles, consuming more fuel than ever before. And a significant proportion of those miles are work related. Telework offers a high tech means of substituting telecommunications for work related travel while also reducing the amount of building space required to house employees.

A study completed by the European Telecommunications Network Operators Association (ETNO) and the World Wildlife Fund concluded that if 20 per cent of business travel in the EU-27 were replaced by audio conferencing video conferencing or telepresence, 25 million tonnes of CO_2 might be saved annually by 2010. The study also noted that if just 10 per cent of EU employees became flexiworkers, another 22 million tonnes of CO_2 might be saved annually.

Intelligent building systems

In the European Union and the United States, buildings are currently responsible for approximately 40 per cent of greenhouse gas emissions or approximately four billion tonnes of CO_2 per year for these two regions alone. These high levels of emissions can be reduced using information technologies – what might be called Intelligent Building Systems – that rely on sensors, transmitters, data acquisition and data processing to optimise and improve building energy use.

These technologies achieve efficiency gains by productively managing a building's chillers, boilers, packaged air conditioners, heat pumps and lights. Since most buildings are not occupied 24 hours per day, space conditioning and lighting services can be reduced or shut down when buildings are unoccupied, reducing energy costs and minimising the wear and tear on equipment. The potential carbon savings provided through the adoption of intelligent building systems have been estimated as high as 20 per cent.

While these case studies suggest that significant reductions are possible for both energy use and carbon dioxide emissions through the adoption of smart technologies, by default they overlook the larger contributions that may be possible from an economy wide versus a case study perspective. The impact of accelerated investments on performance and productivity benefits are illustrated in Figure 2. It shows that especially in the case of high return technologies such as those examined here, the cumulative effort and investments over time would, in effect, provide a discontinuous jump in the performance path as both the advanced technologies and the markets are transformed. This trend would also likely increase overall energy productivity, especially as policy and organisational efforts are catalysing market decisions to move in that direction.

Figure 2. Productivity and performance as a function of cumulative investment.

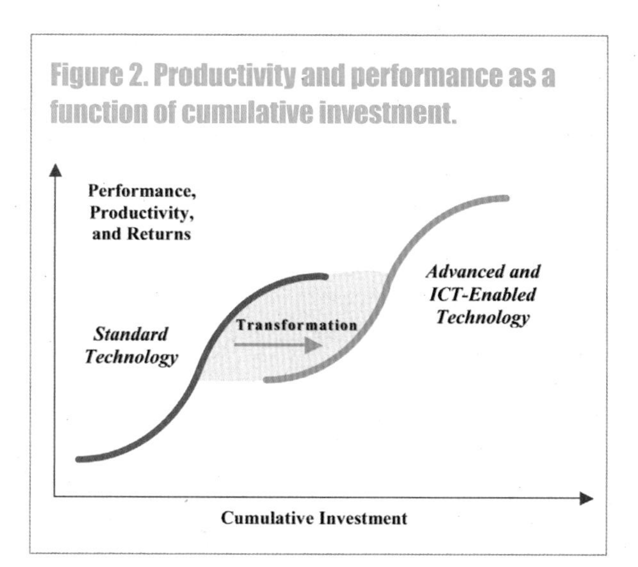

THE MULTIPLE BENEFITS OF ADVANCED TECHNOLOGIES

For businesses, climate change is just one of six major forces driving a renewed focus on energy efficiency. The other five include:

Energy prices

Since the turn of the century, energy prices in many countries have been both high and volatile, upending household and corporate budgets and reducing disposable income and profits. Forecasters generally agree that this pattern is not a temporary aberration, and that the era of cheap energy is over. As such, energy efficiency investments have become a cost management strategy for families and businesses.

The supply straitjacket

Underlying rising energy prices is a set of deeply interrelated energy market problems. The production, processing, and transportation of energy are all experiencing unprecedented constraints that span all major energy markets. This energy straightjacket will not be easily resolved; with supply constrained in so many ways, efficiency has become a near term strategy for balancing energy markets, moderating prices and providing the badly needed headroom to keep energy supply systems reliable.

Consumer and shareholder activism

Consumer, investor and voter groups are increasingly voicing their concerns regarding the environmental

> **"Energy and climate change policies don't have to be about ratcheting down the economy; they can be about technological innovation and leadership."**

and human impacts of corporate behavior in the energy industry. Moreover, socially responsible investing, shareholder activism and public campaigns are being organised to provide real economic incentives and consequences related to corporate action or inaction on efficiency-related environmental issues such as global warming.

Global competition

With the increased globalisation of business, multinational companies are encountering increasingly stringent climate policies and regulations. Efforts to comply with regional policies are driving more widespread investments in energy efficiency. Many companies are also pursuing business opportunities in the 'clean tech' sector. In this context, efficiency is an internal cost management strategy geared towards maintaining competitiveness, as well as an emerging business opportunity.

Better mousetraps

A wealth of energy efficiency technology advancements is nearing market readiness. In the high tech world, microprocessor makers are driving performance per watt to new frontiers. Intel's first (1996) supercomputer capable of a trillion calculations per second consumed 500,000 watts of power. In March 2007, Intel demonstrated a dime-size 80-core chip that used just 62 watts to break the teraflop barrier.

These six factors together have created a fertile environment for energy efficiency investment. However, the required investments and hoped for benefits are by no means guaranteed. Without a sensible framework of government policies, these unfolding technologies might follow any number of less productive paths.

INVESTING IN ENERGY EFFICIENCY: CURRENT AND FUTURE PROSPECTS

How big is the energy efficiency market? And does it make economic sense for businesses to invest in high tech enabled efficiency? Because the efficiency market is fragmented and dispersed, the opportunities are difficult to quantify. This phenomenon is an example of what economists call 'search costs'; a classic market barrier that chronically inhibits investments. Even though industrial energy efficiency investments often have less than a four year payback based on energy savings alone, search costs hinder millions of worthwhile transactions. When other productivity gains are assessed, such as reduced water, feedstock or labour costs, the payback period is often cut by half.

A forthcoming ACEEE analysis estimates that current US investments in energy efficiency are in the order of US$300 billion annually. These investments range from Energy Star® computers and refrigerators to more energy efficient cars, trucks and industrial equipment. These findings are consistent with a recent United Nations Foundation study that identified energy efficiency as both the largest and least expensive energy resource and suggested that the G8 and other nations could double historical rates of efficiency improvement by 2030. A study by McKinsey Global Insight indicates that all future energy service demand growth in North America could be met through cost effective investments in energy efficiency.

Assuming that policies, market forces and new financing mechanisms facilitate substantial movement toward productive investments, assessments by ACEEE and others suggest that energy and climate change policies don't have to be about ratcheting down the economy; they can be about technological innovation and leadership – all in a way that ensures both environmental and economic prosperity, if we choose to move along such a path.

This article draws from an AeA Europe report entitled *Advanced Electronic and Information Technologies: The Innovation-Led Climate Change Solution*. The full report is available at: www.aeanet.org/EUEnergy

Author

Dr Karen Ehrhardt-Martinez is a sociologist and a member of the policy research staff in the National and State Energy Policy program at the American Council for an Energy-Efficient Economy. Her work focuses on the human dimensions of energy and climate change and includes research and analysis of technologies and programmes designed to increase energy efficiency across all social segments and economic sectors of US society.

John 'Skip' Laitner is the Senior Economist for Technology Policy for the American Council for an Energy-Efficient Economy (ACEEE). He previously served for nearly 10 years in a similar capacity at the US Environmental Protection Agency (EPA) where he was awarded EPA's Gold Medal for his work with a team of EPA economists to evaluate the impact of various climate change strategies. Skip has more than 35 years of involvement in the environmental and energy policy arenas.

Organisation

The American Council for an Energy-Efficient Economy (ACEEE) is a nonprofit organisation dedicated to advancing energy efficiency as a means of promoting economic prosperity and environmental protection. ACEEE fulfills its mission by: conducting in-depth technical and policy assessments; advising policymakers and programme managers; working collaboratively with businesses, public interest groups and other organisations. Support for our work comes from a broad range of foundations, governmental organisations, research institutes, utilities and corporations.

Enquiries

Website: www.aceee.org

Compact fluorescent lamps help tackle climate change

Research shows that carbon dioxide emissions, one of the key greenhouse gases responsible for changing the world's climate, are occurring at a rate double the capacity that the Earth can assimilate. At concentrations of 390 ppm, they are at their highest level for over 400,000 years. Lighting is a basic necessity for many of the world's people and greater use of compact fluorescent lamps is widely considered to be one of the simplest options to reduce carbon emissions quickly.

TRUE OR FALSE?

Compact fluorescent lamps (CFLs), also known as energy saving lamps, are widely considered to be the quickest and simplest option for reducing carbon emissions. But misconceptions about them are causing some people to hesitate when considering the switch from conventional lighting.
Below, some of the commonest misconceptions are challenged:

Frequent switching reduces the life of CFLs.

FALSE: in fact a CFL's life is no longer affected by frequent switching. Current standards for 'Energy Saving Recommended' accreditation requires over 3,000 switching cycles per 8,000 hours of tested life, which is many more than would be necessary for normal domestic use.

Megaman produces heavy duty CFLs with a switching cycle capability of up to 600,000 and 15,000 hours' life. This is thanks to the Ingenium technology which combines improved life cycle switching and a shorter preheating time with an integral cooling tube that keeps operating temperatures to a minimum for enhanced safety.

CFLs are too big.

FALSE: the latest generations of CFLs are no longer large. In some cases, they are slightly smaller than regular household bulbs or equivalent and the new classic shapes look almost the same as regular bulbs.

CFLs consume so much energy when first switched on.

FALSE: CFLs do not consume any greater energy during start up and run very efficiently immediately after the first two to three seconds.

CFLs give a colder colour light than ordinary bulbs.

FALSE: CFLs emit the same warm light level as regular bulbs – between 2,600 and 2,800° Kelvin. This is under the Energy Saving Trust's 'Energy Saving Recommended' accreditation.

CFLs flicker with a stroboscopic effect at 50 times on/off per second.

FALSE: CFLs give a constant, flicker-free, non-stroboscopic light. They operate at high frequency through their electronic controller at between 30,000-50,000 hertz (normal mains voltage cycles at just 50 hertz or cycles per second).

You can't dim CFLs.

FALSE: there are new CFLs available right now that you can dim on ordinary domestic dimmer switches or alternately by staged dimming using a standard light switch.

Megaman provides both types of dimmable CFLs, namely the Dimmerable (linear dimming on dimmer switches) and DorS Dimming (4-step dimming on standard light switch).

CFLs are more dangerous.

FALSE: CFLs are much safer to use in confined areas as they produce very little heat when compared to the very hot regular bulbs. They are much safer for use where children are concerned and many are now used in children's light fittings for this reason. Moreover, CFLs burn so much cooler that there is no risk of scorching or discolouration on lampshades or ceilings.

Megaman offers CFL reflectors that can directly replace the blazing hot halogen bulbs, namely the CFL GU10 and CFL GX53, which fit neatly into existing housings.

Energy saving lamps are widely considered to be the quickest and simplest option for reducing carbon omissions.

© Megaman

Green attributes of Megaman lamps

- RoHS Compliance
- Lead-free glass
- Less than 3mg mercury per lamp
- Plastic lamp base – no more rust
- Covered lamps embodied with silicone protection to hold shattered glass and prevent mercury leakage
- Product recovery rate reaches 82 per cent

THE CONVENIENT MOVE

There are no longer any reasons why we should be negative towards CFLs. They will last for a period of between six and 15 years, depending on which product type you choose. Energy savings will be a minimum of 75 per cent of the energy burnt with an old fashioned bulb and the CFL emits much less heat than incandescent bulbs, which is a major contribution in reducing energy consumption from air-conditioners, thus further reducing CO_2 emissions.

> "We should operate profitable business with social responsibility, which means that we have to minimise negative effects to the environment."

A CFL MAKER WITH SUSTAINABLE DEVELOPMENT

Core values, such as environmental concerns and social responsibility, are affecting Megaman's activities on a daily basis. Megaman CFLs are designed and manufactured in accordance with strict ecodesign guidelines that minimise the usage of natural resources and hazardous substances, as well as emissions. To this end, Megaman lamps last longer and achieve an 82 per cent product recovery rate.

Megaman, a leading brand in energy saving lamps, is also an innovative leader in lamp design and the associated technology. Its products are produced in a world class manufacturing plant equipped with state of the art production lines and certified by ISO9001, ISO14001, OHSAS18001 and SA8000. Megaman upholds sustainable development throughout its business activities. According to the Chairman, "Natural resources are getting scarce. Climate change is accelerating. Business operations consume energy and natural resources to make profit. In return, we should operate profitable business with social responsibility, which means that we have to minimise negative effects to the environment and provide a healthy and pleasant working environment for everyone. After all, people are the most important asset of a company. We must take good care of them."

Nothing is more important than taking action to save our planet.

Organisation

Founded in 1994, Megaman is a renowned brand in the compact fluorescent lamp industry. The company was a pioneer that introduced the classic, candle-shaped CFL in the late 1990s. In 2001, it invented the CFL GU10 reflector lamp to replace halogen bulbs and later the dimmable CFLs. Megaman offers a full range of high performance and eco-friendly CFLs to bring a completely new and energy efficient lighting experience to our customers.

Enquiries

Email: info@megaman.cc
Website: www.megaman.cc

© Cezar Serbanescu/iStockphotos

NICLAS SVENNINGSEN, MANAGER,
UNEP SUSTAINABLE BUILDINGS AND CONSTRUCTION INITIATIVE AND
THIERRY BRAINE-BONNAIRE, DIRECTOR FOR CONSTRUCTION R&D,
ARCELORMITTAL GLOBAL R & D

The buildings challenge –

ENTERING THE CLIMATE CHANGE AGENDA

The building industry is responsible for about a third of a country's energy consumption; this is approximately the same across all countries. This puts a special emphasis on the sector's potential to improve efficiency and reduce greenhouse gas (GHG) emissions. In fact, if the building sector potential to reduce GHG emissions could be realised, it would, on its own, hit the GHG reduction goal of Kyoto. So how can such a reduction be realised?

INTRODUCTION

Buildings constitute an essential part of society, providing core functions such as housing, work places, trade centres and leisure. The built environment also forms the character and culture of the local society and has considerable economic impact on private and public economies, typically contributing on average 10 per cent of GDP.

At the same time buildings are responsible for large shares of resource use and waste generation: approximately 40 per cent of materials use, 30 per cent of solid waste generation, and 20 per cent of water use. The materials and technologies used in buildings also have a significant impact on their users health and wellbeing.

Over several years, more attention and effort has been directed at how to improve environmental and health aspects of existing and new buildings. This is a complex issue which has expanded from engineering and architecture to encompass a much wider range of issues, including economics, government policies, standardisation, and links to other development priorities such as social housing, labour markets, and large scale infrastructure coordination. Among all issues, energy use in buildings has climbed to the very top of the agenda in the past five years. Buildings are responsible for about 35 per cent of total energy consumption in society. This not only makes the building sector a prime target for GHG emissions reduction efforts, but makes it highly relevant for national energy security and economic savings through efficiency improvements.

THE CLIMATE CHANGE IMPACT

In 2006, UNEP's Sustainable Buildings and Construction Initiative (SBCI) released *Buildings and Climate Change – Status, Challenges and Opportunities* (www.unepsbci.org/Ressources/ReportsStudies/)

– a report confirming that the high energy use in buildings is common to all countries, regardless of climate and level of economic development. Energy is used for different purposes in different countries, such as heating, cooling, water heating, lighting, ventilation, appliances, and aggregated energy use still constitutes about a third of national energy consumption. The good news is that energy use in buildings can typically be reduced by 50 per cent or more using proven and commercialised technologies and approaches. IPCC's fourth assessment report (2007) also identifies the building sector as having the best potential for achieving drastic cuts in energy use and GHG emissions across all countries. If this full potential was realised, the buildings sector could on its own achieve the full GHG emission reduction goal of the Kyoto Protocol.

"If the building sector potential to reduce GHG emissions could be realised, it would, on its own, hit the GHG reduction goal of Kyoto."

A LACK OF PROGRESS

In spite of the significant potential for reducing energy use, it remains untapped in most buildings. There are a range of reasons why this is so:

- Achieving energy efficiency in buildings requires a coordinated life cycle approach to design, construction, operation and maintenance. In most cases, however, the building sector is very fragmented both horizontally, between different groups of players, eg engineering, architecture, investors, energy suppliers, and vertically, between different stages of the buildings lifespan, eg from preliminary design to end of life. In most cases, there is neither a tradition nor an incentive for groups to work together to optimise building performance over its lifetime.
- There is an economic disconnect between building investors and building users. While opportunities for energy savings can best be realised by the parties responsible for construction and renovation, it is the users who benefit from reduced energy costs. The incentives for building 'energy lean' are therefore very weak, even if stakeholders throughout the building's lifespan would clearly benefit.
- While technologies and approaches for energy efficiency are available, the building sector is by tradition risk averse and generally unwilling to try new approaches. In many countries the majority of relatively small builders also lack know how about how to apply energy saving technologies.
- Governmental policies and support tools for energy efficiency improvements in buildings are often lacking or are not harmonised with other policy instruments affecting the building sector. A common example of obstructive policies is high taxation of renovation works aimed at energy savings.

In the past year, several international organisations, eg UNEP, World Business Council on Sustainable Development, IPCC, have arrived at the same conclusion: the building sector alone is not able to realise the significant energy saving potential in buildings, but requires government intervention in the form of appropriate policy tools.

A RECIPE FOR SUCCESS

The question facing governments is what policy tools are effective and suitable in a local context? According to the UNEP SBCI policy tool database, prepared by the central European University in 2007, Governments have well over 30 types of instruments to chose between (www.unepsbci.org/Ressources/ReportsStudies/). These can broadly be divided into regulatory tools, economic tools, fiscal tools and informative & capacity building tools. The database is based on more than 80 policy tool evaluations from across the world and clearly indicates the most effective policy tools are typically regulatory instruments. Somewhat surprising is that these instruments are normally also the most cost effective, while economic instruments have a more scattered performance.

"The good news is that energy use in buildings can typically be reduced by 50 per cent or more using proven and commercialised technologies and approaches."

Many policy tools, especially regulatory tools, frequently result in substantial economic net savings to society, with savings up to US$192/t CO_2 reported. The study and database also recommends that different types of tools are combined to achieve the best result. In many cases the intended market transformation requires that certain tools, eg economic incentives, are applied only in different stages of the transformation, to be effective and avoid rebound effects.

BUILDINGS AND THE KYOTO PROTOCOL

Of special interest is the extent to which international cooperation can support energy efficiency in buildings. UNEP SBCI is preparing a study of the effect of the Clean Development Mechanism (CDM) on the building sector, due to be released at the end of 2007. Until now, only

© Sang Nguyen/iStockphotos

a handful of CDM projects realise emission reductions through buildings modification. Preliminary results indicate a range of reasons why this is the case:

- Economic benefits that can be achieved through generation of Certified Emissions Reduction Units (CER) under a CDM project are still very small compared to the overall costs for renovating/constructing a building, thereby it is a weak incentive for energy efficiency investments.
- There is failure of CDM to reward other sustainable development benefits than GHG emission reductions. The objective of CDM is in fact to support both emission reductions and sustainable development, but only emission reductions are financially rewarded.
- There is a problem taking into account additional and continuous improvements in energy efficiency after a project has been registered. In the building sector, energy performance is often gradually achieved as an iterative process over time.
- The comparatively small project size of most building projects, makes the overhead cost for registering and developing the project under CDM too high.
- There is a lack of commonly recognised tools and benchmarks which can be used for baseline development in the buildings sector.
- There are difficulties in qualifying and proving the effect of non-technology energy saving options, such as behavioral change of end users.

GOOD IDEAS FOR CHANGE

Clearly CDM in its current set-up is one of the least effective policy tools for encouraging energy efficiency in buildings. There are a number of opportunities for strengthening and revising this set-up to provide better support which include:

- Allowing CDM projects to be credited based on performance indicators (eg energy consumption per square metre), rather than technology based indicators. This avoids costly and cumbersome technology performance verification and monitoring.
- Establishing national policies which provide the additional push and pull for the building sector to pursue energy efficiency at the national level.
- Establishing common baselines for the sector.
- Strengthening the ability of CDM to reward, in economic terms, sustainable development benefits generated from projects.
- Developing rules which take into account the need for low income housing to increase energy use to meet basic needs, without being punished for the associated increased GHG emissions. Finding mechanisms to offset these increased allowances.

A SHARED AGENDA

With or without CDM, there is still an overriding need for governments to support a building sector move towards more energy efficient buildings. This is a requirement that is common for all countries. But special considerations may be warranted for developing countries for two reasons:

- The fastest growing, soon to be the largest, building markets are found in developing countries such as China, India and Brazil. China, for example is building two billion square metres a year which is equivalent to one-third of Japan's existing building area. Rapid urbanisation and the need to provide housing for a growing population presents a unique opportunity and challenge to ensure new builds are efficient from the start.

> **“Many policy tools, especially regulatory tools, frequently result in substantial economic net savings to society, with savings up to US$192/t CO_2 reported.”**

- The sustainable buildings agenda has often been perceived as a developed country issue and the capacity to support sustainable buildings in developing country governments is often weak. But sustainable buildings directly underpin the ability to achieve other

development priorities, such as shelter, safety, health and economic development. Clearly the sustainable buildings agenda is as relevant to developing countries as it is to developed countries, but the tools and data needed may not always be readily available for the former ones.

Among key issues for developing countries when pursuing energy efficient and sustainable buildings are:

- Institutionalising the responsibility for promoting energy efficiency in buildings.
- Removing adverse pricing and fiscal measures.
- Building capacity and know how within responsible institutions.
- Providing financial support to overcome first cost barriers, especially in the poorest countries.
- Developing strategies to pursue overall improved energy efficiency while allowing increased energy consumption needed to overcome energy poverty for individual households.

CONCLUSION

The governments and decision makers of today have an important role to play in ensuring the right signals and support are given to the construction market. The needed market transformation towards energy efficient buildings will not happen without government intervention. And climate change cannot be effectively addressed without this market transformation taking place.

Authors

Niclas Svenningsen is responsible for UNEP's work on sustainable buildings and construction and manages the Sustainable Buildings and Construction Initiative from UNEP's offices in Paris. He has a background in civil engineering and has spent the past 10 years working on various sustainable development issues in the developing world.

Thierry Braine-Bonnaire has chaired the SBCI's board since spring 2007. He is responsible for the Construction portfolio in the ArcelorMittal Global R&D. ArcelorMittal is the world's number one steel company, with 320,000 employees in more than 60 countries and a leader in major global markets, including automotive, construction, household appliances and packaging.

Organisation

The Sustainable Buildings and Construction Initiative (SBCI) is a UNEP coordinated partnership among building sector stakeholders, promoting sustainable buildings in general and energy efficient buildings in particular. SBCI brings together stakeholders from the entire building lifespan, including some building sector heavyweights, such as Arcelor-Mittal, Lafarge and Skanska. SBCI is currently addressing issues related to climate change, government policies, sustainable building baseline development, and sustainable buildings in developing countries.

Enquiries

UNEP, Division of Technology
Industry & Economics
15 Rue de Milan
75441 Paris Cedex 9
France
Tel: +33 1 44 37 14 34
Fax: +33 1 44 37 14 74
E-mail: sbci@unep.fr
Website: www.unep.fr
SBCI website: www.unepsbci.org

The Johnson Controls path to sustainability

It's the road to prosperity. The way to innovation. The avenue to leadership. It's the path to sustainability. Johnson Controls Inc is taking it, and we're helping millions of customers all over the world to follow it, too. The Johnson Controls path to sustainability begins with a common understanding of what sustainability means to businesses and organisations. The goals include financial rewards, more productive employees, and healthy communities. Most important, this route is critical to globally making a difference as sustainability becomes a factor in every person's life.

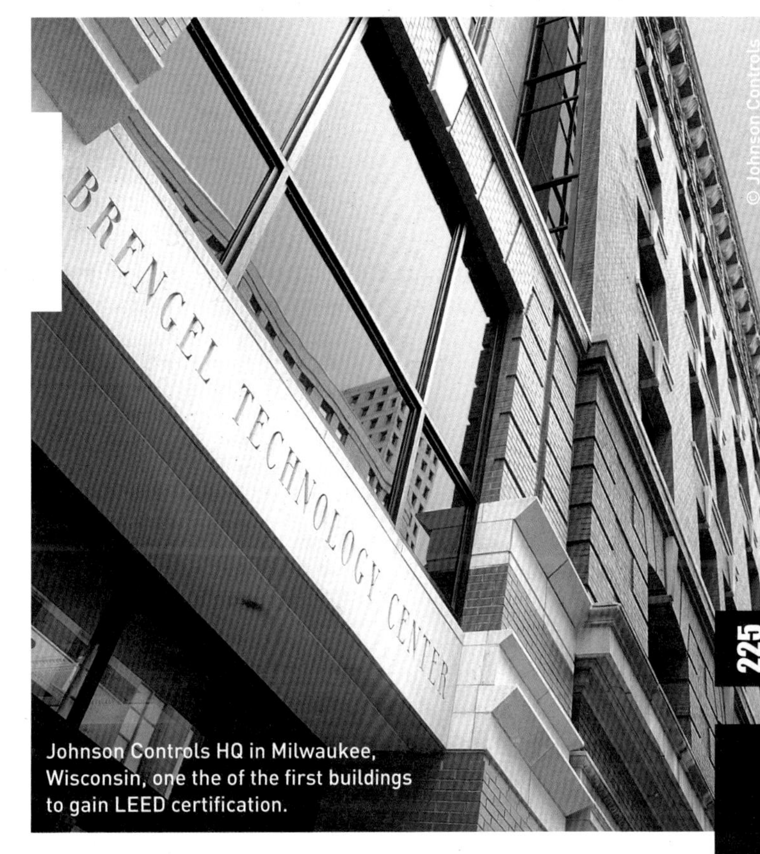

Johnson Controls HQ in Milwaukee, Wisconsin, one the of the first buildings to gain LEED certification.

SUSTAINABLE GREEN BUILDINGS

Since buildings account for 38 per cent of the CO_2 emissions per year, according to the US Green Buildings Council, it makes sense for businesses to begin their path to sustainability with their facilities.

The good news is that high performance facilities, ie those that are environmentally accountable, energy efficient and productive for occupants, are economically possible through an integrated approach to design and construction.

This method means that upfront costs can be the same or less than conventional construction, but the operational costs over the life of the building will be significantly lower. Features of buildings constructed this way include:

- Systems that efficiently use water resources and energy.
- Quality indoor environments that are healthy, secure, pleasing and productive for occupants and owners.
- Wise use of building sites, assets and materials.
- Natural landscaping, efficient material use and recycling.
- Reduced human impact on the natural environment.

A successful integrated building approach requires people, processes and technologies to be brought together early in the design phase. The project team, for example, must be a multidisciplinary group (ie owner, architect, builder, subcontractors and suppliers) and assembled before making any major decisions about the building. This team works together throughout the project, considering each aspect of the project with an eye towards reducing resource use during construction and more efficient long term operation.

Integration technologies should be used to tie together state of the art systems (heating, cooling, ventilation, lighting, fire, security) into one computerised management and automation platform. This approach allows operators to monitor energy and resource use, and to realise significant efficiencies and continuous returns on investment.

As a leading provider of integrated building systems and solutions, Johnson Controls has been at the forefront of the green buildings movement. The company helped develop the Leadership in Energy and Environmental Design (LEED) Green Building Rating System sponsored by the US Green Building Council. We have assessed more than 200 customer buildings for LEED compliance and have achieved certification for many of them including our own headquarters facility in Milwaukee, Wisconsin, one of the first 12 in the world.

LEADING WITH LEED

Johnson Controls offers many solutions that support the LEED rating system and help customers achieve LEED-rated facilities, including:

- Facility Management Systems/Building Automation Systems.
- High efficiency chillers and air handling systems.
- Temperature and humidity monitoring systems.
- Energy measurement and verification.
- Water conservation and measurement systems.
- On site renewable energy.
- Underfloor air distribution systems.
- Personal environments control systems.
- Commissioning services.

Through just a portion of the energy efficiency services it offers, Johnson Controls is helping commercial property owners reduce energy costs by US$18 billion over the next 10 years and eliminating 352 million tonnes of air pollutants.

SUSTAINABILITY EQUALS SUCCESS

Johnson Controls has taken the lessons learned in its own operations and is now helping companies embrace sustainability in all their operations through the following approach:

- **Executive leadership** – An organisation must have a champion with a commitment. Our initial meetings explain why a sustainable approach will help that leader meet the business's goals.
- **Assessment** – Whether a facility is a manufacturing centre, office building or campus, Johnson Controls assesses the triple bottom line risks and the opportunities that may be available through sustainable solutions.
- **Action plan** – Johnson Controls provides clearly defined goals, timelines and guarantees. We can even arrange for funding so that a project's savings pay for the project itself.
- **Implementation** – Our expertise is based on 122 years of experience in applying proven and measurable sustainable strategies.
- **Communications** – Johnson Controls works with all organisational levels to be clear what is being undertaken and provide public visibility to a project's accomplishments.

THE JOHNSON CONTROLS PATH

For Johnson Controls itself, the path to sustainability begins with commitment from our CEO, Steve Roell, who is an absolute advocate. We are developing and investing in areas across all our business lines that help not only the economic bottom line but are also consistent with our environmental and social goals.

Sustainability is fundamental in what we do for our own operations, and it is integrated throughout our customer activities. Using our own processes, we have assembled data on all facilities to establish our true global environmental footprint, recently passing an audit related to the US Environmental Protection Agency Climate Leaders program.

Understanding sustainability

To Johnson Controls, sustainability is today's most significant, but often misunderstood, issue. Some believe sustainability means 'green' and only focus on environmental aspects. Other corporate or government groups will discuss sustainability with regard to social responsibility. Still others look at the dictionary definition, which means able to keep up or keep going, as in sustaining an action or process.

At Johnson Controls, sustainability fully encompasses all of those criteria – and adds a financial component. We know we can better sustain our economic success by integrating environmental and social decisions.

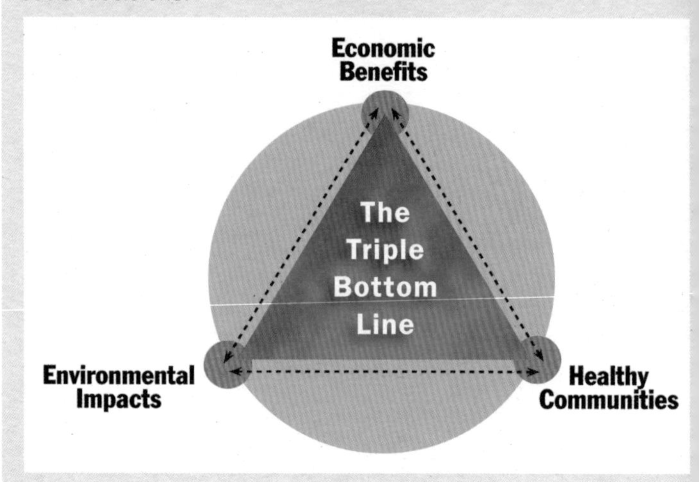

The World Environment Center and the Energy Star Program have both recognised Johnson Controls – factors in our listing with several sustainable financial investment indexes. Through our participation in groups such as the Global Reporting Initiative, Global Environment Management Initiative, World Business Council for Sustainable Development, the UN Global Compact and Clinton Climate Initiative, we're helping others benefit from our experience.

CONTINUOUS IMPROVEMENT

The most interesting thing about the path to sustainability is that it is a process of continuous improvement, and we're living proof. Johnson Controls continues to fully understand and address stakeholder sustainability issues for ourselves and our customers as we progress along the path to financial, social and environmental success.

Organisation

Johnson Controls is the global leader in providing building automation systems and heating, ventilating, air conditioning and refrigeration (HVAC&R) equipment for non-residential buildings. Johnson Controls provides technical services to maintain and repair building systems, and also manages and operates thousands of facilities for Fortune 500 firms around the world.

Our customers include thousands of hospitals, schools, airports, office buildings, data centers, government facilities, high-tech manufacturing sites, pharmaceutical laboratories and other buildings. A dedicated team of 56,000 employees is creating a more comfortable, safe and sustainable world through our products and services.

Enquiries

Website: www.johnsoncontrols.com

CHRISTIAN KORNEVALL, PROJECT DIRECTOR,
ENERGY EFFICIENCY IN BUILDING PROJECT (EEB), WORLD BUSINESS
COUNCIL FOR SUSTAINABLE DEVELOPMENT (WBCSD)

A more effective policy framework for the building sector

This article reports on a recent study of the Energy Efficiency in Buildings project of the World Business Council for Sustainable Development that highlights the need for a more effective policy framework for the building sector. This will be vital in order to kick start the industry into action to slash the energy use in buildings. The technology and knowledge already exist, yet action has been woefully slow.

INTRODUCTION

Buildings can make a major contribution to tackling climate change and energy use, but the lifespan of buildings means we need to act now. We can start to act now because we have the knowledge and technology to slash the energy buildings use. Why is this not happening?

The Energy Efficiency in Buildings project of the World Business Council for Sustainable Development tries to answer that question. We published a Facts & Trends Summary Report in 2007 which identified three kinds of barriers preventing energy reductions in buildings: behavioural, organisational and financial. The report also outlined three approaches to overcome them:

- Encourage interdependence by adopting holistic, integrated approaches to whole communities and individual buildings.
- Make energy more valued by those involved in the development, operation and use of buildings.
- Transform behaviour by educating and motivating professionals involved in building transactions.

COMPLEXITY IN THE BUILDING INDUSTRY

The building market is diverse and complex. The commercial relationships between the many specialists involved are intricate and critical in sparking action on energy efficiency. The sector is characterised by fragmentation within sections of the value chain and non-integration among them. Even the largest players are small and relatively local by international business standards, with the exception of materials and equipment suppliers. Incentives to reduce energy

THE EEB VISION

The EEB vision is a world in which buildings consume zero net energy. That is ambitious, but we believe a high ambition is necessary to achieve the progress needed to address climate change and energy use. Zero net energy means buildings as a whole (but not every individual building) would generate as much energy as they use over the course a year.

use are usually split between different players and not matched to those who can save the most through energy efficiency. The complexity of interaction among these participants is one of the greatest barriers to energy efficient buildings.

KNOWLEDGE GAPS IN THE INDUSTRY

The EEB project commissioned research that revealed serious gaps in knowledge about energy efficiency among building professionals, as well as a lack of leadership throughout the industry. We found that people believe financiers and developers are the main barriers to more sustainable approaches.

In general, we found that people recognise sustainable buildings are important for the environment but underestimate buildings' contribution to greenhouse gas emissions. They also generally overestimate the cost premium, which is likely to be under five per cent in developed countries.

Awareness of environmental building issues is relatively high in all countries we researched, but few people have been directly involved. Our research identified four factors that are the main barriers to greater adoption of sustainable building by building professionals:

- Personal know-how – whether people understand how to improve a building's environmental performance and where to go for good advice.
- Business community acceptance – whether people think the business community in their market sees sustainable buildings as a priority.
- A supportive corporate environment – whether people think their company's leaders will support them in decisions to build sustainably.
- Personal commitment – whether action on the environment is important to them as individuals.

ACHIEVING CHANGE

The knowledge, technology and skills are already available but are not being widely used to achieve dramatically lower energy use in buildings. Progress is hampered by barriers in the form of industry structure and practices, professionals' lack of know-how and support and a lack of leadership.

The EEB project believes that a more effective policy framework would help to stimulate activity. Businesses in the building industry need a supportive policy and regulatory framework to achieve dramatic improvements in energy efficiency. This is supported by the project's research findings on industry leadership, which reveal that many building industry professionals only adopt new practices if they are required by regulation.

Governments need to concentrate on the most efficient and cost-effective approaches. Research for the UNEP Sustainable Buildings and Construction Initiative (SBCI) found that the most effective instruments achieve net savings for society and that packages of measures combining different elements are desirable. The study identified policies that were both cost effective and successful in reducing emissions.

A better policy framework would cover:

- Urban planning.
- More effective building codes to enforce minimum required technical standards.
- Information and communication, to overcome the lack of know-how and to highlight the energy performance of individual buildings. A combination of voluntary and mandatory schemes is already emerging, for example: voluntary labeling schemes such as CASBEE, Japan, and LEED, US, and the building 'energy passport', EU.
- Tax and other incentives to encourage energy efficiency in building equipment, materials and occupation.
- Energy pricing to make energy more valued by users, to decouple utilities' revenues from the volume of energy supplied and to encourage local and renewable generation. For example, electricity consumers in Germany receive credit for power fed into the grid from local generation at a rate four times the cost of the electricity they use from the grid.
- Enforcement, measurement and verification to make sure policies and regulations (including building codes) are effective and support market measures such as trading.

A HOLISTIC APPROACH

While an effective policy framework is important, it will not be enough on its own. EEB research also identified the need for a holistic approach to construction, new financial approaches and behaviour changes.

A holistic approach begins with master planning and considers energy use over the whole life cycle and embraces integrated building design processes. Master planning considers the community in its entirety as well as single buildings. Some new urban centres are being created from scratch with an entirely sustainable plan, such as Dongtan near Shanghai, China, and Songdo, Korea. But many existing and rapidly growing cities have little room to manoeuvre due to existing constraints. In that case, master planning has to be implemented within the existing urban environment.

EEB CORE GROUP COMPANIES

The EEB project has brought together 10 leading companies (CEMEX, DuPont, EdF, GdF, Kansai, Lafarge, Philips, Sonae Sierra, Tepco and UTC) in the global building industry to tackle this vitally important subject. The project is led by Lafarge, and United Technologies. The EEB project covers six countries or regions that are together responsible for two-thirds of world energy demand, including developed and developing countries and a range of climates: Brazil, China, Europe, India, Japan and the United States.

ENERGY USE IN BUILDINGS

Energy efficiency factors in buildings vary according to geography, climate, building and location. The distinction between developed and developing countries is important, as is the contrast between retrofitting existing buildings and new construction. In some cases, there are different standards of building quality. It is vital that energy efficiency permeates all levels and is not restricted to high end properties.

This complexity means it is impossible to develop a single solution for all markets and cultures. Instead, the EEB project aims to identify approaches, market factors and policy initiatives that will together achieve the needed results.

Climate change will increase site energy demand as people seek to maintain comfort levels in more extreme conditions. Other main drivers pushing energy use up include demographics, economic development, lifestyle changes and technology.

Integrated Design Process

Many professionals are involved at different stages of a design project, and many factors need to be taken into account: climate, building shape, comfort levels, materials and systems, occupant health and security. Most projects follow a sequential approach, finalising one stage before moving to the next, with fee structures aligned to this linear approach and compartmentalisation. Designers need to be able to carry out extra iterations, revisiting earlier stages, to optimise the many factors and introduce cost-effective innovations at an early stage.

An Integrated Design Process (IDP) involves all participants in the early design phase of the project. Multidisciplinary workshops bring together owners, architects, engineers and others. They cooperate across the different specialties rather than working in the traditional 'silo' approach that involves little communication between specialists and results in buildings with sub-optimal performance.

IDP can achieve improved building performance with lower costs and fewer disruptive changes during the later project stages. The earlier in the process that IDP occurs, the greater the impact on building performance and the lower the impact on costs.

FINANCE

Financial considerations are critical to property development and investment, but they appear to be limiting the advance of energy efficiency. Financial pressures have become more powerful, especially in the US, because of the rise of real estate as an investment class and a decline in the number of owner-occupied buildings. Investors' time horizons are likely to be short. This increases the importance for their investment calculations of the property's residual value when they sell compared with operational returns during their ownership.

There is some evidence that an energy efficient building can command a premium, and this may increase as awareness of climate change and expectations of rising energy costs leads people and organisations to attach more value to energy efficiency. In the US, buildings with high energy performance are becoming more attractive financially because of markets for renewable energy (in 20 states as of mid-2007) and energy efficiency credits (10 states).

THE COST OF ENERGY AND ENERGY EFFICIENCY

Energy is typically a small proportion of total occupancy costs for buildings and this limits the incentive to invest in energy efficiency. For example, in a high quality office building in Germany, heating and electricity made up less than five per cent of the total operating cost of the building, including rent and maintenance.

This is particularly disappointing given that the cost of energy efficiency can be quite low. For commercial properties, the Fraunhofer Institute has shown that the energy demand of new office buildings can be reduced by 50 per cent compared with the existing building stock without increasing construction costs. The US Green Building Council has performed numerous studies and concluded that the cost of reaching certification under its Leadership in Energy and Environmental Design (LEED) standards system is between zero and three per cent, while the cost of reaching the highest level of LEED (platinum) comes at a cost premium of less than 10 per cent.

CONCLUSION

Design and technology available today can achieve dramatic improvements in building energy efficiency, but market failures and behavioural barriers are blocking progress toward the EEB vision of zero net energy.

It is the responsibility of Governments to set the global long term pathway to a low carbon future. Building is a key to reduce CO_2 and an effective and robust policy framework that helps the building sector to overcome its barriers could mainstream the market towards green buildings.

Author

Christian Kornevall joined the WBCSD in October 2005 to head the Energy Efficiency in Buildings project. Previously, he was Senior Vice President, Group Function Sustainability Affairs at ABB. He has also worked at WWF, the International Committee of the Red Cross, the International Labour Union (ILO) and the Swedish International Development Authority (SIDA). Originally from Sweden, Christian studied Economics, Business Administration and Statistics at Stockholm University before exploring the world on various assignments in South America, Asia and Europe with the above mentioned employers.

Organisation

The World Business Council for Sustainable Development (WBCSD) is a CEO led, global association of some 200 companies dealing exclusively with business and sustainable development. The Council provides a platform for companies to explore sustainable development, share knowledge, experiences and best practices, and to advocate business positions on these issues in a variety of forums, working withgovernments, non-governmental and intergovernmental organisations.

Enquiries

Christian Kornevall, Project Director
Tel: +41 (22) 839 3102 | Fax: +41 (22) 839 3131
E-mail: kornevall@wbcsd.org
Website: www.wbcsd.org

1984
Light bulbs

1995
Refrigerators

2005
Circulator pumps

GRUNDFOS

Grundfos circulator pumps

ENERGY EFFICIENCY THROUGH INNOVATION

Not many people know that pumps in the heating system, hidden in the basement, take up a large share of the electricity consumption in our homes. Even fewer know that substantial savings and a significant contribution to reducing greenhouse gases can be achieved simply by upgrading the circulator to a more energy efficient alternative.

For many years Grundfos has been the preferred supplier of some of the most energy efficient circulator pumps on the market. With the introduction of the European Energy Labelling Scheme for circulators in 2005, Grundfos committed to labelling all its circulator pumps for heating systems with a standardised declaration specifying how efficiently they use energy. It allows private homeowners as well as commercial building customers, ie consultant engineers, installers and distributors alike, to directly compare the energy efficiency of different circulator pumps and to make an informed choice.

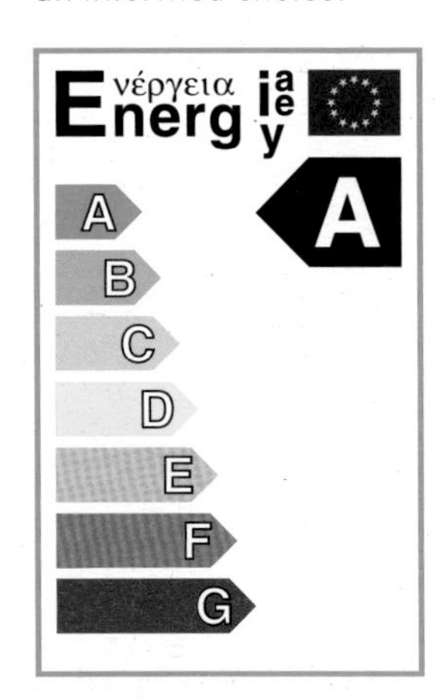

At present only a few European countries have legislation that demands the change to energy efficient circulators in new domestic and commercial building projects. In cooperation with its partners, Grundfos has chosen to make large investments in developing the technology for innovative pumps and in bringing out the message to the European countries through a large scale marketing campaign called the Energy Project.

CIRCULATOR PUMPS

In simplest terms, circulator pumps are electrically powered pumps used for circulating water in heating, air conditioning, and hot water systems. Grundfos offers two A-labelled variants: the Grundfos ALPHA2 range for private homes and the larger Grundfos MAGNA for commercial buildings.

STANDARD CIRCULATOR PUMPS

Regardless of heat requirement, a standard 3-speed pump will run at a fixed speed. If the radiator valves are closed (manually or by thermostat), the pump will continue to build up a higher pressure than required. The result is unnecessary electricity consumption: up to 100W, and often noise in the system. You could compare this to driving a car at constant full throttle and using only the brake to control your speed.

INTELLIGENT SPEED CONTROLLED PUMPS

A speed controlled Grundfos pump will automatically adjust its output to the changing heat demand regulated by the (manual or thermostatic) radiator valves. Therefore, speed controlled pumps never produce unnecessary pressure. Rather, they reduce their speed to the level needed in the heating system and save electricity as well as reduce noise in the system.

SAVE UP TO 80 PER CENT

A traditional D-labelled circulator, which is the average already installed in European households, runs at full speed day and night, summer and winter. Consequently, it is one of the big entries on the electricity bill, responsible for up to 10 per cent of the total electricity consumption of a private household.

The intelligent A-labelled Grundfos ALPHA2 circulator automatically adjusts its pressure to demand, and runs on down to as little as 5W, or typically 90 kWh a year. The power input of a standard pump is 60-100W, meaning that the yearly power consumption is 4-500 kWh. Replacing a standard pump with an A-labelled circulator can result in savings of up to 80 per cent.

Average annual energy consumption in European households in kWh

Circulation pump	
D	550
A	115
Washing machine	
G	398
A	236
Refrigerator	
G	305
A	115

Payback time of the extra initial investment from a standard to the A-rated alternative is just under two years. The next approximately 10-15 years of operation offer substantial savings year after year, after year.

AUTO*ADAPT* TECHNOLOGY

Combining unique compact design and revolutionary technology, Grundfos ALPHA2 and MAGNA circulators lead the way to the future of A-labelled pumps.

The intelligent proportional pressure control, AUTO*ADAPT*, is the heart of the MAGNA and ALPHA2 designs. The innovative Grundfos technology allows the pump to automatically recognise and adjust itself to the specific system in which it is installed and saves even more.

HOMEOWNERS SURVEY

To get an insight into the energy habits and behaviour of its customers, Grundfos has just finished a consumer survey. The survey focused on three European countries, Germany, France and England and involved more than 3,300 houseowners.

The conclusion of the survey was hardly surprising: The vast majority confirm a great interest in energy savings. Figures also show that more than 90 per cent of respondents are very conscious that the effort for a better environment is a common responsibility.

Although homeowners are conscious about saving energy, most believe energy savings primarily concern remembering to turn off the lights and other electric devices when not in use. Although many know the energy labelling scheme, very few are aware of the energy rating of the electrical products in their home.

And although we hate to admit it, very few of the respondents knew that a standard D-labelled circulator pump is one of the biggest energy thieves in the house. Most were convinced that the washing machine, the refrigerator, or the tumbler use more energy than the circulator. Had they installed an A-labelled ALPHA2 they would be absolutely right...

SAVINGS THAT MAKE A DIFFERENCE

The survey also shows that it is not so easy to take the decisive steps alone. The majority of respondents do things to reduce the power consumption, but most only turn off or disconnect devices when they are not in use and are not very confident when it comes to buying A-labelled products.

Together the European customers who in 2005 and 2006 chose one of the energy efficient Grundfos circulators each year save the equivalent of the total electricity consumption of approximately 200,000 average European households.

If the 120 million circulators already installed in Europe were all replaced by A-labelled pumps, savings would be approximately 44 billion kWh per year – or equal to the annual power production of five nuclear power stations or 10 conventional power plants.

REDUCE CO_2 EMISSIONS

These significant savings have a direct impact on the amount of CO_2 emitted from European households. By making the A-rated choice, consumers save on their electricity bill, save energy, and make a personal contribution to reducing greenhouse gas emissions.

CIRCULATOR FACTS

- 120 million circulators run in houses in Europe.
- Each year approximately 10 per cent are replaced.
- A-labelled Grundfos pumps reduce energy consumption by up to 80 per cent.
- The Grundfos ALPHA2 and MAGNA A-rated circulators with integrated AUTO*ADAPT* can replace virtually any old circulator without extra efforts.

GRUNDFOS ALPHA2 LEADS THE WAY

The new ALPHA2 is a revolution within circulator technology. The pump offers a wide range of new functions and innovative technology that will save the homeowners a lot of money and the installers a lot of trouble.

The display makes the energy consumption very visible to homeowners allowing them to make the informed choice and see the difference an A-labelled circulator makes. In addition, its innovative, energy saving technology contributes significantly to reducing CO_2 emissions from private homes.

Organisation

Grundfos is a leading manufacturer of centrifugal pumps and systems for water applications in industry, irrigation, heating and wastewater treatment. The Grundfos Group is represented by 72 companies in 41 countries.

Enquiries

Grundfos Management A/S
Poul Due Jensens Vej 7, 8850 Bjerringbro, Denmark
Tel: +45 87501400 | Website: www.grundfos.com

Agriculture and climate change –

CHALLENGES AND OPPORTUNITIES

PARVIZ KOOHAFKAN, DIRECTOR,
ENVIRONMENT, CLIMATE CHANGE AND BIOENERGY DIVISION,
UN FAO

There is concern that climate change could increase food insecurity due to the myriad interactions between climate variability and food systems, of which little is known. It is therefore critical that implications of climate change for food security are explored and understood, in order to respond efficiently and effectively.

INTRODUCTION

According to estimates from the UN Food and Agriculture Organization (FAO), the number of undernourished people in the world in 2000–2002 was 852 million. Of this, 815 million came from the developing countries, 28 million in the countries in transition and nine million in the developed market economies. According to the State Of the World Food and Agriculture report (SOFA), today's estimated 820 million undernourished people in developing countries represent a marginal reduction of three million as against the 1990-1992 baselines of 823 million used by the World Food Summit.

CLIMATE CHANGE ASSESSMENT REPORT

The Fourth Assessment Report of the Intergovernmental Panel on Climate Change (IPCC) on Climate Change Impacts, Adaptation and Vulnerability highlights observed changes in the natural and human environment as a result of climate change. The changes show that natural systems have been affected by climate change and temperature increases in particular. Changes include those in hydrological, terrestrial, biological and marine and freshwater systems. These changes have also impacted human systems, although specific impacts are hard to discern due to adaptation and non-climatic drivers. Some of the impacts include changes in agricultural and forestry management, impacts on human health and the reduced ability to undertake certain activities. Other observations, that cannot be verified as trends yet, include increased flooding risk due to glacial lake outbursts, warmer and drier conditions in parts of Africa and sea level rise impacting coastal development. The food, fibre and forest products section makes a number of important points. These are:

- Crop productivity is projected to increase slightly at mid to high latitudes for local mean temperature increases of up to 1-3°C depending on the crop, and then decrease beyond that in some regions.
- At lower latitudes, especially seasonally dry and tropical regions, crop productivity is projected to decrease for even small local temperature increases of 1-2°C, which would increase the risk of hunger.
- Globally, the potential for food production is projected to increase with increases in local average temperature over a range of 1-3°C, but above this it is projected to decrease.
- Adaptations, such as altered cultivars and planting times, allow low and mid to high latitude cereal yields

to be maintained at, or above, baseline yields for modest warming.

- Increases in the frequency of droughts and floods are projected to affect local production negatively, especially in subsistence sectors at low latitudes.
- Globally, commercial timber productivity rises modestly with climate change in the short to medium term, with large regional variability from the global trend.
- Regional changes in the distribution and production of particular fish species are expected due to continued warming, with adverse effects projected for aquaculture and fisheries.

THE NEED FOR ADAPTATION

The continuing impact of climate on many aspects of food systems is expected to grow in the coming century. The outcome of climate change occurs at the regional, national and local scale and requires an integrated response. Natural systems can only react to climate change, while human systems can adapt before the change has negative consequences. Adaptation depends on the adaptive capacity of the individual, systems, institutions and governments but also global community. This adaptive capacity or ability to make changes can depend on many things including resources, knowledge about change and options for adapting, policies, norms and social capital among other things. In order to reduce the negative impacts of change on food security it is necessary to adapt, although the specific impact of climate change on regions and systems should first be established.

When assessing opportunities for adapting food systems to climate change, it is also critical to acknowledge how food systems contribute to emissions and how elements of the food system could mitigate these emissions by changing agronomic, management and processing practices.

CLIMATE CHANGE IMPACTS ON FOOD AND AGRICULTURE

The social and economic costs of not responding to climate change are generally seen to be much higher than the costs of taking immediate corrective action. There is much better understanding today about the regional and continental impacts, although there remain uncertainties as to when, where and how climate change will affect specific countries. Changes in temperature and precipitation and an increase in extreme weather events are likely to change food production potential in many areas of the world, especially Africa and Asia. There is the potential to disrupt food distribution systems and infrastructure and to change the purchasing power of, for example, the rural poor.

FAO, in collaboration with the International Institute of Applied Systems Analysis (IIASA), has developed the Agro-Ecological Zones (AEZ) methodology, a worldwide spatial soil and climate suitability database. The AEZ approach has been used by IIASA to quantify regional impacts and geographical shifts in agricultural land and productivity potentials and the implications for food security resulting from climate change and variability. The analysis indicates that, on average, industrialised countries could gain in production potential, while developing countries may lose. Findings show that the potential impact of a changing distribution of water availability for food and agricultural production and food security include:

- Global agricultural production potential is likely to rise with increases in global average temperature up to about 3°C, but above this it is very likely to decrease.
- Cold climates would benefit from higher temperatures and new agricultural land may become available at high latitudes and high elevations.
- There is significant potential for expansion of suitable land and increased production potential for cereals only when considering the use of 'new land' made available by the warming of these cold climates at high latitudes.
- At lower latitudes, especially the seasonally dry tropics, crop yield potential is likely to decrease for even small global temperature increases, which would increase risk of hunger.
- Increased frequency of droughts and floods would affect local production negatively, especially in subsistence sectors at low latitudes. This will have much more serious consequences for chronic and transitory food insecurity and for sustainable development than shifts in the patterns of average temperature and precipitation.

> **"Changes in temperature and precipitation and an increase in extreme weather events are likely to change food production potential in many areas of the world, especially Africa and Asia."**

EMERGING CHALLENGES TO ADDRESS CLIMATE CHANGE IN AGRICULTURE

It is clear that the countries which are most vulnerable to climate change impacts also tend to be the poorest ones. In addition, they do not have the means, eg data, observations, methods, tools, technical and institutional infrastructure capacity building, to deal with this new situation.

Even with effective mitigation and reduction of greenhouse gases there will be a need to equally promote adaptation. Developed countries are aiming to reduce their greenhouse gas (GHG) emissions by five per cent below 1990 levels, but much higher reductions in emissions (60 per cent for carbon dioxide, 15-20 per cent for methane and 70-80 per cent for nitrous oxide) are needed to stabilise GHG concentrations at current levels. More efficient strategies for mitigation are a key challenge for developed countries in particular.

Figure: Impact of climate change in the 2080s based on currently cultivated land (million tonnes).

ECHAM4 (European Centre Hamburg Model - global climate model)

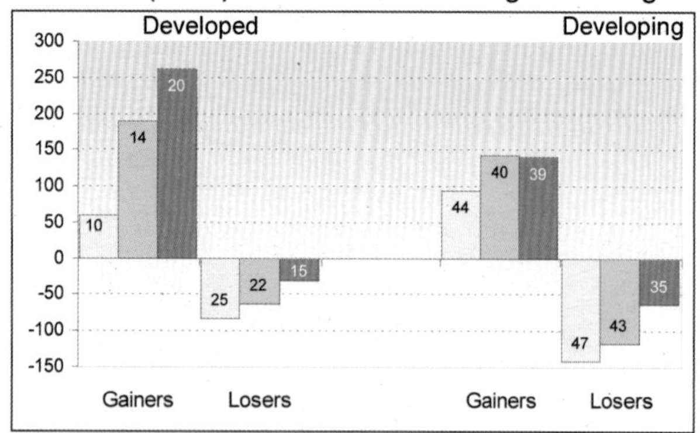

CGCM1 (Canadian Global Coupled Model)

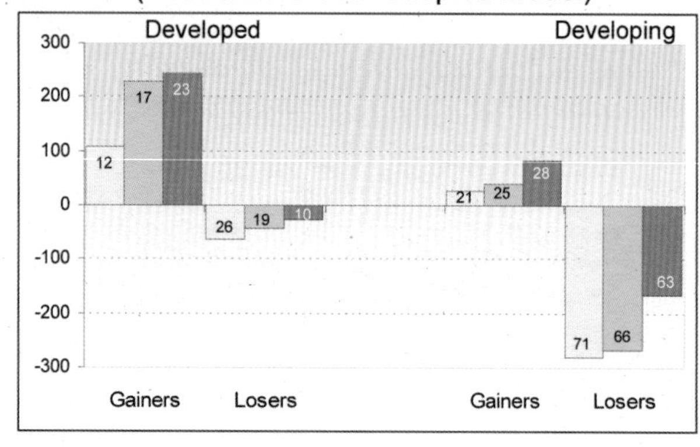

- Aggregate impacts of projected climate change on the global food system are relatively small.
- Developed countries tend to benefit and have effective options for adaptation.
- Less developed countries bear the brunt of negative impacts.

Rain-fed, single
Rain-fed, multiple
Rain-fed and/or irrigated, multiple

Source: Fischer, G, Shah, M and van Velthuizen, H (2002) Climate change and agricultural vulnerability. Laxenburg, Austria: IIASA.

Adaptation will be crucial in both developed as well as developing countries but will be a key challenge for the least developed countries.

People living in the least developing countries (LDC) where economic reliance on natural resources and rain-fed agriculture is high, in low altitude small island developing states (SIDS) or in marginal areas such as drylands or mountains, face additional challenges through climate change due to limited access to management options which could reduce impacts. In addition, they will be affected indirectly through prices, food availability on markets and job opportunities. Climate adaptation strategies must reflect such circumstances in terms of the speed of the response and the choice of options.

In the context of adaptation, assistance should be focused on helping poor and vulnerable people and their countries in particular, to strengthen their livelihoods and improve their capacity to adapt. Climate change adaptation should be a specific area within development policy/assistance and within poverty alleviation strategies in particular, with its own criterion of adaptive capacity to measure success among vulnerable groups.

In terms of agricultural production there will be winners and losers. The losers will most likely be those who are least responsible for greenhouse gas emissions. The geographical areas of anticipated losses are notably located in the tropics and in developing countries where the adaptation potential is limited. There is a strong demand to establish a global responsibility system, functioning mechanisms for compensation, including Clean Development Mechanism (CDM) and fair trade agreements which consider the impacts of climate change and production patterns and payment for environmental services.

A framework for climate change adaptation

Key pillars for a framework to address climate change adaptation in agriculture in LDCs and SIDs countries need to be defined. Such a framework needs to be directed simultaneously along several interrelated lines:

- Legal and institutional elements – decision making, institutional mechanisms, legislation, tenure and ownership, regulatory tools, legal principles, governance and coordination arrangements, resource allocation, networking civil society; strengthening local institutions and demand led rural service provision.
- Policy and planning elements – risk assessment and monitoring, analysis, strategy formulation, sectoral measures; involving poor people in natural resource governance, planning and policy.
- Livelihood elements – food security, hunger, poverty, promotion of indigenous good practices for risk management and development; strengthening through livelihood diversification social resilience, access to productive assets; offering exit strategies, where needed, from agriculture sectors.
- Cropping livestock, forestry, fisheries and integrated farming system elements – food crops, cash crops, growing season, crop suitability, livestock fodder and grazing, non-timber forest products, aquaculture, water management, land use planning, soil fertility, soil organisms; new technologies in the agriculture forestry and fisheries sectors and natural resource use under variable rainfall regimes.
- Tailored strategies and decision support tools for risk aware management (including small scale assets) in agriculture, livestock, fisheries and agroforestry.

- Ecosystem elements – species composition, biodiversity, resilience, ecosystem goods and services.
- Linking climate change adaptation processes and technologies for promoting carbon sequestration, substitution of fossil fields and use of bioenergy.

Ways of working may need to change. Since it is likely that agro-ecological zones and farming systems may slowly change, there is a need to promote more holistic approaches and systems thinking when addressing climate change. Partnerships building, at and between all levels, on research, development and policy are crucial. Systems approaches would need to address a variety of factors in an integrated way and with more political commitment, such as:

- Combining risk reduction with productivity and natural resource management objectives.
- Democratic accountability, governance and planning.
- Better linking the local, where much of the adaptation and research is conducted, to the global, where policy debates are and instruments, such as conventions, are formed.

A wide range of basic tools and methods to interact and communicate with local people and decision makers have been developed and used successfully in the context of our ongoing development work. These will be as suitable in the context of a more systems-oriented way of approaching climate change issues. There is no need to reinvent the wheel but rather to build on what we have.

New research strategies

New research strategies may be needed to provide focused and demand responsive scientific and external support. The capabilities of poor people, rather than their vulnerability *per se*, provide a starting point for demand led research for development that can moderate the negative effects of climate change and empower people to take hold of opportunities. Research strategies should evolve beyond 'more of the same' to respond flexibly to new challenges. In line with production and sustainability oriented research portfolios, they need to also focus on the emerging knowledge gaps such as:

- How to address longer term ecosystem change.
- The risk of changing disease and pest patterns.
- Better understanding of related institutional issues; who to do what, incentive systems, and how to reinforce adaptation.
- Better understanding of location and time specific dimensions of the political-economic contexts of existing development and expected change.
- Better understanding of how poor people can optimise new knowledge in a context of competing demands; what encourages the uptake of innovations more effectively.
- Incorporating economic drivers better in modeling work related to climate change.
- How to better link livelihood diversification strategies with agriculture, natural resource management and environmental services.

Author

Dr Parviz Koohafkan is Director of the Environment, Climate Change and Bioenergy Division (NRC) in the Natural Resources Management and Environment Department (NRD) of the UN Food and Agriculture Organization. Dr Koohafkan worked as Assistant Professor in Ecole Nationale des Genies Rurales des Eaux et Forêts (ENGREF), Montpellier, France, from 1982 to 1985 and joined FAO in 1985 as team leader in the natural resources management programme. He joined FAO Headquarters in 1991 as Senior Officer in sustainable development and was appointed as Service Chief in the Land and Water Development Division of FAO in 1996. From 2005-2006, he was Director of the Rural Development Division in the Sustainable Development Department.

Organisation

The Food and Agriculture Organization of the United Nations leads international efforts to defeat hunger. Serving both developed and developing countries, FAO acts as a neutral forum where all nations meet as equals to negotiate agreements and debate policy. FAO is also a source of knowledge and information. We help developing countries and countries in transition modernise and improve agriculture, forestry and fisheries practices and ensure good nutrition for all.

Enquiries

FAO Headquarters
Viale delle Terme di Caracalla
00153 Rome, Italy
Tel: +39 06 57051
Fax: +39 06 570 53152
E-mail: FAO-HQ@fao.org
Website: www.fao.org

Keeping the greenhouse green

YARA

A response to climate change is needed fast, and the political and commercial sectors must work together as a catalyst for change. Yara International intends to be part of the solution to the complex task of providing enough food and energy for the world, while addressing the global climate challenge. As a knowledge leader within plant nutrition and environmental applications for industry, we have started the process by committing to a five million tonnes reduction in our greenhouse gas emissions before 2009. Our aim must be to play a defining role in making the planet a safe and healthy place to live.

FOOD, ENERGY AND CLIMATE CHANGE

Yara International intends to lead the quest for a solution to the complex task of providing enough food and energy for the world, while addressing the global climate challenge. As a knowledge leader within plant nutrition and environmental applications for industry, we are committed to reducing our total greenhouse gas emissions by five million tonnes of CO_2 equivalents before 2009.

Fertiliser management tools allow farmers to apply optimal types and amounts of fertiliser.

But we need to push further. We must set standards and support the fertiliser industry, our agricultural customers and partners so they can meet their environmental obligations. A response to climate change is needed fast and the political and commercial sectors must work together as a catalyst for change. Our aim must be to play a defining role in making the planet a safe and healthy place to live.

AGRICULTURE – A BALANCED VIEW

Until the 19th century, area expansion was largely responsible for increases in food production. With the advent of modern high yield farming practices in the 20th century, growers were able to produce more food per acre each year. This trend towards harvesting more from every acre is perhaps the greatest development within food security and environmental protection during the past 50 years. Without high yield farming, massive additional land areas would have to be tilled for agriculture to avoid widespread starvation.

Fertilisers play an important role because they energise biomass. Nearly all life depends on a plant's ability to provide energy in the food we eat and capture oxygen in the air we breathe. Plants need energy to grow and any energy that is surplus to growth, they store. The optimal use of fertiliser significantly enhances this process. Life cycle analysis studies show that the energy yield from optimally fertilised crops is 10 to 15 times higher than the total energy consumed in fertiliser manufacture, transport and application.

Agriculture today faces the global challenge of increasing biomass production while reducing its environmental footprint. The production of food has to keep pace with a world population that expands every year at a rate equal to the combined populations of France and Belgium and all the while the average consumption per person is also increasing. At the same time, rising energy prices are pushing demand for bioenergy and creating a growing market for agricultural products, and this puts added pressure on scarce agricultural resources.

Moreover, only a fraction of the planet's non-developed land can be utilised for agricultural expansion. The remainder is needed to preserve forest cover or support infrastructure development, or is inaccessible. In addition, the pool of unused suitable cropland is very unevenly distributed. Sub Saharan Africa, the CIS countries, and Latin America are the last frontiers. Unless we are willing to level entire forests, and this is clearly not environmentally sustainable, no other sizeable tracts of potentially arable land are available. All these factors explain the need to increase the intensity of agricultural production.

INTENSIVE SUSTAINABLE FOOD PRODUCTION

Yara works with a broad coalition of partners, from farmers and agricultural advisors to the UN Millennium Project and local NGOs, to support sustainable high yield agriculture. Fertilisers used correctly can be environmentally benign and consist of naturally occurring plant nutrients necessary for maintaining soil fertility and adequate food production.

The main environmental concern is the irresponsible use of fertilisers. Excessive application is a waste of money and can result in nutrient leaching into groundwater/rivers and soil acidification, while inadequate fertiliser application leads to suboptimal yield and can cause erosion and loss of fertile soil.

Yara promotes crop-specific nutrition plans and supplies fertiliser management tools that allow farmers to apply optimal types and amounts of fertiliser – taking into account crop and soil type, weather conditions and tillage/irrigation practices. This service is provided via dedicated teams of agronomists who issue recommendations, via published guidelines and expert systems for farmers such as the N-Tester, N-Sensor and Yara Plan. Traditionally, our efforts have been aimed at maximising yield and minimising ecological impact, but now we also advise growers on how to mitigate their greenhouse gas emissions. Nitrate-based fertilisers such as ammonium nitrate, calcium ammonium nitrate and complex fertilisers lose very little nitrogen in this manner, while nitrate fertilisers like calcium nitrate have no ammonia emissions.

GHG EMISSIONS

Yara is among the most efficient and environmentally friendly fertiliser producers. However, the production of nitrogen fertiliser is extremely energy intensive and Yara still emits considerable amounts of greenhouse gases. Even though fertilisers contribute positively to society from the value chain perspective, Yara recognises that it has an environmental responsibility to reduce emissions. Yara has therefore implemented measures that will reduce GHG emissions by 25 per cent over a four year period. This corresponds to a total reduction of five million tonnes of greenhouse gases before 2009.

Yara's greenhouse gas emissions emanate from two main sources: CO_2 from ammonia plants and N_2O from nitric acid plants. Per unit of weight, N_2O produces approximately 300 times more global warming than CO_2. If we convert N_2O emissions to CO_2 equivalents, Yara emits an equal amount of these two greenhouse gases. In the production of ammonia, CO_2 emissions are kept to a minimum by operating at the highest efficiency levels. Yara's ammonia plants are among the most energy efficient. On average, Yara's plants in Europe are 10 to 20 per cent more efficient than competing facilities. Ongoing revamps of Yara facilities in Europe and overseas will result in further improvements.

The energy yield in optimally fertilised crops is 10 to 15 times higher than the total energy consumed in the manufacture, transport and application of fertilisers.

If yields had stayed at 1960 levels, an additional 1.4 billion hectares of farmland, would be needed to produce today's food supply.

MEET THE CATALYST

In 2005, after 10 years of research and development, Yara introduced a new catalyst technology for the reduction of N_2O from nitric acid plants. This technology, which has already been installed at several of our own facilities as well as at external plants, has demonstrated a 70 to 90 per cent reduction of greenhouse gas emissions from nitric acid plants. Competitive technologies exist but Yara's is the most effective and can be implemented in existing plants. Yara's catalyst technology is now used in 50 per cent of all the Clean Development Mechanism (CDM) projects worldwide, including projects in emission hotspots, such as Russia and China. The potential is truly huge. If Yara's catalyst technology were adopted throughout Europe it would result in a 30 million tonnes reduction of GHG emissions annually. If adopted globally, the outcome would be a cut of 75 million tonnes. This would be a significant reduction of GHG emissions at a relatively low cost. But the industry is slow to implement best available technology without proper regulatory incentives.

Yara favours regulatory initiatives to stimulate emission reductions. Such initiatives should naturally be fair, cost effective and encourage early compliance. They should be composed in such a way that national or regional initiatives do not prompt the transfer of production capacity to countries with poor emission standards, which would only lead to increased global emissions. The EU is moving to introduce a permit system, but experience shows that industry is slow to adopt new technologies that increase costs and may result in a competitive disadvantage. Yara has therefore proposed that the EU includes nitrous oxide in its emissions trading framework. This would be a substantial and effective incentive to reduce global emissions from the agricultural value chain.

© Yara International ASA

Nitrogen ferilisers provide about half of the nutrients in harvested crops and roughly 40 percent of the world's dietary protein supply.

Organisation

Yara International ASA is a leading chemical company that converts energy and nitrogen from the air into essential products for farmers and industrial customers. As the number one global supplier of mineral fertilisers and agronomic solutions, we help provide food for a growing world population. Our industrial product portfolio includes environmental protection agents that safeguard air and water purity and preserve food quality. Yara's global workforce of 7,000 employees represents great diversity and talent enabling Yara to remain a leading performer in its industry.

Enquiries

Yara International ASA, Bygdøy allé 2
PO Box 2464, Solli, N-0202 Oslo
Norway
Tel: +47 24 15 70 00
Fax: +47 24 15 70 01
E-mail: Yara@yara.com
Website: www.yara.com

New paper markets mean opportunity for measurable climate action

JOSHUA MARTIN
DIRECTOR OF THE
ENVIRONMENTAL PAPER NETWORK

Today governments and businesses are looking to reduce their impact on climate change and adopt practices that are environmentally and socially responsible. A critical, effective, and often overlooked, avenue for significant, measurable results is paper usage. In fact, purchasing and using paper thoughtfully and responsibly is a must have component of any institution's comprehensive action plan for reducing its impact on climate change. Today, there are many opportunities to purchase paper made with recycled content and third party certified, sustainably harvested forest fibre.

Despite predictions that the digital revolution would make paper as obsolete as the typewriter, paper remains central to our lives. In fact, according to industry forecasters, global demand for paper and paperboard is projected to increase nearly 60 per cent, from 368 million tonnes in 2005 to 579 million tonnes in 2021.

PAPER PUTS PRESSURE ON FORESTS AND CLIMATE

Paper production has traditionally been a major contributor to climate change due to significant environmental impacts throughout its entire life cycle: from consumption of forests globally, to energy intensive mill processing, to its disposal and decomposition in landfills. However, increasing market demand for environmentally responsible papers has spurred innovation and new product development that allows a wide adoption of effective climate action through positive paper choices. Today there are papers produced with climate and forest sustainability as a key distinguishing attribute available in almost every grade that perform exceptionally for all manner of applications. Leading companies are now working with suppliers to develop new environmentally improved papers. In doing so they achieve measurable net climate benefits they can report.

> “Global demand for paper and paperboard is projected to increase nearly 60 per cent, from 368 million tonnes in 2005 to 579 million tonnes in 2021.”

THE CASE OF HARRY POTTER

In a demonstration of the availability of environmentally improved papers and how benefits can be quantified, the final edition of the Harry Potter series, published in 2007, has been touted as the greenest book in publishing history and is a model for all paper users. Globally, 16 editions of the book were printed on ecopapers and the climate and forest benefits were significant: 137,609 BTU's (British thermal unit) of electricity conserved;

> Today there are papers produced with climate and forest sustainability as a key distinguishing attribute available in almost every grade that perform exceptionally for all manner of applications.

7,876 tonnes of carbon dioxide emissions avoided; 20,404,246 kg of solid waste diverted from landfills and savings of 197,685 trees which did not need to be harvested. None of these book papers even existed just a few years ago. Harry Potter's publishers worked with their suppliers to achieve this switch with help from experts at organisations including Markets Initiative, Green Press Initiative and others.

Any institution developing a strategy for analysing and reducing their carbon footprint will benefit greatly from adopting a responsible paper purchasing policy. The Environmental Paper Network, an international coalition of leading environmental NGOs advancing social and environmental responsibility in the pulp and paper industry, advocates developing a policy which embraces the four pillars of its members' 'Common Vision for the Transformation of the Pulp and Paper Industry':

- Increasing paper use efficiency.
- Maximising recycled content.
- Sourcing fibre responsibly.
- Employing cleaner production practices.

INCREASING PAPER USE EFFICIENCY

When any business or organisation begins to address its paper usage, the volume of waste and inefficiency becomes starkly visible and it's easy to see that increasing paper use efficiency saves money. Just as a unit of conservation/efficiency is the most economically valuable unit of alternative/clean energy, the same applies to paper use. Lower paper volumes benefit your bottom line directly by reducing your purchase costs. This also has indirect cost benefits that can be 10 times the cost of the paper alone. These include: reducing the costs of technology such as photocopy toner and printer ink; paying for less storage space and filing equipment; slashing postage costs and saving time. Many companies find that paper reduction strategies have significant additional benefits by introducing new systems of information efficiency which improve the quality of services by speeding up information flows. These savings can then be applied to balance other aspects of the paper policy if necessary.

> The final edition of the Harry Potter series, published in 2007, has been touted as the greenest book in publishing history.

The climate benefits of reducing paper consumption are significant. If, for example, the United States cut its office paper use by roughly 10 per cent, or 540,000 tonnes, greenhouse gas emissions would fall by 1.6 million tonnes. This is the equivalent of taking 280,000 cars off the road for a year.

Using less paper is also about equitable use of the Earth's resources. Think how much better the world would be if current levels of paper production were used to make books for schools in impoverished nations instead of wasted office print outs and junk mail. Paper usage volumes vary enormously around the world: North Americans and Europeans use more than 200kg each per year, while the average African uses just 6.5kg.

MAXIMISING RECYCLED CONTENT

Compared to copy paper made from 100 per cent virgin forest fibre, a copy paper made from 100 per cent recycled content reduces total energy consumption by 44 per cent, net greenhouse gas emissions by 41per cent and wood use by 100 per cent. In the same comparison, 21 million BTU's of energy and 2,108 lb of CO_2 are saved for every tonne of paper replaced with recycled fibre.

Purchasing recycled paper also helps keep waste paper out of landfill. If paper is landfilled rather than recycled, it decomposes and produces methane, a greenhouse gas with 23 times the heat trapping power of carbon dioxide. In the United States, more than one-third of municipal solid waste is paper, and municipal landfills account for 34 per cent of human related methane emissions to the atmosphere, making landfills the single largest source of such emissions. The US

© AVTG/iStockphoto

Environmental Protection Agency has identified the decomposition of paper as among the most significant sources of landfill methane.
In 2007, the United States House of Representatives changed its office copy paper purchasing to 100 per cent recycled paper. In doing so, their action will save approximately 84,000 trees per year, enough energy to power 643 homes per year, and save the greenhouse gas equivalent of taking 670 cars per year off the road.

" Deforestation accounts for 25 per cent of the annual carbon emissions caused by human activity. "

SOURCING FIBRE RESPONSIBLY

Most paper still comes from trees sourced from the world's forests. But knowing where your paper comes from, where in the world – what forest, can help you determine whether it was harvested in a manner that does not negatively impact the global climate. According to the UN Food and Agriculture Organization, deforestation accounts for 25 per cent of the annual carbon emissions caused by human activity. A growing body of evidence suggests that avoided deforestation, the protection of standing intact forests and their natural functions, will be a key strategy for protection against climate change.
Of course, not all logging for paper results in deforestation and loss of natural carbon sinks, but is there any way to know? Is there a way to make an active, positive choice to support a forest industry rooted in sustainability? There is, thanks to the independent, third party certification of well managed forests by the Forest Stewardship Council (FSC). Its label, a tree and check mark, is appearing on a rapidly increasing number of paper products and in the credentials of forest and paper users worldwide.
FSC is an independent, nonprofit, non-governmental organisation. It is an association of members from environmental and social groups, the timber trade and the forestry profession, community forestry groups and other organisations from around the world. FSC runs a global forest certification system that includes two key aspects: Forest Management and Chain of Custody certification. This ensures that timber produced in certified forests has been traced from the forest to the end user.
It is estimated that over 40 per cent, of the world's industrial wood harvest is used for paper production.

Recycled paper logo.

© Sava Miokovic/iStockphoto

The paper and printing industry has the power to have a massive impact on how our forests are managed. Through using and specifying FSC paper from well managed forests, purchasers that seek to take climate action give value to responsible forestry practices and ensure that purchases are not contributing to forestry practices that exacerbate global warming.

EMPLOYING CLEAN PRODUCTION METHODS

The biggest greenhouse gas releases in pulp and paper manufacturing come from the energy production needed to power the pulp and paper mill. The pulp and paper industry is the fourth largest emitter of greenhouse gases among United States manufacturing industries, and contributes nine per cent of total manufacturing carbon dioxide emissions. Worldwide, the paper industry uses six per cent of all energy.

As stated, purchasing recycled paper reduces total energy consumed. Secondly, purchasers of paper should determine what energy source was used in the manufacturing process and find paper that is produced using clean, renewable energy sources, such as wind and solar. Increasingly, these products are coming to market bearing certifications and logos that announce they are produced in this way. Be sure to achieve confidence that the claims are genuine, logos are credible, and that offsets are achieved through reliable institutions.

WHAT'S IN YOUR PAPER?

Purchasing and using paper thoughtfully and responsibly can be a key component of any institution's comprehensive action plan for reducing its impact on climate change. By embracing the four pillars of the Common Vision, paper manufacturers, suppliers and purchasers can dramatically reduce the climate change impacts of paper use.

Thanks to a rapidly increasing market demand, product innovation by the industry, and tools and information for carbon conscious decision making, climate action via responsible paper choices is an opportunity that is here today, not tomorrow. It is a tremendous opportunity to take voluntary, positive actions in the commercial marketplace which lead us firmly in the direction of sustainability.

THREE STEPS TO CLIMATE ACTION VIA RESPONSIBLE PAPER CHOICES

1 Get information and assistance

The Environmental Paper Network (EPN) can refer purchasers to the most appropriate organisation to help you learn what's in your paper now and how to take climate action via responsible paper choices.

2 Develop a formal policy

Find step by step guidance for developing a comprehensive responsible paper purchasing policy that will meet your specific goals on EPN's website. Be sure to set goals and to track and report progress towards them.

3 Calculate the benefits and celebrate them

The Paper Calculator, www.papercalculator.org, developed by United States NGO Environmental Defense, quantifies the impact of a change to environmental papers and can be used to produce professional reports to communicate the accomplishment.

Author

Joshua Martin is the Director of the Environmental Paper Network, located in Asheville, North Carolina, USA.

Organisation

The Environmental Paper Network links together over 100 organisations worldwide, working collaboratively to advance social and environmental transformation in the paper industry.

Enquiries

Environmental Paper Network
16 Eagle Street
Suite 200
Asheville, NC 28801
USA
Tel: +1 (828) 251-8558
E-mail: info@environmentalpaper.org
Website: www.environmentalpaper.org

The Mosaic Plantation Concept (MPC) and our commitment to High Conservation Value Forest (HCVF) protection is believed to be the most viable option to stop forest degradation and loss in Indonesia.

Sustainability in action

APRIL

Asia Pacific Resources International Holdings Limited (APRIL) offers a precedent-setting example of a sustainable business model that successfully balances economic benefits to the wider community and conservation of globally and nationally recognised ecological and social values.

SUSTAINABLE FOREST MANAGEMENT

APRIL rigorously pursues a rational sustainable business model, in direct opposition to the unfortunate but incontrovertible trajectory of forest loss and often unmanaged social development in Indonesia. We believe that leaving the forests without active management and protection is not a wise option amid constant threats from illegal logging, forest fires, shifting agriculture and poor forestry practices.

We manage forest resources sustainably, maintaining the efficient infrastructure needed to ensure healthy growing forests, the provision of valuable ecosystem services and the responsible use of forest resources. This is achieved through our Mosaic Plantation Concept (MPC), a landscape based approach to manage the forest resources to meet the conservation, social and economic needs of today's society without compromising the needs of the future generations.

At APRIL we have publicly committed to identify, protect and manage the High Conservation Value Forests (HCVF) in our operations, becoming one of only two pulp and paper companies in the world to do so. The HCVF mechanism provides a framework for responsible companies to proactively exceed legal conservation requirements. Presently, about 200,000 hectares are being protected and managed as conservation areas in APRIL's operations. This makes APRIL the second largest conservation agency in the province of Riau (Sumatra, Indonesia).

We are also the first to achieve certification for Sustainable Plantation Forest Management under the Lembaga Ekolabel Indonesia (LEI, Indonesian Eco-labeling Institute). The LEI certification system was created by a multi-stakeholder body that included leading NGOs to promote fair, just and sustainable forest management in Indonesia. It is supported by the European Union, ITTO, and the International Conservation Union among others, and is an acknowledged standard by the Japanese government.

Our forest plantations are also certified under OHSAS 18001:1999 for Health and Safety Management.

MITIGATING CLIMATE CHANGE THROUGH RESPONSIBLE FORESTRY

According to a 2006 report by the Food and Agricultural Organization, about 20 per cent of CO_2 emissions have been caused by land use change with the loss of about seven million hectares per year between 2000 and 2005.

> **"We have committed to identify, protect and manage the High Conservation Value Forests (HCVF) in our operations, becoming one of only two pulp and paper companies in the world to do so."**

The report, however, also said that this amount of forest loss would have been almost twice as large, were it not for the reforestation and conservation efforts of plantation forestry companies like APRIL.

The National Council for Air and Stream Improvement reported that carbon stored in forest products is increasing by about 150 million tonnes per year. This is

APRIL's Community Development programme helps alleviate poverty, as well as provide an alternative source of income to illegal logging.

equivalent to removing 540 million tonnes of CO_2 from the atmosphere per year.
Moreover, the Intergovernmental Panel on Climate Change (IPCC) in its Fourth Assessment Report in 2007 noted that, "in the long term, a sustainable forest management strategy aimed at maintaining or increasing forest carbon stocks, while producing an annual sustained yield of fiber, (timber or energy) from the forest will generate the largest sustained mitigation benefit."
As the sole member in Indonesia of the World Business Council for Sustainable Development (WBCSD), we are committed to maximising our contribution to climate change mitigation through the adoption and promotion of sustainable forest management, including the science based management of the peatlands, as important mitigation strategies.
Our Science Based Management Support Program, a consortium of experts on carbon and peatland development, enables us to practice optimal water level management, prevent soil subsidence and significantly minimise CO_2 emissions in our peatland operations. In addition, the consortium's inputs also guide our conservation and forest protection practices.
Presently, indicators show that CO_2 emission from degrading peatlands is cut by almost 80 per cent when placed under APRIL active management as proven by our present operations.
We are committed to track, manage and report on our carbon impact as part of our responsibility as a member of the WBCSD, a signatory to the UN Global Compact and to the CEO Climate Change Statement launched in the 2007 UN Global Leaders' Summit by the UN Environment Programme, UNGC and the WBCSD.

APRIL's proposed project in the Kampar Peninsula (Riau, Sumatra, Indonesia) offers a model for the protection of forest carbon stock that helps mitigate climate change, while also protecting the conservation values in the area and potentially generating financing incentives for the government and the communities. If implemented, the project will achieve avoided deforestation in 270,000 hectares of conservation areas that store about 1,100 Mt of CO_2 or roughly the equivalent of 50 per cent of Indonesia's present total emissions.

STAKEHOLDER ENGAGEMENT

At APRIL, we incorporate the values of our stakeholders, both government and civil society, into our operations and actively seek opportunities for partnerships that address common environmental and social concerns.
Our company publishes an independently verified sustainability report, with the third edition released in 2006. Our report follows the Global Reporting Initiative and the UN Global Compact guidelines on Communications of Progress.
We co-hosted The Forest Dialogue in 2007 and have been the UNEP's corporate partner for the Champions of the Earth Award since 2005.

ENSURING A CLEAN SUPPLY CHAIN

APRIL is the only pulp and paper company in Indonesia with a wood tracking system that is annually verified by independent third parties, and observed by WWF. Results of these audits confirm that no questionable logs enter our mill.
To secure the integrity of our fibre source and to assure our customers, we continue to enforce and strengthen our Acacia Chain of Custody (CoC) and Pulpwood Tracking Systems. The Acacia CoC, regularly audited by SGS, ensures that the flow of Acacia fibre into our mill operations is monitored, tracked, documented and independently verified.
To ensure all pulpwood coming from outside suppliers has come from an approved source, we conduct documentary and field inspections to determine that suppliers have complied with both their legal requirements and our own policy.

COMBATING ILLEGAL LOGGING

More than any other plantation company in Indonesia, our company has been leading the fight against illegal logging in Riau, together with the local and provincial governments and WWF Indonesia. Over the past years, we have had demonstrations against us, our security personnel killed and trucks and machinery destroyed because of our fight against illegal logging.
We know that we cannot halt illegal logging on our own and need support from law enforcement authorities. We also continue to develop relationships with other key stakeholders and further strengthen our collaboration with WWF. Working together with the Task Force that includes WWF Indonesia and concerned government authorities, we help organise strategic checkpoints and composite patrols, reinforced by the blocking or cutting off of access roads used by illegal loggers.
We understand that economic necessity generally underlies the proliferation of this illegal activity. We have undertaken major community development initiatives in

In collaboration with WWF, APRIL supports the Elephant Flying Squad which keeps wild elephants within the forest and away from village farms.

order to provide local people with alternative means of employment. Through our operations, we generate 35 direct and indirect jobs for every 100 hectares managed. These provide a legitimate livelihood option to people who might otherwise engage in illegal logging.

PROTECTING THE SUMATRAN ELEPHANTS' HABITAT

Together with the Indonesian Government through the Ministry of Forestry, the Riau Provincial Government, concerned NGOs including WWF Indonesia, and private forestry companies, we established the Tesso Nilo National Park Foundation in April 2006. The immediate objective of the multisectoral initiative is to expand the Tesso Nilo National Park from 38,000 hectares to about 100,000 hectares.

In support of the human-elephant conflict mitigation protocol for Riau signed in 2005 by the Indonesian Ministry of Forestry, WWF Indonesia and other NGOs, the Elephant Flying Squad Project was established. This project uses trained elephants and staff to undertake patrols along the border of the National Park and drive back wild elephants into the Park. In 1996 we adopted four elephants and are now cooperating with the Riau Ministry of Forestry (BKSDA) and WWF Indonesia in training them and eight mahouts to become the second Flying Squad.

PREVENTING FIRE AND HAZE

The occurrence of haze is largely due to slash and burn practices that are still undertaken as the traditional and cheapest method to clear land. APRIL does not use fire to clear land as it makes no business or environmental sense. We have employed a 'No Burn' policy from the very outset of our operations and use only mechanical methods to prepare the land for planting.

Our fire fighting team, which is equipped with air, water and land based capabilities, carry out daily patrols to spot fires within and around the concessions and extinguish them at the quickest possible time. In addition to daily ground and aerial patrols, we also conduct daily calculations of the Fire Danger Rating in each fibre estate and monitor the NOAA (US National Oceanic Atmospheric Administration) satellite data on fire hotspots.

In addition, we also actively engage the local communities in fire prevention initiatives including teaching the local communities about fire and haze prevention and sustainable agricultural methods instead of the slash and burn approach.

ENSURING CLEAN AND GREEN PRODUCTION

APRIL's integrated pulp and paper mills are equipped with the best available technology and benchmarked against the world's best. All emissions and effluent produced by the mill are monitored regularly and measured against external standards. Our modern treatment processes enables us to meet and exceed Indonesian regulations and the US Cluster Rule for New Mills and World Bank Pollution Prevention Guidelines. Wastewater quality, air emission levels and ambient air quality more than satisfy national and international standards.

The mill is largely self sufficient in energy generation with approximately 97 per cent of total energy produced coming from biofuels which are a byproduct of the production process – black liquor, wood bark and rejected wood chips.

EMPOWERING THE COMMUNITY

In 2006, APRIL became a signatory to the United Nations Global Compact, the world's largest corporate citizenship initiative. The Compact, established in July 2000, offers a framework for businesses to follow in response to the challenges of globalisation. It embraces 10 principles covering human rights protection, fair labour, environment conservation, and anti-corruption.

APRIL has established several social development initiatives aimed at fostering an economically viable community, empowering local people and establishing a thriving social infrastructure. APRIL's Community Development programme encompasses three main areas: integrated farming, community fibre farms and small and medium enterprises. These programmes serve to help alleviate poverty, as well as provide an alternative source of income to illegal logging.

At APRIL, we continue to believe that sustainability is an imperative and only with a responsible, balanced approach to commercial, environmental and social values can a business achieve long term profitability and meaningful contribution to the wider natural and social landscapes.

Organisation

APRIL, with offices in Singapore and operations in Indonesia and China, is a leading producer of fibre, pulp and paper. APRIL operates one of the world's largest pulp mills with an annual production capacity of 2,000,000 tonnes in Indonesia. The company is committed to protecting the natural resources in its care through sustainable management of its mills and plantation operations to benefit our stakeholders, both now and in the future.

Enquiries

Lucy Jasmin, Head of APRIL's Corporate Communications
Tel: +62 761 499113
E-mail: lucy_jasmin@aprilasia.com
Website: www.aprilasia.com

REGIONE ABRUZZO
PARCHI TERRITORIO AMBIENTE ENERGIA
PARKS TERRITORY ENVIRONMENT ENERGY

HOW CAN LOCAL GOVERNMENTS HELP TO REDUCE CARBON FOOTPRINT?

When designing sustainable energy strategies it is very important that Regions have very clear the concepct of reduction of energy consumption. Starting from here, Regions must decide specific targets leading to a tangible result and work with IGOs, NGOs, universities and local businesses. Regione Abruzzo goal is the reduction of energy consumption by 30% by 2010, as established by Regione Abruzzo *Regional Energy Plan* (REP) and the control of the emissions deriving from puntual or widespread sources and local traffic as described by the *Regional Quality Air Plan* (RAP).

2010- Expected Energy Production

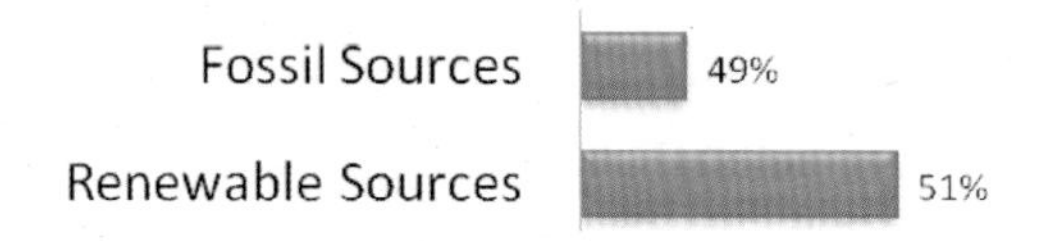

Energy Production from renewable sources: expected target at 2010

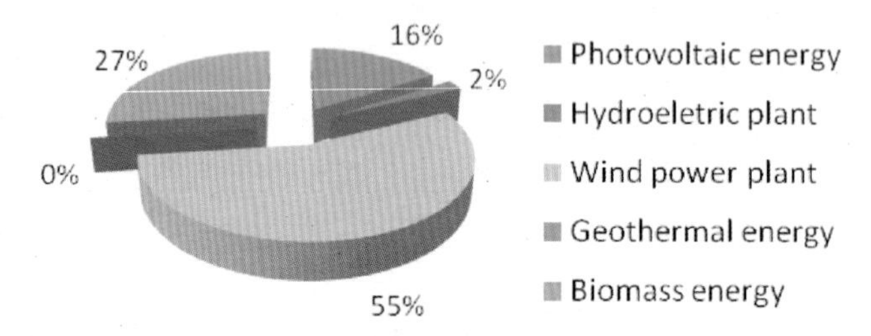

In order to reach these objectives Regions must implement 3 different strategies:

- *PUBLIC AWARENESS CAMPAIGN* among citizens on the correct use and management of energy
- *REGIONAL LAWS FOR THE USE OF ENERGY SAVING MATERIALS AND BUILDING TECHNIQUES*
- *REGIONAL PUBLIC CALLS FOR TENDER* to boost the implementation of renewable energy sources.

Achieving the 2010 goals

BIOMASS ENERGY

Regione Abruzzo signed an agreement with the Ministry for the Environment to start a programme of biomass enhancement involving public and private stakeholders. The objective is to realize projects to implement the use of biomass creating local systems at sustainable costs as well as new employment.

WIND POWER PLANTS

Regione Abruzzo has recently produced guidelines for the creation of wind power plants, focusing on the environmental impact and at the same time encouraging private companies to implement this technology. Actually the energy production is around 160 MW bringing Abruzzo on the National Top-Ten.

HYDROGEN

The road to the "zero-emissions" entices Regione Abruzzo to create a new generation fuel station network to produce and commercialize hydrogen locally in order to use it immediatly in a natural gas mixture.

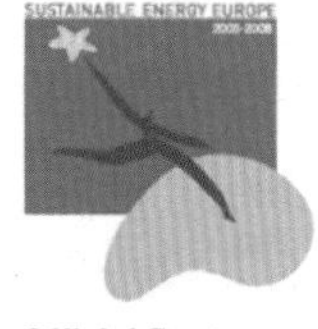

Official Partner

CONTACTS
Regione Abruzzo - Direzione Parchi, Territorio, Ambiente, Energia
ARAEN - Abruzzo Regional Energy Agency
Via Passolanciano, 75, 65124, Pescara, ITALY - Tel: +39 85 767 2570/71 Fax: +39 85 767 2545
e-mail: araen@regione.abruzzo.it
www.regione.abruzzo.it/xAmbiente
www.araen.org

The role of cities in tackling climate change

NICKY GAVRON
DEPUTY MAYOR OF LONDON

Cooperating with other cities worldwide is a key initiative for the mayoral office in London because the battle against climate change will be won or lost in cities. The role of national governments is, of course, widely debated, analysed and understood. Yet the challenge is so huge that cross-cutting action at all levels will be needed. The central role of city leaders in our rapidly urbanising world will be key to reducing the world's greenhouse gas emissions. The leaders of large cities have a particular responsibility to act, and national and subnational governments must empower and enable city governments to take on this role.

© Greenpeace/Reynaers

The Copenhagen district heating system supplies 97 per cent of the city with clean, reliable and affordable heating.

WHY ARE CITIES SO IMPORTANT?

By 2030, two-thirds of humanity will live in cities or urban areas. Half already do. Even now, cities consume 75 per cent of the world's energy and are responsible for 80 per cent of carbon dioxide emissions. Moreover, all cities are highly vulnerable to the impacts of climate change, and none more so than fast growing cities in developing countries. About 20 of the 30 largest cities of the world, London included, are situated on low lying coasts. Rising sea levels of a few metres would have catastrophic implications. So there's an extraordinary responsibility and motivation for cities to act.

But there are also great opportunities for cities. With their concentrations of people and activities at high densities, cities can use resources such as energy, materials and land efficiently. They are the places where high level, knowledge-based activities congregate, with the expertise to tackle climate change. Many cities are the drivers of their national economies, generating a large part of their country's GDP. Five US cities: New York, Los Angeles, Chicago, Boston and Philadelphia, together, make the world's fourth largest economy. Bangkok and Sao Paulo are home to 10 per cent of their countries populations, but generate 40 per cent of the national wealth. It could be argued that cities, not nations, are the engines of development.

> "Cities consume 75 per cent of the world's energy and are responsible for 80 per cent of carbon dioxide emissions."

Mayors and their municipalities have the powers and levers to reduce carbon emissions. They control the development of land, have housing powers, regulate transport and often manage public transport systems.

© TFL

Towards a low carbon London

The Greater London Authority was set up in 2000 with a new brand of political leadership: a directly elected mayor. From the very start, the Mayor set an overarching vision "to develop London as an exemplary sustainable world city". All strategies and policies – transport, housing, energy, waste and the London Plan – have taken that vision as their starting point, together with initiatives such as the introduction of congestion charging. In February 2007, all policies and implementation programmes were pulled together into the London Climate Change Action Plan, setting the ambitious target of reducing London's carbon emissions by 60 per cent by 2025. The Action Plan is a comprehensive, holistic approach addressing transport, new and existing buildings, energy supply and aviation as well as seeking fundamental changes in behaviour.

They have varying degrees of responsibility for the collection and processing of waste and have responsibility for other environmental infrastructure such as energy and water. They own and manage buildings and vehicle fleets. Significantly, they have huge purchasing power. They are able to form partnerships with private interests as well as mobilising and coordinating community action. Mayors have responsibilities in areas key to taking swift action to reduce emissions, and can show leadership in taking decisive and radical action. It is at city level that innovation and progress on climate change action is most likely to be achieved.

Although leadership from national governments is crucial in negotiating international agreements, setting frameworks and standards and for providing fiscal and financial incentives, when it comes to practical action on the ground, city leadership must take centre stage.

THE C40 LARGE CITIES CLIMATE LEADERSHIP GROUP

All over the world, city governments are taking their own initiatives, recognising the need to cooperate across national and international boundaries. Substantial carbon reductions have been achieved by 670 municipalities through ICLEI's 'Cities for Climate Protection' campaign. (ICLEI is an international association of local governments and national and regional local government organisations that have made a commitment to sustainable development.) Hundreds of mayors in the USA are mobilising to meet or beat Kyoto targets. While it is often the smaller cities that lead the way, it is in the larger cities where huge reductions in emissions can be achieved. But it is the very largest cities that pose the greatest challenge.

That is why, in October 2005, many of the world's largest cities met at the C20 summit in London Its aim was to develop long term international collaboration among large cities to drive down their emissions, act as champions and stimulate business and national governments. A set of practical actions were agreed, including creating municipal procurement alliances – buyers clubs – jump-starting the supply and demand for climate change technologies and measurably influencing the markets. The establishment of the Large Cities Climate Leadership Group and these commitments were set out in a short communiqué to the UN climate negotiations in Montreal.

In August 2006, as chair of the Large cities group the Mayor of London, Ken Livingstone, joined with former US President, Bill Clinton, to announce a partnership with the Clinton Foundation Climate Initiative (CCI). The CCI was to become the operational arm of the Cities Group and work on an accelerated programme of carbon reductions in each of the cities.

> **"It is at city level that innovation and progress on climate change action is most likely to be achieved."**

Participation in what has now become the C40 spans the globe: including Berlin, Buenos Aires, Cairo, Caracas, Chicago, Delhi, Dhaka, Houston, Istanbul, Johannesburg, Karachi, Lagos, London, Los Angeles, Madrid, Melbourne, Mexico City, Moscow, New York, Paris, Rome, Sao Paulo, Seoul and Toronto.This global partnership also includes an affiliated group of smaller cities that are exemplars of innovative practice, such as Curitiba and Copenhagen.

The C40 is not about exclusivity it is about delivery! C40 cities are expected to act as catalysts for change within their country or region. Any initiative or procurement package developed within the C40 cities will become available to other cities, once they are up and running.

SHARING BEST PRACTICE

Retrofits for large buildings – Berlin

We have plenty of best practice to share. The city of Berlin in partnership with Berlin Energy Agency (BEA) has pioneered an excellent model for improving energy efficiency in buildings. The BEA, a leading energy consultancy, partly owned by the government of Berlin, organises retrofits for large government and commercial buildings. Contracts between the building owners and energy systems companies agree to make energy efficiencies of around 24 per cent. They do this

In 2006 the C40 announced a partnership with the Clinton Foundation Climate Initiative to accelerate carbon reductions in cities.

© Dan Avila for the Clinton Foundation

by installing hardware such as heating control systems, lights and insulation. So far, 1,400 buildings have been upgraded, delivering CO_2 reductions of more than 60,400 tonnes annually.

These retrofits cost the building owners nothing and the buildings make immediate savings. Energy bills are slashed by an average of 26 per cent. It is these savings that fund the retrofit. Average payback periods are between 8 and 12 years.

District heating, Copenhagen

The Copenhagen district heating system is one of the world's largest, oldest and most successful, supplying 97 per cent of the city with clean, reliable and affordable heating. Set up by five mayors in 1984, the system simply captures waste heat from electricity production, normally released into the sea, and channels it back through pipes into peoples' homes. The system cuts household bills by €1,400 annually and has saved the Copenhagen district the equivalent of 203,000 tonnes of oil every year: that's 665,000 tonnes of CO_2.

Other best practice examples

Other examples of best practice include the city of Seattle, which sets the standard for US Green Buildings, Shanghai's carbon neutral quarter, being built in Dongtan, lake water air conditioning in Toronto and Tokyo's decentralised energy programme.

It is these extremely effective means of reducing emissions that can encourage others to implement change.

THE NEW YORK SUMMIT

The C40 held its second summit in May 2007 hosted by Mayor Bloomberg in New York. Thirty two Mayors attended with representatives from over 50 cities as well as senior executives from the world's largest companies and financial institutions.

Many delegations declared it the most productive conference they had ever attended. The range of discussion topics included:

- Transport – congestion charging, bus rapid transport, car free days, hybrid and hydrogen bus engines
- Waste – recycling and renewable gas production
- Water – wastage and energy used in delivery of water
- Energy – use of renewables, decentralised energy and combined cooling, heat and power systems
- Buildings – retrofitting of existing buildings and low and zero emission homes for the future.

On almost every topic covered, the world's leading city in that field was on hand to enable other cities to learn from its experiences.

The power of procurement

The Summit also saw the launch of the first C40 procurement package, the Energy Efficiency Buildings Retrofit programme, developed by the Clinton Climate Initiative. This exemplifies the approach of negotiating deals between customers, suppliers and financial institutions to establish economies of scale, reduce costs and accelerate the introduction of technologies. In this instance, 16 of the world's largest cities, four of the world's largest energy services companies and five of the largest banks come together to offer city authorities and the owners of public and private buildings an energy audit, with recommendations for energy and emission reductions; a comprehensive discount on the goods and services to achieve these reductions and the financing to pay for them, paid from energy savings and underwritten by the banks. This procurement package will be followed by others on transport and waste with more to come. This is a new way of doing business that scales up and catalyses markets for public goods and services.

THE 2007 TRANSPORT AND CONGESTION WORKSHOP

Summit attendees agreed to hold workshops between summits. The first, on transport and congestion, will be held in London in December 2007. The workshop, co-hosted with Stockholm is discussing and comparing best practices among the transportation and congestion experts from C40 municipal governments.

In particular, the workshop will draw on experiences of specific cities. Bogotá recently introduced a Bus Rapid Transit (BRT) system through the city, consisting of 850 buses used daily by 1,400,000 passengers. The system has reduced travelling time by 32 per cent, taken 2,109 public service vehicles off the road, and reduced greenhouse gas emissions by 40 per cent. Such is the success of the 'Transmilenio' system that it is being expanded to 130 km of dedicated lanes to transport 1.8 million passengers a day.

" C40 has an important role to play in ensuring that best practise is shared between cities and that adaptation measures are consistent with reducing emissions. "

Also speaking will be representatives from Seoul, drawing upon the success of their car free day's initiative. This is a voluntary scheme, encouraging residents to participate as often as they can, and yet already two million cars are staying off the road every year, decreasing traffic volume by 3.7 per cent. The initiative delivers a reduction of about 243,000 tonnes of CO_2 emissions, a reduction rate of 9.3 per cent and equivalent to about six per cent of total CO_2 emissions per year in Seoul.

A number of Chinese cities have also been developing the car free day initiative including Shanghai and Beijing. Beijing recently tested the impact of emissions reductions by taking 1.3 million of the city's three million vehicles off the road for several days.

Paris's Deputy Mayor will also discuss recent experiences of introducing a bike scheme that has revolutionised the streets and the way people get around the city. Over 10,000 brand new self-service bicycles went up for rental at 750 ranks across the city in July. One survey has recorded that in its first two months, the 'Vélib' bikes (short for vélo-liberté) were used five million times.

LOOKING FORWARD

Further workshops are planned for 2008 on port and airport emissions. Four of the world's largest container ports are in Asia, with three of them in China (Shanghai, Hong Kong and Shenzhen), while Asian airports such as Hong Kong, Beijing and Shanghai are growing rapidly.

Adaptation to climate change is also a crucial area of engagement. Many cities are already hit hard by the climate change. C40 has an important role to play in ensuring that best practise is shared between cities and that adaptation measures are consistent with reducing emissions. Asia is under threat from flooding, storm surges and sea level rise. Cities in the low lying areas along the east and south China coast, and the delta areas of South and South East Asia are particularly vulnerable. The economic benefits of wise adaptation strategies that dovetail with mitigation measures should be a focus for cities in 2008

CONCLUSION

As the climate debate shifts from whether the scientific evidence demands global mandatory targets to what level those targets should be and the measures to meet them, the need to work with different spheres of government will become more apparent, and the role of cities will become critical. The potential of the C40 to make deep cuts in carbon emissions and to reconfigure global markets for cutting edge technologies, will take on global significance.

To achieve this potential and reach the highest reduction targets, national and subnational governments need to put cities in the driving seat. They need to open up a new era of municipal enterprise by collaborating with and empowering cities. In the spirit of this, it would be a step forward if cities could be included with national delegations at the COP and MOP meetings.

Author

Nicky Gavron has been the Deputy Mayor of London since 2000, responsible for Strategic Planning, Under 18s, Climate Change and the Environment. Nicky initiated the C40 and brokered the partnership with the Clinton Foundation. As Chair of the London Planning Advisory Committee in the 1990s, she commissioned the research and formulated policies, including congestion charging, which underpin much of the Greater London Authority's work today. At a national level, she was a member of the Government's Commissions for Integrated Transport and Sustainable Development and was an Advisor to the Urban Task Force. She was the first Chair of the Local Government Association's Planning Committee.

Organisation

The Greater London Authority is a unique form of strategic citywide government for London. It is made up of a directly elected Mayor and a separately elected Assembly. There are around 600 staff to help the Mayor and Assembly in their duties. The GLA's main areas of responsibility are: transport, policing, fire and emergency planning, economic development, planning, culture, environment and health.

Enquiries

Mayor of London, Greater London Authority
City Hall, The Queen's Walk, More London
London SE1 2AA
Tel: +44 (0)20 7983 4100 | Fax: +44 (0)20 7983 4057
E-mail: mayor@london.gov.uk
Website: www.london.gov.uk

Energy efficiency in an imaginary megacity

SIEMENS

Numerous energy efficient solutions are already available for use today with the potential to make considerable energy savings. These solutions are demonstrated here by presenting an imaginary megacity, incorporating real energy consumption figures for Germany as its basis. The megacity is a world champion in energy efficiency, reducing its primary energy consumption and carbon dioxide emissions by 50 per cent.

New York's new subway monitoring and train control system enables more trains to travel faster through the network and thus use the infrastructure more efficiently.

The combustion of raw materials such as coal, gas, and oil results in the emission of 26 billion tonnes of carbon dioxide (CO_2) annually. But what would happen if we used the most energy efficient technology? If we could start afresh, how much energy would an imaginary city of 10 million people require? This thought experiment leads to surprising conclusions.

ENERGY REQUIREMENTS FOR AN IMAGINARY MEGACITY

Germany, the sixth biggest energy consumer after the US, China, Russia, Japan, and India, consumes about 14,200 petajoules of primary energy (1 PJ equals 10^{15} joules). Its population is 82 million, which means an imaginary city of 10 million would consume around 1,750 PJ. The energy mix for the megacity consists of:

- Petroleum – 36 per cent
- Natural gas – 23 per cent
- Hard coal – 13 per cent and brown coal – 11 per cent
- Nuclear power – 12 per cent
- Renewables (water, wind, solar cells or biomass) – five per cent.

Converting this primary energy into usable energy will inevitably lead to losses. As a result, consumers only obtain 1,120 PJ of so-called 'site energy', of which industry and commerce consume 42 per cent, households 29.5 per cent, and transport 28.5 per cent.

ENERGY SAVINGS IN INDUSTRY AND BUILDINGS

In the imaginary city, the first issue to address is heat production, one of the biggest drains on energy. In Germany, 58 per cent of site energy is used to generate heat for buildings, hot water, and process heat for industry, eg in metal production. In private households, heating accounts for 80 per cent of energy use. Focusing on the imaginary city, some immediate savings can be exploited. Building renovation as well as intelligent building automation will reduce energy consumption by 40 per cent or more.

In industry, electric motors for drive systems, conveyor belts, and pumps account for more than two-thirds of its power consumption. Up to 60 per cent of the power used by an electric motor will be saved by using frequency converters and more efficient motors. Waste heat, for example in the glass, metal, and cement industries, will also be used for boiling a liquid to drive a turbine for generating electricity.

One way to finance efficiency-boosting measures ideal for municipalities such as the imaginary megacity is energy-performance contracting. The local government pays for the investment in instalments financed by the energy savings achieved. This puts no additional burden on local budgets, and once the contract expires after around 10 years, all savings flow directly into the municipal budget.

ENERGY EFFICIENT TRANSPORTATION

The megacity's second biggest energy guzzler is traffic. In Germany, 28.5 per cent of site energy is consumed in the form of fuels. The imaginary city's residents therefore often use the extensive public transport network, especially since taxes and toll fees have made it expensive to drive vehicles with high CO_2 emissions. The new buses and trains are not only comfortable and travel at short intervals; they also consume 30 per cent less energy than before, thanks to lightweight materials and braking energy regeneration systems.

The world's largest and most efficient gas turbine produces enough electricity to supply the population of a city the size of Hamburg or Barcelona.

Those who still need a car drive hybrid vehicles, or use piezo technology to optimise fuel injection reducing fuel consumption by around 20 per cent. Internet-based information systems and traffic management systems prevent congestion and make it easier to find a parking space. This also considerably reduces CO_2 emissions, as studies have shown that the search for a parking space accounts for up to 40 per cent of city driving.

RESOURCE EFFICIENCY IN POWER GENERATION

The world champions of energy efficiency wouldn't be known as such if they hadn't also cut electricity consumption. Although electricity only accounts for 20 per cent of all site energy consumed in Germany, the electricity must first be produced in power plants, which convert only an average of about 40 per cent of primary energy into electricity. The megacity efficiency champions make better use of primary energy in facilities, such as combined cycle power plants, which convert more than 60 per cent of the energy contained in gas into electricity. By exploiting the heat produced, they bring the rating to over 80 per cent. Here, process steam and heat is sent via pipes to nearby factories and apartment buildings.

ONE GAS TURBINE FOR 1.8 M INHABITANTS

The imaginary megacity has invested in a 530 megawatt combined cycle power plant which will radically boost efficiency. At 444 tonnes, its gas turbine is as heavy as six diesel locomotives but has 100 times the output. Its output of 340 megawatts produces enough electricity to supply 1.8 million of the megacity's population. Replacing all coal-fired plants worldwide with such high efficiency plants would result in over four billion tonnes less of CO_2 being released into the atmosphere each year, equivalent to the total annual emissions from the whole of Europe. (Siemens is currently building such a plant for the energy supplier E.ON, due to go on line in 2008.)

Of course, renewables also reduce CO_2 emissions in the imaginary city. Solar cells can be found on top of public and private buildings and windmills provide their share of electricity, as do geothermal and biomass power plants. In addition, a large portion of household waste is converted into fuel for power plants. The city is also installing a fuel cell power plant in the megawatt class which, when combined with a gas turbine, will convert around 70 per cent of the energy into electricity. (Siemens and Energie Baden-Württemberg (EnBW) are developing such a fuel cell power plant which they plan to complete by 2012.)

THE POTENTIAL OF HOUSEHOLD APPLIANCES AND LIGHTING

Today, almost half of all electricity consumed in households is used by refrigerators, stoves, washing machines, and dishwashers. Purchasing new appliances whose consumption has been cut by 30-75 per cent since 1990 is the best investment here. Replacing all household appliances in the fictitious megacity would reduce electricity consumption by the amount used today by 600,000 people.

Lighting systems have been revamped in this city as well. Lighting accounts for more than 10 per cent of total electricity consumption in Germany and 19 per cent worldwide. Energy-saving lamps as well as LED lamps reduce electricity consumption by 80 per cent, compared to conventional lightbulbs. Some real cities are already making the switch. Budapest, for example, is replacing the lightbulbs in its traffic lights with LEDs. The financing model is similar to energy-performance contracting, as the monthly installments the city has to pay are lower than the savings generated from reduced power consumption and the elimination of traffic light servicing. This investment in sustainability pays for itself.

CONCLUSION

The combined potential for energy conservation in households, industry, transportation, and power plant technology enables the imaginary megacity to reduce both the consumption of primary energy and CO_2 emissions by 50 per cent. Nearly all of the described solutions already exist today. They don't have to be developed; they could be implemented immediately.

Author

Dr Ulrich Eberl is head of the Technology Press and Innovation Communications department of Siemens. Eberl studied physics at the Technical University of Munich and wrote his dissertation about converting solar energy in the process of photosynthesis. In 1988, he became a science and technology correspondent for various newspapers and magazines. After working for Daimler's technology publications, he joined Siemens in 1996. Since 2001, Eberl has been editor-in-chief of the magazine *Pictures of the Future* (www.siemens.com/pof), which has been the recipient of several international awards.

Organisation

Siemens (Berlin and Munich) is a global powerhouse in electrical engineering and electronics. The company has around 475,000 employees and provides innovative technologies to benefit customers in over 190 countries. Founded 160 years ago, the company focuses on fields such as energy and environment, industry and infrastructures and healthcare. In fiscal 2006, Siemens had sales of €87.3 billion and a net income of €3.033 billion.

Enquiries

E-mail: Ulrich.eberl@siemens.com
Website: www.siemens.com/innovation

The power of green public procurement in the EU

Energy-efficient public buildings.

MOGENS PETER CARL
DIRECTOR-GENERAL FOR THE ENVIRONMENT,
EUROPEAN COMMISSION

Our current consumption in the EU causes environmental damage at unsustainable rates. In addition, rapidly developing economies are further increasing the pressure on the environment. If the world as a whole were to follow these same patterns of consumption, global resource use could quadruple within 20 years. So we must find ways to reduce the negative impacts of our production and consumption patterns on our environment, health and natural resources. One essential way to do this is to stimulate the faster development and wider use of environmentally beneficial products. And where better to start than with our public authorities.

GREEN PUBLIC PROCUREMENT

Europe's public authorities are influential consumers. They have an annual budget equivalent to 16 per cent of the EU's GDP, around €1,800 billion. There are many areas with major environmental impacts where public authorities, because of their relatively large spending power, have the capacity to trigger the supply of greener products and reduce their own environmental impact. Examples include office machinery, transport, food and catering, energy, construction and waste management.

This process is green public procurement (GPP). In other words, public authorities seek to procure goods, services and works with a reduced environmental impact throughout their life cycle, as alternatives to the ones they would otherwise have procured. GPP procedures can bring a number of different benefits.

Cutting costs

Where products consume resources over their lifetime, the purchase of more resource-efficient products can significantly cut costs for users, even though the initial expenditure may be higher. Examples are office equipment, lighting, vehicles and buildings.

Stimulating innovation

GPP rewards firms that develop goods and services with a reduced environmental impact. It also encourages the development of new technologies, promoting innovation that supports the EU economy. Environmentally improved goods are often niche products for which increased demand will lead to economies of scale, allowing products to move into mainstream markets. The pull of public procurement may stimulate greater and faster technological innovation or breakthroughs that will ultimately lead to lower unit costs and mass market availability. This will help consolidate the international position of EU industry.

Achieving environmental goals

Where GPP leads to the purchase of greener products, the reduced environmental impact from those products will help achieve environmental goals, and could do so more cheaply than other available policy instruments. This would reduce the cost of achieving those goals.

A study conducted for the European Commission, the EU's executive body, between 2001 and 2003 shows just how important this role could be. Each year the public

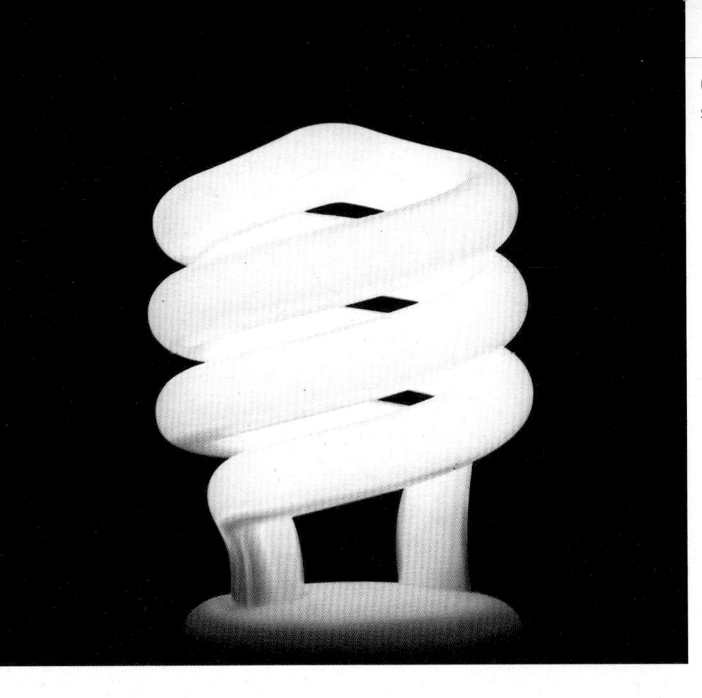
© Robyn Mackenzie/Fotolia

Greener office supplies could contribute significantly to reducing carbon emissions.

sector buys more than 2.8 million PCs. If it started purchasing energy efficient desktop computers, 830,000 tonnes of CO_2 would be saved, which would bring the EU 0.25 per cent closer to its Kyoto goal.

According to an expert study by Hans Nilsson in 2003 on technology procurement, the US Department of Energy set out in early 1998 to use procurement to develop the market for a new generation of smaller, brighter and less expensive compact fluorescent lamps. The initial sales goal of one million lamps was exceeded by more than 50 per cent. And as a result of the programme, 16 new lamp models have been brought onto the US market at reasonable prices.

A recent study carried out for the Commission on the costs and benefits of GPP refers to the procurement of public railcars with particle filters for the *Taunusbahn* in Germany (2004-2005). The filter ecotechnology used comes from recent technological efforts to reduce emissions by improving engine technologies so as to meet the high emission standards that will come into force in 2012. This procurement approach not only stimulated R&D to develop economically efficient ecosolutions, but also resulted in the breakthrough of this ecotechnology onto international markets.

In the bus sector, public procurement represents about 33 per cent of the market. Specifying requirements aimed at reducing CO_2 and pollutant emissions could significantly improve the offer of cleaner buses on the mass market as a result of the economies of scale achieved within this market segment.

STATE OF PLAY

A recent survey shows that only seven of the 27 EU Member States practise a significant amount of green purchasing, ie green criteria appearing in more than 40 per cent of tenders published in the year. Even in those front running countries, ie Austria, Denmark, Finland, Germany, the Netherlands, Sweden and the UK, there is scope for improvement, because many of the green tenders were not formulated in a clear, non-discriminatory way.

Despite strong recommendation by the Commission in its 2003 Communication on Integrated Product Policy calling for Member States to adopt national action plans on GPP by the end of 2006, only 11 Member States have so far adopted such plans.

Outside Europe, green or sustainable procurement policies have been launched by most of our important trading partners, such as the USA, Japan, Canada and Australia, followed by rapidly developing countries, such as China, South Korea, Thailand and the Philippines. Back in 2002 the OECD Council adopted a recommendation on green public procurement. As a follow up to the Johannesburg summit on Sustainable Development in September 2002, a Marrakech Task force on sustainable procurement was created with the aim of spreading sustainable (green) public procurement practices, in particular in developing countries.

MISCONCEPTIONS AND BARRIERS HINDERING FURTHER UPTAKE OF GREEN PUBLIC PROCUREMENT

The most common arguments against green public procurement are that it is more expensive than conventional procurement and problematic from a legal point of view. Both are incorrect. Many green alternatives, such as energy efficient computers, printers and photocopiers, are no more expensive than conventional products. And even if some of these products are more expensive in terms of their initial purchase price, they save money over their life cycle, which offsets the higher cost. Typical examples are energy saving bulbs, which are around four times more energy efficient than the alternatives and last up to 10 times longer, and printers and copiers that can print double sided, thus saving paper.

There is a need to further clarify the legal framework on public procurement. The EU's new Public Procurement Directives of 2004 clearly allow public purchasers to include environmental criteria in their tender documents, provided those criteria are: relevant for the final product purchased; sufficiently detailed; and transparent and non-discriminatory. Governments are even allowed to go one step further and decide, for example, to make green purchasing mandatory for certain parts of their administration, or set targets for certain product groups. However, a problem raised by many stakeholders is the need to clarify whether it is legal to include environmental criteria which relate to the production process.

Member States do not do enough to help implementation, eg by providing freely accessible environmental databases for criteria and guidance, helpdesks and training. This results in poor quality GPP in terms of both legal compliance and quality of environmental specifications. This leads to added costs and less GPP.

Current GPP activity tends to be decided locally with different environmental criteria applying in different geographical markets. This increases the administrative costs for companies that reply to different procurement requests, which may involve remeasurement, or restatement of environmental attributes. This could reduce the rewards from innovation for suppliers whose greener products may only meet some of a wide variety of environmental criteria and reduce the clarity of public sectors requirements, and hence the incentive to innovate.

Finally, in most Member States there is still a general lack of high level political support, resulting in a shortage of resources required to implement GPP. If there is no strong centralised political message on GPP, individual purchasing organisations will fail to give management support for GPP, thus exacerbating the problem.

THE POLITICAL CONTEXT FOR GREEN PUBLIC PROCUREMENT AT EUROPEAN LEVEL

In recent years, the strategic importance of public procurement for both the environment and competitiveness in Europe has been receiving growing attention in EU Member States and at European level. The benefits are two-fold: new products, services and technologies are kind to the environment, but they also have the potential to open up new markets and create new jobs. The EU eco-industries sector already accounts for one-third of the global market, which is estimated at €550 billion per year, and it has enjoyed a healthy annual growth rate of around five per cent since the mid 1990s. Green public procurement plays an important role in many of our policy initiatives, from our climate change policies, to the Environmental Technologies Action Plan (which includes a range of measures to overcome barriers to the development and wider use of ecotechnologies) and Integrated Product Policy (which seeks to reduce the negative environmental impacts of a product throughout its life cycle). EU leaders have repeatedly stressed the importance of eco-innovation for EU competitiveness in the context of the EU Strategy for Growth and Jobs. By promoting energy efficiency and renewable energy sources, green public procurement can also help create a more sustainable energy system in Europe, as set out in the recent Green Paper on a European Strategy for Sustainable, Competitive and Secure Energy. Other benefits include more sustainable use of natural resources, waste prevention and recycling and more sustainable cities.

It is necessary to maximise the effectiveness of GPP by raising political support and identifying common criteria for its application. This will pull the latest environmental technologies onto the market and reduce the environmental impact of public consumption. EU leaders recognised this need by including an EU wide GPP target in the renewed EU Sustainable Development Strategy. The aim is that by 2010 the average level of EU GPP should meet the standard currently achieved by the best performing Member States.

EUROPEAN INITIATIVES TO STIMULATE GREEN PUBLIC PROCUREMENT

The European Commission has published a handbook entitled Buying Green in all EU languages. It explains in clear, non legal terms how to include environmental criteria at the different stages of a public procurement procedure. The new website on green public procurement also has some useful background information including links to databases of environmental criteria for products, to national websites on green public procurement, and to environmental management schemes and eco-label websites. The Commission promotes the use of eco label criteria in tender documents, since they are an easily accessible source of environmental information for many product groups.

The Commission has financed various workshops on GPP and is developing a web based training toolkit on GPP, including examples of environmental tender specifications for 11 product and service groups.

To reach the ambitious goal set down in the renewed Sustainable Development Strategy, the Commission is stepping up its efforts and early in 2008 will publish a Communication that will set targets for GPP and provide legal and operational guidance. The Communication will also kick off a process of cooperation with Member States and stakeholders to identify criteria for priority product and service groups. Finally, it will stimulate awareness raising and training and encourage the application of GPP to EU funded projects, focusing in particular on local authorities.

CONCLUSION

The European Commission will step up its cooperation with EU governments to implement green public procurement throughout Europe and reach the EU wide target set by the EU leaders in the renewed Sustainable Development Strategy. Efforts will particularly focus on local and regional authorities, as it is ultimately each individual purchasing officer who must be persuaded to give GPP a try and help to achieve a more ecofriendly world. By doing this, public authorities will give a strong signal to the supply side and stimulate EU industry to develop new environmental technologies. This will help ensure competitiveness in a global world which is increasingly concerned with the environmental aspects of production and consumption.

Because of the huge purchasing power of the public sector, we can achieve immense environmental benefits by buying green. We must not miss this chance.

Author

Prior to his appointment as the European Commission's Director-General for Environment in November 2005, Mogens Peter Carl served as Director-General for Trade for over five years. He joined the Commission in 1974, specialising in international maritime issues, covering transport, law of the sea and fisheries. In 1979, he was posted to the Commission's delegation in Washington, and was then seconded to the World Bank. He returned to Brussels in 1985 to concentrate on the Commission's international trade policy. Mr Carl holds a Masters degree in Economics from the University of Cambridge, and an MBA from INSEAD at Fontainebleau.

Organisation

The European Commission is the executive arm of the European Union. The mission of its Environment Directorate-General is to protect, preserve and improve the environment for present and future generations, and to promote sustainable development. This includes, of course, developing and implementing policies on climate change, biodiversity and air, water and soil pollution. Part of this mission is to encourage the private and public sectors to implement sustainable policies and voluntary approaches by taking account of environmental requirements in their activities.

Enquiries

Website: http://ec.europa.eu/environment/gpp/index_en.htm

Flood-related costs associated with climate change for a city such as Bilbao (pictured past and present) are estimated at between €130 and 160 million.

EUSKO JAURLARITZA
GOBIERNO VASCO
INGURUMEN ETA LURRALDE ANTOLAMENDU SAILA
DEPARTAMENTO DE MEDIO AMBIENTE Y ORDENACIÓN DEL TERRITORIO

© Basque Government

The Basque commitment to combat climate change

The Basque Country has a surface area of 7,000 square kilometres and just over two million inhabitants. Although its contribution to global climate change is small in quantitative terms, the Basque Government works to the basic principle that governs international policies on emissions reduction: shared responsibility, especially among industrialised countries. The Government has set up a policy based on two strategic axes: act against climate change and prepare for its consequences; and encourage a culture of innovation to work towards a sustainable Basque economy. In 2006, the Basque Office on Climate Change was set up and in autumn 2007, the Basque plan to combat climate change was completed, deploying specific actions to reduce and adapt to climate change in areas such as energy, industry, transport and the residential sector. Here, Esther Larrañga Galdos, Minister for the Environment and Land Use, highlights the issues and outlines the Basque Government's plan to combat climate change.

The Basque environmental strategy for sustainable development has been in place since 2002. Based on a timeframe running to 2020, it is aligned with the premises and coordinates set by the European Union and the United Nations. Two particular priorities of the strategy are the struggle against climate change and the conservation of biodiversity.

BASQUE SOCIETY'S SHARE OF RESPONSIBILITY

The environmental strategy also envisages two major commitments on the part of the public authorities. The first addresses developing countries and assumes Basque society's share of responsibility at international level. This is in the context of our position as joint president of the Network of Regional Governments for Sustainable Development, collaboration agreements with Latin America and initiatives to offset CO_2 emissions through reforestation in countries such as Kenya. The Government has, for example, agreed to plant a total of 232,475 trees in Kenya in the next three years through its agreement signed with the Green Belt Movement. This is to offset an estimated 23,400 tonnes of CO_2 emitted as a result of car and plane journeys.

Our second commitment as public authorities is to strive for excellence in governance, ie the "good governance" that Professor Wangari Maathai, Nobel Prize winner and Chair of the Green Belt Movement, has spoken of. This envisages environmental policy and the struggle against climate change as a contract with society based on transparency, participation, orientation towards results and accountability.

Figure: Trends in GHG emissions in the Basque Country, the EU and Spain 1990-2006.

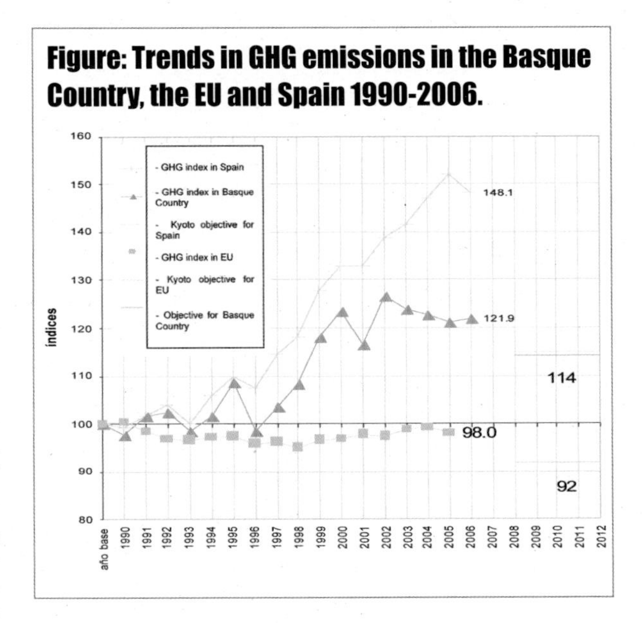

© Basque Government

OBJECTIVES OF THE BASQUE OFFICE ON CLIMATE CHANGE

If asked to identify the biggest difference between the 2005 UN climate change conference in Montreal and the 2006 UN conference in Nairobi, I would say that while Montreal focused on whether climate change was real, Nairobi looked at how to tackle climate change and its consequences. Bali, in 2007, must look at this last question in greater depth.

The Stern Report (*The Economics of Climate Change: The Stern Review*) has contributed substantially to this change of attitude by revealing that the cost of early action to adapt to climate change (around one per cent per annum of GDP by 2050) is substantially less than the costs arising from the adverse effects of that change, which are put at between five per cent and 20 per cent of GDP. This is precisely the philosophy underlying climate change work in the Basque Country. The Basque Office on Climate Change was set up in January 2006 with the remit of reducing greenhouse gas emissions (GHGs), meeting the requirements of the Kyoto Protocol and minimising the effects of climate change.

The Basque Country is currently emitting 24 per cent more CO_2 than in 1990. Following the completion in autumn 2007 of the Basque Plan to Combat Climate Change, the goal is to bring that figure down to 14 per cent, one point below the limit set by Kyoto for 2012. The sectors in which most effort will be required are energy and transport, which account for 65 per cent of emissions, with emissions increasing by 199 per cent and 101 per cent respectively since 1990.

In this context we must stress our government's backing for and patronage of the Global Action Plan (GAP) set up under the United Nations Environment Programme (UNEP). Over 13,000 families in the Basque Country have taken part in this programme, which sets out to encourage responsible consumption on a daily basis in the home.

© J.M. Gonzalo

ADAPTING TO CLIMATE CHANGE

We must adapt to the consequences of climate change, and understand how vulnerable the Basque Country is to its impacts. We must then design the most suitable policy to deal with the challenge accordingly.

The Basque Government is promoting the establishment of a multi-disciplinary working group to perform research on the most highly vulnerable areas (marine & land ecosystems, soil & agricultural resources and impact on energy infrastructures). Flood-related costs associated with climate change for a city such as Bilbao are estimated at between €130 and 160 million. This figure is for damage per annum envisaged up to 2080, though of course such damage is not expected to occur every year. The potential cost of flooding in such a city is expected to increase by 56 per cent as a result of climate change.

Reduction and adaptation is vital from the local to the global, or as the slogan for the Nairobi Summit so aptly put it: "Pulling together for our planet". We must do this for our own good, for our immediate surroundings, and out of solidarity and justice with other peoples and individuals.

Author

Esther Larrañaga was born in Gipuzkoa (Basque Country) in 1959. She holds a Bachelor's Degree in Law from the University of the Basque Country. She has been Basque Government Minister for the Environment and Land Use since June 2005, having previously worked in the Basque Government as Junior Minister for Justice at the Department of Justice, Employment and Social Security (November 2001 – June 2005) and Junior Minister for the Environment at the Department of Land Use, Housing and the Environment (1995 – November 2001).

Organisation

The Basque Country has a population of 2,082,587 and its own Parliament. The three governing councils of the historic provinces of Araba, Bizkaia and Gipuzkoa are in charge of collecting the main taxes and the high level of autonomy allows for direct government and administration of such matters as: treasury and tax collection, industry and economic promotion, research and innovation, transport, housing, the environment, education, public health and law and order. A multi-annual economic agreement regulates financial relations with the Spanish State.

Enquiries

Ministry of the Environment and Land Use,
Basque Government, Calle Donostia-San Sebastian,
101010 Vitoria-Gasteiz, Spain
Tel: +34 94 5019811
Fax: +34 94 5019849
Websites: www.euskadi.net I www.ingurumena.net

China's circular economy:

A NEW ECONOMIC PATTERN FOR FUTURE DEVELOPMENT

QIAN YI
ENVIRONMENTAL SCIENTIST,
TSINGHUA UNIVERSITY, CHINA

Rapid economic development and population growth worldwide have inevitably resulted in increases in pollution, waste and carbon emissions, putting further pressure on the planet and contributing to global warming. Closed loop processes and cleaner production techniques may be able to offer solutions to reduce these environmental impacts and, hence the impacts of climate change. Such processes combine to offer a circular economy approach to reducing environmental impact, while at the same time furthering significant country development. China is embracing the circular economy as a potential new economic pattern for its future development.

INTRODUCTION

Rapid economic development and population growth in the 20th century have caused serious environmental problems that are now affecting our climate. Experiences and lessons from practices around the world have proven that end-of-pipe treatments are not a cost-effective way of reducing these impacts. Instead, cleaner production, first developed in the 1970s, has obtained both environmental and economical benefits in different countries. The circular economy, closed loop concept was first developed in the 1990s and, together, these approaches can offer ways to minimise resource consumption and pollution production, thereby reducing overall emissions and mitigating, to some extent, the impacts of climate change.

CLEANER PRODUCTION

The cleaner production concept was developed in the 1980s with the Cleaner Production Program launched at the UNEP in 1989. For production processes, cleaner production involves conserving raw materials and energy and eliminating toxic raw materials, reducing the quantity and toxicity of all emissions and wastes before they leave the process. For products, cleaner production involves reducing the negative environmental impacts of a product along its life cycle, from raw materials extraction to its ultimate disposal. For services, cleaner production involves incorporating environmental concerns into designing and delivering services. Most importantly, cleaner production requires a change of attitudes and technology improvements.

In China, a strategy to implement cleaner production was proposed and endorsed by the Chinese Government at the Second National Conference on Industrial Pollution Prevention and Control in 1993. Promotion of cleaner production was listed as one of the nine top priorities in China's Agenda 21 issued in the same year. Since then, a series of efforts have been made for its implementation:

- Cleaner production implementation is being encouraged in laws and regulations related to environmental protection when they are modified and a new law, the Law of Promoting Cleaner Production of PRC, was issued in 2002 and put into effect in Jan 1, 2003.
- Cleaner production centres have been set up at national and provincial levels and industrial sectors to provide technical support to implement the cleaner production programme.
- Cleaner production implementation projects have been set up in China with the help of UNEP, the World Bank,

UNIDO, UNDP, the EU, and other foreign countries, including Canada, the US, Norway, Australia and the UK. These include: developing the Chinese cleaner production audit manual for enterprises; raising awareness on cleaner production; training of cleaner production auditors; executing cleaner production audits in demonstration plants and implementing cleaner production options requiring low cost or no cost;

Experiences of those who adopt cleaner production, show that both significant environmental and economic benefits are possible. Most organisations, for example, recoup their investment within two to three years, while significantly reducing pollutant discharges and emissions.

CIRCULAR ECONOMY

Closed loop processes or the circular economy concept have evolved from decades of worldwide effort to work towards a sustainable economic development model.

> "In the circular economy, however, a development pattern is promoted that is harmonious with the earth. The circular economy organises economic activities towards a closed loop process of 'resource – production – consumption – regenerated resource'."

Circular economy, not linear economy

Traditional industrial economy is a one way linear economy consisting of 'resource – production – consumption – disposal'. In this kind of open loop system, energy and materials are 'drained' from the planet while also releasing greenhouse gas (GHG) emissions and other pollutants to land, sea and air. Economic activities are typically 'high exploitation, low utilisation, and severe pollution'.

In the circular economy, however, a development pattern is promoted that is harmonious with the earth. The circular economy organises economic activities towards a closed loop process of 'resource – production – consumption – regenerated resource'. All materials and energy are used rationally and continuously in sustained economy cycles, and the harmful effects of emissions and pollutants can be reduced to a minimal level.

The three 'Rs' as a working principle of a circular economy

The waste mantra of the three Rs: 'reduce, reuse and recycle' form the principles of a circular economy:

- **Reduce**: reducing the input energy and material flows into the production and consumption process; producing essential products with as few resources as possible.
- **Reuse**: using natural resources and products in every possible ways.
- **Recycle**: turning wastes to secondary resources, reducing wastes for final disposal and decreasing the consumption of natural resources.

Some may simply regard circular economy as waste recycling, yet the fundamental goal is to systematically prevent and reduce wastes in the industrial process. And according to the China Council for International Cooperation on Environmental Protection and Sustainable Development (CCICED)Task Force on Circular Economic and Cleaner Production, cleaner production is the cornerstone of circular economy.

Circular economy as the economic pattern for sustainable development

Traditionally, economic growth and environment protection do not always go together. They are separated instead of integrated in solving economic, social and environmental problems. Traditional economy, for example, solely pursues GDP growth. It grows at the cost of natural capital consumption and leads to conflict between economy and environment. It reduces employment positions by increasing the level of automation, and leads to conflict between economy and society.

Unlike a traditional economy, the circular economy is a 'triple-win' economy. It combines economic growth, environmental protection (which includes those actions needed to reduce the impacts of climate change) and social employment. In every aspect of development, a circular economy means a revolutionary change. In promoting economic growth, a circular economy brings a change: from material growth in volume to service growth in quality. In solving environmental problems, the circular economy brings change from open-loop terminal recovery to closed-loop process control. And in promoting social employment, a circular economy brings change from an employment-downwards society to an employment-upwards society.

We have long followed a linear growth pattern and as a result economic growth has accelerated at the cost of ecological deterioration with profound changes to the world's climate. We must find a new economic pattern for China's future development, ie the organic integration of economic growth, environment and resource protection, and social employment achieved by a circular economy.

Some iron and steel complexes in China, for example, have made efforts to follow a circular, closed loop strategy. All emissions from the production process are collected and used to produce electricity and all solid waste is used to produce cement. In addition, water recycling occurs as much as possible for reducing water consumption. As a result, an iron and steel complex with the capacity of 8-10 X 10^6 tonnes of steel can support a power station with 1.2 X 10^6 KW of electricity and a

cement manufacture with 3 X 10^6 tonnes of cement, greatly reducing overall emissions.

Some municipalities and provinces in China have also launched programmes promoting the circular economy. Ecological industrial parks have developed rapidly in China, linking several enterprises in order to make efficient use of materials and energy. In Liao Ning Province, Jiangsu Province and GuiYang city and Shanghai Municipality, strategic overview plans have been developed to carry out activities based on a circular economy at municipality or provincial level. Industrial, agricultural and social aspects are included. It is expected that a law for promoting circular economy will be drafted and issued in the near future.

> "We have long followed a linear growth pattern and as a result economic growth has accelerated at the cost of ecological deterioration with profound changes to the world's climate."

CONCLUSION

The circular economy is a new mode of economic development which not only concerns the change of production patterns but also the change of consumption patterns. The urgent need now, if we are at all serious in our efforts to curb the impacts of climate change, is implementation. To do this, the coordination of different institutes and the combination of different measures: educational, logistical, instructional, financial and technological, will be vital.

Author

Professor Qian is an environmental scientist and educator in China. Ms Qian has been involved in education and research on water pollution control at Tsinghua University as an associate professor. She has conducted research on: appropriate technologies for wastewater treatment in China, mechanisms and technologies for treating refractory organic pollutants, the treatment process for wastewater reuse, strategy for sustainable management of water resources. She is an elected member of the China Academy of Engineering and serves as Vice Chairman of the Committee of Environmental and Resources Protection, the National People's Congress of China. Professor Qian is also a member of the Board of Directors of the World Resources Institute. She is actively engaged in the implementation of cleaner production and sustainable development strategy in China.

Organisation

Tsinghua University is the leading university within science and technology in China with a long and distinguished record in environmental sciences and engineering education and research. Courses in municipal engineering have been offered here since 1928. At present, the Department of Environmental Sciences and Engineering has over 80 faculty members, including three members of the Chinese Academy of Engineering, over 360 undergraduates, and over 400 graduate students. Main research focuses on water pollution, atmospheric pollution, solid waste, remediation of polluted environments, environmental chemistry, environmental biology, environmental management and policy, environmental information systems.

Enquiries

Yi Qian
Department of Environmental Science & Engineering
Tsinghua University, Beijing 100084, China
E-mail: qiany@tsinghua.edu.cn
Website: www.tsinghua.edu.cn

© Zoran Mrdjenovic/Fotolia

Progress in climate change adaptation in the Caribbean community

DONNA MCRAE SMITH
PROJECT OFFICER,
CARICOM

The Caribbean community is deeply committed to the implementation of adaptation and mitigation measures to minimise the effects of climate change on the region's sustainable development. Governments, as well as the private sector, are involved in such efforts. The region is making commendable progress through the delivery of training programmes, the practice of sustainable tourism and initiatives in renewable energy.

A REGION EXPOSED

While Small Island Developing States (SIDS) produce only a tiny fraction of global greenhouse gas (GHG) emissions, many, because of their location barely above sea level, are among the most vulnerable to the effects of climate change. Climate change is therefore a matter of critical importance for the very survival of the Caribbean community (CARICOM).

Over the past decade, the Caribbean has experienced extensive social, economic and environmental impacts of intensive storms and hurricanes, with consequent significant loss of life, and major set backs in the national and regional development programmes. Ignoring the spectre of climate change and its anticipated impacts, would truly expose the region to unprecedented consequences, and display poor stewardship in our responsibility for the natural and built environment and the regional patrimony.

> "More than 95 per cent of the tourism infrastructure in many CARICOM Member States is located within 10km of the coastline, thus rendering the industry highly susceptible to the effects of exaggerated wave action and tidal surges."

The scope of the required adjustments varies widely since each CARICOM Member State has its own special adaptation needs and resource constraints. These states have for some time been engaged in gathering information and are now arriving at a position where the risks can be better understood and adaptation measures can be identified.

Source: www.tiamoresorts.com/tiamo/power.cfm

IMPACTS OF CLIMATE CHANGE

The anticipated impacts of climate change on the sustainable development of the region will affect all sectors, but most critically, tourism, water, agriculture, human habitat and economic infrastructure. More than 95 per cent of the tourism infrastructure in many CARICOM Member States is located within 10km of the coastline, thus rendering the industry highly susceptible to the effects of exaggerated wave action and tidal surges.

Climate variability, as expressed by changing and unpredictable weather patterns, already represents a major challenge for the agriculture sector as well as for planners generally in the community. Both the tourism and agriculture sectors will be severely affected by declining water availability and by extended periods of drought.

Another significant threat is the projected impact of climate change on human health, through an increase in the presence of vectors of tropical diseases, such as malaria and dengue, and the prevalence of respiratory illnesses. These diseases will affect the well being and productivity of the workforce of the Caribbean and compromise the region's economic growth and development potential.

THE CARIBBEAN COMMUNITY CLIMATE CHANGE CENTRE

In 2002 the CARICOM Heads of Government decided to establish the Caribbean Community Climate Change Centre and to place the issue of climate change at the highest level on their agenda. The Caribbean community coordinates its climate change activities, through the Climate Change Centre, which is based in Belize. The centre provides a range of services and products relating to research, impact assessment, response strategies and systematic observation of climate change in the region.

The Caribbean community recognises the need to deal with climate change through international agreements and joint implementation efforts, but is also committed to action at national level. Such action is aimed at implementing adaptation and mitigation plans and programmes.

ADAPTATION EFFORTS IN THE CARIBBEAN COMMUNITY

There are many opportunities for a wide range of adaptation options to be adopted throughout the Caribbean community. At national level, governments as well as the private sector are now taking action to address climate change issues with noticeable progress in a number of areas.

It is to their credit that these governments contributed to, and supported, for over five years commencing in 1997, the foundation project, Caribbean Planning for Adaptation to Global Climate Change (CPACC) with the support of the Global Environment Fund (GEF), through the World Bank, and in association with the University of the West Indies and the CARICOM Secretariat. Presently, the GEF funded Mainstreaming Adaptation to Climate Change (MACC) Project is being implemented by the Caribbean Community Climate Change Centre.

The project's objectives include:

- Build capacity to identify climate change vulnerability and risk.
- Build capacity to reduce vulnerability to climate change.
- Build capacity to access and effectively utilise resources to minimise the costs of climate change.
- Public education and outreach.

Sustainable tourism

In CARICOM, the tourism sector makes the greatest use of the coastal and marine resources and many of the region's adaptation efforts can be seen through the response of this major economic sector. The tourism sector is highly dependent on infrastructure. This includes airports, coastal protection structures, natural amenities, beaches, reefs and wetlands, access to clean and adequate supply of water, reliable energy and food. Even before climate change issues were on the regional and global agenda, the tourism sector was turning its attention to sustainable tourism. While much emphasis has been placed on training, there are also many practical examples of adaptation measures being implemented.

The Caribbean Environment Network (CEN) Project, an initiative of the Caribbean Environmental Programme (CEP) was implemented during 1996-99 to improve environmental quality and coastal and marine natural resource protection, by promoting the use of environmentally sound practices in the tourism industry. This project conducted training workshops and developed training manuals for a number of areas including Improved Training and Public Awareness on Caribbean Coastal Tourism and Manual for Sand Dune Management in the Wider Caribbean. The effects of these training workshops will become more evident in the near future.

Water management

With respect to water management, desalination plants have been constructed in Antigua and Barbuda and Barbados. In some CARICOM states, such as St Vincent and the Grenadines, building codes require cisterns to be constructed for water capture and storage. At the regional level, the Caribbean Environmental Health Institute (CEHI) is encouraging Member countries to practice Integrated Water Resource Management. The Integrated Watershed and Coastal Area Management (IWCAM) Project was approved by the Global Environment Facility (GEF) in May 2004.

> **"The Caribbean community recognises the need to deal with climate change through international agreements and joint implementation efforts, but is also committed to action at national level."**

Sustainable energy

The climate change challenge of small island nations is compounded by their struggle with expensive fossil fuel imports and an inability to supply electricity in rural areas. It is recognised however, that these nations are especially suited to utilise combinations of modern renewable energy technologies and the implementation of energy efficiency measures. The development and use of renewable energy is gaining momentum across the region.

Solar energy initiatives are largely private sector based and focus primarily on the provision of water heaters, but in Barbados the government has offered incentives to homeowners for the installation of these units. Solar companies in Barbados and St Lucia have expanded their services to several other CARICOM States. Households as well as hotels across the region are now utilising these services. Solar driers have also been profitably used in drying onions, hay and plastics in Antigua and Barbuda, Barbados and Trinidad and Tobago respectively. In Barbados, sugar factories are now engaged in cogeneration using bagasse to produce electricity and are contributing approximately 15-18 per cent of that country's primary energy supply. The Government of Belize recently invested US$62 million in a cogeneration plant, Belize Co-Generation Energy Limited (BECOGEN), which will use bagasse to generate 31.5 megawatts of electricity most of which will be sold to the national grid.

The private sector is also very actively involved in energy conservation schemes. The Caribbean Alliance for Sustainable Tourism (CAST) is assisting hotels by conducting energy audits, and is a leading advocate of incentives for the promotion of energy efficient technology, the development of standards and building codes and the provision of attractive interest rates for the purchase of energy efficient technology.

Other sustainable energy initiatives are ongoing in the region. With assistance from the Global Sustainable Energy Islands Initiative (GSEII), St Lucia has adopted a National Sustainable Energy Plan, establishing aggressive targets for renewables and energy efficiency, and setting the stage for significant changes in the energy sector. Grenada and Dominica are developing similar plans.

Food and agriculture

Another major initiative is the Regional Special Programme for Food Security (RSPFS) which is spearheaded by the Food and Agricultural Organization (FAO) in collaboration with the CARICOM Secretariat with support from the Government of Italy. This project involves the CARICOM Member States, and the Dominican Republic. A key aim of the Programme is to address constraints related to increasing trade, increasing small farmer productivity, and improving food policy, food insecurity information and linkages between food product development and promotion and food health related issues.

GOING FORWARD

In October 2007, the Climate Change Centre signed a Memorandum of Understanding with the Government of the United Kingdom to facilitate funding for the development of the comprehensive Caribbean Regional Climate Change Strategy. This project will provide the Caribbean community with a coherent and cohesive policy framework and implementation plan to guide its efforts.

Adaptation efforts in the region so far, have focused primarily on capacity building for adaptation. However, it should be pointed out that the adaptation activities highlighted herein are in keeping with the Region's Sustainable Development agenda, namely the implementation of the Barbados Programme of Action (BPoA) and the more recent Mauritius Strategy of Implementation. There is now a need however, for developing a mechanism which would provide a more deliberate feed of climate change adaptation issues into the sustainable development agenda.

Author

Donna McRae Smith is the Project Officer in the Sustainable Development and Environment Unit of the Caribbean Community (CARICOM) Secretariat since 2003. She has several years professional experience in the fields of environmental management, agriculture, industrial development and mining. In her present capacity, she assists in the coordination of CARICOM's activities particularly with regard to the areas of climate change, marine resource management, and biodiversity. She holds a B.A. in Applied Geography and an LLM in Environmental Law and Management.

Organisation

The objectives of the Caribbean Community (CARICOM), formerly the Caribbean Free Trade Association, are to improve standards of living and work; the full employment of labour and other factors of production; accelerated, coordinated and sustained economic development and convergence; expansion of trade and economic relations with third states; enhanced levels of international competitiveness; organisation for increased production and productivity; and achievement of a greater measure of economic leverage.

Enquiries

Caribbean Community (CARICOM) Secretariat
PO Box 10827, Georgetown, Guyana
Tel: +592 222 0001 75
Fax: +592 222 0171
E-mail: info@caricom.org
Website: www.caricom.org

Steps...

There are a number of steps that you can take to reduce your organisation's carbon footprint. This section of Climate Action is dedicated to these positive actions. Over the next 26 pages, you will see how easy it can be to reduce your carbon footprint – and how, by doing so, you can reduce costs and increase profits. Some of the steps require little investment in time or money, while others require substantial time and capital. What they all require is a commitment to succeed.

... to shrink your carbon footprint

Business is at the forefront of efforts to reduce climate change. Without the support of business, efforts to reduce greenhouse gas emissions will be at best only partially successful and not nearly enough to avoid the worst impacts of global warming. Most business leaders inherently know this and realise that climate change is a serious challenge – maybe the most serious challenge human societies collectively face today.

What is often missed, however, is that climate change has another side. Reducing the amount of carbon pollution that your business creates and releases into the atmosphere is an opportunity to reduce costs and increase profits. It also offers the possibility of taking new business and market opportunities. To start with, it's an opportunity to engage your staff in a team effort that will surprise you in the ways in which you can achieve results.

Of course, there are other drivers besides the immediate bottom line: mainly your customers, an increasing number of which want to reduce their carbon footprints. Like many others in business, however, you may be a little stumped on where to begin to reduce your carbon emissions, the total of which is often described as a carbon footprint. However, if you are reading this publication, you have already begun.

There are a number of actions you can take. Some seem almost too simple and some are more challenging. Some require little or no investment in time or money, while others require substantial time and capital. Some will succeed in the short term and others in a longer horizon. However, they all necessitate action, first requiring a commitment to succeed and improve continuously, the basis for any successful business.

Fortunately, you have help in the climate action publication and website (www.climateactionprogramme.org). You are not alone and, importantly, you are not the first one out of the foxhole. The following outlines several actions you can consider to help your company shrink its carbon footprint. Now is the time to start to change your business culture, not the climate.

© Sławomir Jastrzebski/Fotolia

These Actions have been compiled by **Peter Fries**, a journalist specialising in energy and environmental issues, and **Cornis van der Lugt**, who is responsible for corporate environmental and social responsibility at UNEP's Division of Technology, Industry and Economics (DTIE).

"EVERYTHING CAN ALWAYS BE DONE BETTER THAN IT IS."
Henry Ford

"Climate change is shaping global markets and global consumer attitudes. There will be winners and losers. Companies who seize the opportunities, who adopt strategic environmental, social and governance policies and who evolve, innovate and respond to these challenges are likely to be the pioneers and industry leaders of the 21st century."

Achim Steiner, Executive Director,
United Nations Environment Programme

CLIMATE SCEPTIC?

Are you someone who thinks the statement by climate scientists that "there is a 90 per cent probability that humans are changing the climate" is wrong? What if there was only a 10 per cent chance that humans are changing the climate? What's it worth to insure against that risk?

Think of the insurance you and your business already carry. For most insurance products, the probability of an insurable event is small, much less than one per cent. In Australia, the cost to insure a vehicle is about three per cent of the capital cost. Most families in OECD countries would spend at least one per cent of their net income on insurances of some type. The average for OECD is €2,700 *per capita* spending on insurance for an average income of €40,000 – that's about six per cent.

With this logic, an investment of one per cent of income to mitigate the risk is reasonable. In business, however, this spending can often lead to savings that are much, much greater. So even if you do not believe climate change is a serious threat, you can still reduce your carbon pollution in ways that add to your bottom line.

1 Make a commitment and form an action team

Leadership, focus and commitment are success factors in creating a climate for change and a business that strives towards a zero carbon footprint. Reducing carbon is no different from many other management tasks to improve a company's performance and reputation, and prudent companies embed such thinking into their core business management systems.

A comprehensive climate change strategy can affect almost every aspect of the operations of a company. As with all organizational change, a phased approach with strong leadership, commitment to continual improvement, effective employee engagement, ongoing monitoring and regular evaluation is necessary.

However, the signal to staff, shareholders and other stakeholders of the company's intentions is also important. Signals move markets and the signal that management intends to reduce emissions cannot be overstated. Telling your staff that your company will reduce carbon emissions may seem simplistic, but even simple actions, such as a notice on the company's intranet and in the common room, can be effective. When the US Environmental Protection Agency wanted to promote the Green Lights programme for energy efficiency lighting (see www.greenlights.org), for example, they wrote directly to CEOs. Many companies reported that the simple action of posting the letter to company bulletin boards motivated staff, with positive results following. The simple act of asking for ideas leads to some innovative solutions.

Forming a climate action team can include a diverse staff, including representatives from operations management, design, product/service development, procurement, marketing, communications and accounting. The team can then produce a strategy with specific goals. Specific targets will help to provide focus and goalposts against which to track progress.

Some companies have used a continual improvement process called 'energy kaizen' (from *kai* meaning 'change,' and *zen* meaning 'for the better'), a process of implementing change and investing in thinking and changing processes. In the case of Kodak, an energy kaizen led to immediate changes in climate and lighting controls and a reduction in unnecessary ventilation exhaust. This helped the company stop losing US$1 million a year in unnecessary electricity costs at no capital outlay.

The team may also have to engage your board members to approve investments that have varying returns. As mentioned above, some carbon reductions can be made just by simple changes to existing processes using only human capital as well as the cost of coordination and control. Others however, require financial capital that may have varying returns and payback periods. Even with lucrative returns, however, businesses can be reluctant to part with reserve or borrowed cash. This is where leadership and a strategic commitment can really make a difference.

NOT ALWAYS AN EASY SELL

In 2003, the world's largest shipper, FedEx, announced a programme to replace the company's 30,000 medium duty trucks with diesel-electric hybrids over the next 10 years. Four years later, FedEx has purchased fewer than 100 hybrid trucks, or less than one-third of one per cent of its fleet, although fuel savings should pay for the higher costs of hybrids over the 10 year lifespan of the vehicles. FedEx, which reported record profits of US$2 billion for the fiscal year that ended May 31, decided that breaking even over a decade wasn't the best use of company capital. According to *Business Week*, the company said its fiduciary responsibility to shareholders meant it couldn't subsidise the development of this technology for their competitors.

2 Assess where you stand

An assessment of your company's climate impacts needs to consider both the opportunities to profitably reduce emissions as well as the risks of not reducing them. These can vary substantially between different companies and sectors.

Knowing where and how your company generates greenhouse gases is the first step to reducing them. This has to be determined through the use of methods such as energy audits and environmental technology assessments. As the business adage goes: 'what gets measured, gets done'. For small businesses, online calculators and internal assessments can help start the process. Larger organisations may need specialised advice and use of tools, such as the new ISO 14064 standard for greenhouse gas accounting and verification.

> "Knowing where and how your company generates greenhouse gases is the first step to reducing them."

The Greenhouse Gas Protocol of the World Resources Institute (www.wri.org) and the World Business Council for Sustainable Development (www.wbcsd.org) provides an accounting tool for government and business managers to understand, quantify, manage and report greenhouse gas emissions. Its guidance notes five important factors for a company's assessment:

- **Relevance**, ensuring the greenhouse gas (GHG) inventory appropriately reflects the company's amount of emissions and serves the decision making needs of users – both internal and external to the company.
- **Completeness**, accounting for and reporting all GHG emission sources and activities within the chosen inventory boundary, and justifying any specific exclusions.
- **Consistency**, using consistent methodologies to allow for meaningful comparisons of emissions over time.
- **Transparency**, addressing all relevant issues in a factual and coherent manner, based on a clear audit trail.
- **Accuracy**, achieving sufficient accuracy to enable users to make decisions with reasonable assurance as to the integrity of the reported information.

Compiling and maintaining a comprehensive inventory of greenhouse gas emissions improves a company's understanding of its emissions profile and any potential liability or exposure. Such risks are increasingly a

Intro | Step 1: Electricity | Step 2: Heating | Step 3: Auto Travel | Step 4: Air Travel | Offset Now

Introduction

How much CO2 and other greenhouse gasses (CO2e) do your activities create? Fill in the blanks in our CO2 emissions calculator to find out.

Each section in our calculator will ask you to fill-in information about your energy use. You may complete each of the sections in order or jump ahead as you wish by using the tabs at the top of the calculator. If you do not complete all of the sections, the calculator will indicate your CO2 emissions for the sections that you choose to complete.

Note: you will need cookies enabled in your browser to use the carbon calculator.

Start >>

Source: www.greentagsusa.org

"Significant GHG emissions in a company's value chain may increase costs or reduce sales, even if the company itself is not directly subject to regulations."

management issue due to heightened scrutiny by the insurance industry and shareholders and increasing moves by political leaders to reduce greenhouse gas emissions through regulations. Although the time frame is uncertain, like many of the impacts of climate change, there is little doubt that carbon will eventually be judged as an atmospheric pollutant and regulated accordingly. In the context of future greenhouse gas regulations, significant GHG emissions in a company's value chain may increase costs or reduce sales, even if the company itself is not directly subject to regulations. Thus investors may view significant indirect emissions upstream or downstream of a company's operations as potential liabilities that need to be managed and reduced. A limited focus on direct emissions from a company's own operations may miss major GHG risks and opportunities, while leading to a misinterpretation of the company's actual GHG exposure.

CALCULATING THE FOOTPRINT

For a basic idea of the carbon footprint of a business, a number of online calculators can help. Estimating the carbon footprint of commercial buildings can be done via, for example, the Portfolio Manager, US Environmental Protection Agency's online energy rating system for commercial buildings.

Webconferencing business Webex has an online calculator on its website: www.webex.com.au

Other calculators:
www.clevel.co.uk/business-calc.htm.
(UK/EU specific)

www.puretrust.org.uk/Business/Calculator.aspx
(UK/EU specific)

www.greentagsusa.org/GreenTags/calculator_intro.cfm (North America specific)

www.seegreennow.com/calculate.aspx

"WHAT GETS MEASURED, GETS DONE."

© Anton Foltin/iStockphoto

3 Decide and plan where you want to go

Based on your company's assessment of its climate related risks and opportunities, a strategy and action plan can be developed. This may involve pursuing a new business model. It also needs to be linked with, and build on, your company's environmental management systems (EMSs). Targeted objectives will help focus efforts and also provide a benchmark for measuring success. Most businesses can reduce energy use by 10 per cent, almost always resulting in a 10 per cent reduction in greenhouse gas emissions, with a one year payback or less.

The next step is to make a plan and allocate responsibilities and resources. For most organisations, a plan to reduce carbon emissions will first focus on the type and way energy is used. This energy includes the electricity for buildings and manufacturing, the gasoline/diesel/gas for fleets and the fuel for other types of travel.

Reducing this energy creates savings that go straight to the bottom line. A US$1,000 reduction in a company's energy bill equals US$1,000 more profit. This type of efficiency, however, can also increase productivity, see action 'Get efficient', see page 276. But there is also another way to think of this other than saving energy and emissions. Your company is actually losing tens of thousands of dollars, perhaps millions, by purchasing unnecessary energy and carbon.

One of the most effective tools for plugging this hole is an energy audit referred to earlier. This can be done internally with the help of an outside expert. Many electric utilities and government energy offices now will offer an audit as part of their efforts to reduce carbon emissions.

"Your company is actually throwing tens of thousands of dollars away by purchasing unnecessary energy and carbon."

© Kutay Tanir/iStockphoto

4 De-carbon your company DNA

There is a broader way to think about carbon and climate. Everything in a business embodies some form of carbon, either in the product itself or in the energy and materials it takes to make the product. Buildings, fittings and equipment are all proxies for carbon, 'carbon copies' that can be chosen based on the least amount of impact they will have on the climate and the bottom line.

As carbon is already part of a company's DNA, engrained in its systems and business model, it's important to take actions to reduce it. Integrating climate friendly criteria into R&D, product design and development, supply chain management, manufacturing, distribution, marketing and communications creates a systematic company response that can trigger a ripple effect by helping to 'unchain' suppliers from their climate impacts and multiplying both the savings in carbon emissions and unnecessary costs.

The supply chain

For some organisatons, the biggest carbon emissions may come from suppliers. Reducing those emissions are more challenging, but many companies are setting procurement guidelines that include environmental requirements that encourage low or no carbon products and services. Unbundling the carbon from your supplies can also create multiple benefits and is itself a process embedded in the larger moves to sustainable development. Sometimes this may cost more, but often the cost can be the same or less. A partnering effort with suppliers and/or downstream clients can result in efficiency gains that provide a win-win for all involved. More intelligent application of information and communications technologies can help to drastically cut material and energy waste in the value chain.

Sourcing materials, such as paper from companies adhering to internationally certified standards can mitigate some supply chain risks. The Forestry Stewardship Council (www.fsc.org), for example, is an international nonprofit organisation promoting responsible management of the world's forests. The FSC trademark is increasingly recognised as an international standard for responsible forest management. More than 90 million hectares in more than 70 countries have been certified according to FSC standards while several thousand products are produced using FSC certified wood and carrying the FSC trademark.

PACKAGING SMART WAL-MART

US retail giant Wal-Mart worked with one of their toy suppliers to help them reduce packaging on just 16 items. The toy suppliers saved on packaging costs while Wal-Mart used 230 fewer shipping containers to distribute their products, saving about 356 barrels of oil and 1,300 trees. By broadening this initiative to 255 items, the company believes it can save 1,000 barrels of oil, 3,800 trees, and millions of dollars in transportation costs.

There is one supplier that is actually critical to a business – staff. The time and energy they use getting to and from their place of employment can produce significant carbon emissions. Helping them to reduce their emissions at the same time they are helping the company to reduce its emissions can produce multiple benefits, and a more engaged staff.

Reduce the waste

In the area of recycled material use, switching to recycled or sustainably sourced paper can lead to considerable savings, reducing both waste directed to landfill and carbon emissions. The Waste & Resources Action Programme (www.wrap.org.uk/) estimates the average greenhouse gas savings from using recycled paper instead of it going to landfill is 1.4 tonnes of CO_2 for every tonne of paper and cardboard. A WRAP study of 112 companies, including BT, Carillion, Ernst and Young, KPMG, Sainsburys, Tate & Lyle and all London's boroughs, found these organisations bought 23,000 tonnes of recycled paper and board in 2005, saving 53,000 m^3 of landfill space, 690,000,000 litres of water, 621,000 kg of air pollutants and 32,000 tonnes of CO_2.

© Olga D. van de Veer/Fotolia.com

Buildings

As a business periodically needs to replace existing assets, new infrastructure presents an excellent opportunity to reduce emissions and costs. This involves office buildings (including leased buildings), factories and plants, vehicle fleets and equipment.

Conventional buildings can account for almost 40 per cent of CO_2 emissions per year, but high performance facilities, ie those that are environmentally accountable, energy efficient and productive for occupants, are economically possible through an integrated approach to design and construction. This approach can produce buildings with capital costs the same or less than conventional construction, and significantly lower operational costs over the life of the building, which also lowers the impact on climate.

Features of buildings constructed in this way include:

- Systems that efficiently use water resources and energy.
- Quality indoor environments that are healthy, secure, pleasing and productive for occupants and owners.
- Natural landscaping, efficient material use and recycling.

Such designs require people, processes and technologies to be brought together early in the design phase. The project team includes the owner, architect, builder, subcontractors and suppliers and is assembled before making any major decisions about the building. The team will be together throughout the project, considering each aspect with an eye towards reducing resource use during construction.

For many businesses, an office retrofit can produce significant reductions in pollution while increasing productivity. There are many good examples of using the approach above to create new offices that are more productive and cheaper to run, all while reducing environmental impacts. Most retrofits address the key physical assets: lighting, space conditioning, furniture and floor coverings.

WHOLE SYSTEM ENGINEERING

There is substantial experience with retrofits where a 'whole system engineering' approach, ie looking at the entire building as a system, produces surprising, multiple benefits. In an analysis of one Chicago 20,000 m^2 office tower retrofit, for example, super efficient windows, lights and office equipment could reduce cooling load by 85 per cent compared to the original building. This made the replacement cooling equipment 75 per cent smaller and US$200,000 cheaper. This was enough savings to pay for all the other renovations and reduced the annual energy bill by 75 per cent.

Source: Niclas Svenningsen and Thierry Braine-Bonnaire, (2007) 'The buildings challenge, entering the climate change agenda.' in Climate action

How many office employees does it take to change a light bulb?

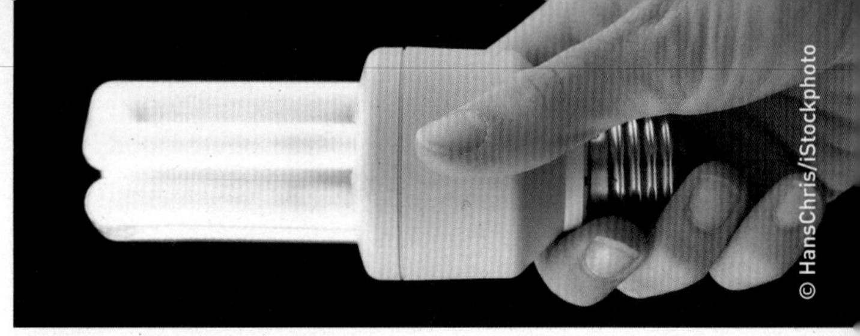
© HansChris/iStockphoto

5 Get energy efficient

Improving the efficiency of your buildings, computers, cars and products is the fastest and most lucrative way to save money, energy and carbon emissions. This does not mean going without. Energy efficiency is about increasing productivity but doing more with less. More efficient buildings, cars and products are a direct and lasting contribution to limiting carbon emissions and many energy efficient products cost no more than inefficient ones.

Simple is effective

It's important to realise that very simple measures can lead to immediate savings. Just turning off unused lights, motors, computers and space heating can substantially reduce wasted energy. In one of their distribution facilities, Staples, the world's biggest suppliers of office products, replaced conventional conveyor belt motors with high efficiency soft start motors and sensors to turn off lighting and fans when parts of the building were unoccupied. The result was a US$100,000/yr saving in electricity – a pay back of less than 12 months.

Energy efficiency improves the bottom line but it also yields even more valuable side benefits:

- Higher quality and reliability in energy efficient factories.
- Six to 16 per cent higher labour productivity in efficient offices.
- Forty per cent higher sales in stores skilfully designed to be illuminated primarily by daylight.

Revenue generated from increased productivity can be 10 times as high as the energy cost savings received from performing upgrades. Every dollar invested in an energy efficient upgrade can produce between US$2 and US$3 in increased asset value, which can make commercial properties more attractive to buyers and lenders.

Office equipment and appliances

Because of their rapid turnover, office equipment and other appliances, such as refrigerators, can be a potent source of wasted money from unnecessary energy use and consequently greenhouse gas emissions. Computers, copiers and printers come in many different configurations that can either reduce or increase operating expenses. Generally, laptop computers use less energy than desktop computers and LCD monitors less energy than CRT screens. Any large organisation that relies on computing can substantially reduce electricity bills and carbon emissions by simply using the energy saving mode when the computer is not in use. This is generally a very quick and easy action with an almost immediate payback.

There is also the issue of what to do with equipment when it is finished with its useful life, referred to as 'e-waste'. Procurement guidelines that stipulate energy efficiency and product take back or recycling create the most resource savings and reductions in greenhouse gas emissions.

For appliances, the Energy Star rating is a way to describe the efficiency of products; the more stars, the more efficient they are. For many brands now, the highest energy efficiency rating – four or five stars – does not cost any more than products with few stars. Originally from the US, Energy Star is now applicable in Europe (www.eu-energystar.org).

In 2007, the Energy Star guidelines were upgraded. Called Energy Star 4.0, the standard raises the bar for manufacturers of desktops, notebooks, tablets, workstations and low end servers. The new standard requires computing devices to switch their displays into sleep mode and reduces the amount of electricity needed for standby and sleep modes.

MOVE ELECTRONS, NOT PEOPLE

Telecommuting

Travel to attend meetings and conferences is expensive, time consuming and often unpleasant. Modern communications equipment has relieved some of this burden. Emails have replaced letters and the mobile phone has saved time, money and hassle. Modern advances in technology, particularly web and video conferencing, mean the time is quickly approaching when the need to travel will be substantially diminished. Consider a two day business trip to attend a meeting 1,000 km (600 miles) away. Physical travel costs about US$2,000 per person when accommodation, travel and meals are included, while a video conference costs about US$200. The savings are US$1,800 and about half a tonne of carbon. For medium size company, Lawson, the savings from employing web conferencing via WebEx in its Global Support Center saved the company US$600,000 per year, for an ROI in excess of 700 per cent.

Telecommuting can also be both a winner for the bottom line and the environment. With the average commute of nearly 12,000 kilometres a year, a study by the Telework Coalition (www.telcoa.org) found that if 32 million Americans who could telecommute did so one day a week, they would not:

- Drive around two billion kilometres – a distance equal to 51,000 times round the Earth.
- Waste US$100 million from buying an unnecessary 300 million litres of petrol.

TELECONFERENCE SUPER SAVER

Bell Canada, has focused on climate change to drive teleconferencing, saving 142,000 tonnes of carbon dioxide emissions annually from more than 2.7 million teleconferences of their customers. The company has also conducted a Climate Protection Programme for several years, including a programme called Everyday Kyoto for employees that aims to reduce the emissions of company operations. By allowing more than 15,000 of their employees to telecommute, the company has saved 6,000 tonnes of carbon dioxide emissions annually.

The more energy stars a product has, the more efficient it is.

Additionally, telecommuters gain the equivalent of 32 million extra hours every week for leisure, family or work, which would provide a US$300 billion bottom line benefit to the economy every year.

Whether it is environmental improvements or productivity gains as the principle outcome of teleworking and teleconferencing, it seems one supports the other. AT&T has found, for example, its teleworking programme has helped staff save time, petrol and pollution, and helped the company to slash its annual real estate costs by US$30 million and gain US$150 million in extra hours of productive work from teleworkers. An important secondary benefit is that the teleworking programme also lowers training and staff costs, as teleworkers are much less likely to switch jobs than in office staff.

The German data processing company, SVI, evaluated the benefits and costs of a teleworking pilot and found that for a set-up cost of around €5,500, plus monthly running costs of €120-225 per person, the benefits were at least three to four times greater than these costs and included:

- Technology savings of €1,200-1,300 per person at work.
- Increased productivity of 2-5 per cent per year.
- Reduction of sick leave by an average of two days annually.
- Savings in recruitment and training costs because of increased staff retention.

British Telecom used video conferencing technology to replace nearly 900,000 face to face meetings in one year. The company has saved £135 million for travel and travel allowances and £103 million for time saved. This represents a 2,500 per cent return in just 12 months. They also saved 97,000 tonnes of CO_2. That's good business, and good for the environment. The company is working with others under the Global e-Sustainability Initiative (www.gesi.org) of UNEP and industry partners to further examine opportunities related to climate action and improved use of information and communications technology.

Lighten up

For an average office building or factory, lighting makes up about 15-20 per cent of total electricity use. Converting coal at the power plant into incandescent light in your facility, however, is only three per cent efficient.

Compact fluorescent lights, also called simply CFLs, have evolved rapidly in the past decade. They now last between six and 15 years, depending on which product you choose, and reduce electricity use by a minimum of 75 per cent compared to a standard incandescent bulb. What's more, CFL emits much less heat than incandescent bulbs, creating a double benefit by reducing energy consumption from air-conditioners. Because they emit less heat, they are also safer to use in confined areas. New designs can replace hot halogen bulbs and be operated on dimmers.

The advantages of CFLs and other high efficiency lighting have prompted legislation to ban incandescent bulbs. In 2007, Australia is the first country to mandate that no incandescent bulbs will be sold in Australia by 2012, a move that will reduce emissions by four million tonnes and cut power bills for lighting by up to 66 per cent.

EFFICIENCY PAYS

Over the decade from 1995-2005, chemical manufacturer DuPont boosted production by nearly 30 per cent but cut energy use by seven per cent and greenhouse gas emissions by 72 per cent (measured in terms of their carbon dioxide equivalent), saving more than US$2 billion. Five other major firms, IBM, British Telecom, Alcan, NorskeCanada and Bayer, have collectively saved at least another US$2 billion since the early 1990s by reducing their carbon emissions by more than 60 per cent. In 2001, oil giant BP met its 2010 goal of reducing carbon dioxide emissions 10 per cent below the company's 1990 level, thereby cutting its energy bills US$650 million over 10 years. General Electric has vowed to raise its energy efficiency 30 per cent by 2012 to enhance the company's shareholder value.

CFLs certainly use less energy than traditional incandescent lights, but free sunlight is even better. Prince Street Technologies, a subdivision of Interface Carpets, relocated its factory into a new building with 32 skylights providing excellent natural lighting. The energy bills were less, but more interestingly, after three years, workers' compensation cases had dropped from 20 per year to less than one, saving between US$100,000 and US$200,000 each year.

Money in the tank

For many businesses, the vehicle fleet is a major capital outlay and operating expense. Despite more than a century of refinement, however, the modern car remains astonishingly inefficient. Just six per cent of the fuel energy actually accelerates the car, and all this energy converts to brake heating when the car stops. And, because 95 per cent of the accelerated mass is the car itself, less than one per cent of the fuel ends up moving the driver. It's important to choose the most efficient vehicle that meets your needs. Smaller and lighter vehicles not only cost less to buy and run, they do so without compromising safety. Already auto manufacturers are voluntarily producing or planning to produce cars with tailpipe emissions significantly lower than conventional vehicles and there are moves to tax fuels based on their carbon content.

New vehicle designs, such as hybrids and advanced diesel designs, offer excellent fuel economy and life cycle costs. With oil prices increasing, there is already some market pressure to increase efficiency, but often tax codes work against this through higher deductions for larger engines. This confusing signal can be countered by company procurement guidelines that stipulate either fuel economy or tailpipe emissions of carbon dioxide.

BETTER BUILDINGS THROUGH PERFORMANCE CONTRACTS

Upgrading the energy performance of buildings can be done using performance contracts. Siemens, for example, will upgrade buildings, promising prescribed amounts of energy and monetary savings. If there is a shortfall, the difference is made up by Siemens, while better than expected results are shared with the building operator. Siemens has reported total worldwide savings of nine million tonnes of CO_2 and one billion Euros.

"green" is cheaper and smarter

A "green" IT and network approach to business can also be easily extended to hardware. Our expertise in infrastructure consolidation means our customers can halve the number of servers they need and reduce their data centres by tenfold, resulting in significantly less energy consumption and fewer CO_2 emissions. Further energy savings can be achieved by using communication technologies to build paper-free workflows: electronic billing, ordering and reporting; and using web portals to manage the whole customer experience. E-government applications for online administrative procedures can also help: according to the WWF and ETNO study, 100 million tax declarations made over the Internet would result in a saving of 100,000 tonnes of CO_2.
New communication technologies can also help reduce road traffic and optimise the real-time use of thousands of vehicles. Thanks to embedded GPS, "Machine-to-Machine" fleet management tools, journeys can be streamlined and shortened, and at the same time allow improvements in the preventive maintenance of those vehicles, minimising still further their environmental impact. This is why, for example, the Singapore bus company uses the Orange Fleet Performance system. Elsewhere, "Machine-to-Machine" services can provide solutions to help to manage environmental risks: we are currently developing monitoring systems for flood and pollution detection, as well as noise control.

we're our own case study...

The good news about these "green" approaches to communication technologies is that, beyond their obvious environmental benefits, they also happen to be very profitable for their users. At the France Telecom group, we are convinced of that, and in order to reduce our own ecological footprint, we practise what we preach and use the same products and services that we recommend to our customers. For example, some 30,000 persons are using our "thin-client" e-bureau solution. In 2006, we held 600,000 videoconference meetings, so that our teams avoided making 1.2 million business trips, a saving of several tens of millions of tonnes of CO_2. We are also developing electronic billing – already adopted by 80% of our Spanish customers – saving 200 tonnes of paper a year, or 12 million litres of water. Our R&D and purchasing policies enable us to reduce the number of servers and data centres we need, and by optimising ventilation they use less and less power. In 2006, we bought 6,000 "green" cars emitting CO_2 below 130 g/km, and we are including electric and hybrid vehicles in our fleet.

recycling as part of design

We are putting into place in each country where we operate an environmental management system that meets the ISO 14001 standard. We are also including ecodesign in our product development, to integrate recycling at an early stage and to optimise resources. Finally, we place particular importance on accurately measuring the ecological impact of our activities, using the Global Reporting Initiative indicators amongst other ones. We believe that even the simplest solutions can have a lasting impact, such as making it possible for customers to manage billing online without having to print. Of course, we also run more ambitious projects. For example, in Senegal we are testing a 100% solar-powered mobile phone network which could one day provide coverage to rural areas around the world, and promote economic development without hurting the environment. To summarise, Orange believes that together we can make the future "green".

France Telecom group in brief...

Orange is the flagship brand of France Telecom group, one of the world's leading telecommunications services providers. We serve more than 163 million customers on 5 continents. In 2005, we adopted the New Experience in Telecommunications (NExT) program, which is aimed at transforming us into the benchmark operator of integrated services in Europe. With a 2006 turnover of €51.7 billion, we are the number three mobile phone operator and the number one ADSL service provider in Europe, and amongst the world leaders in providing telecommunications services to multinational companies.

For more information:

www.francetelecom.com/en/group/development/index.html

orange™

6 Switch to low carbon energy

There are several ways to switch to energy sources that emit less carbon and can reduce costs and emissions. Generally, coal produces twice the emissions of gas, six times the amount of solar, 40 times the amount of wind and 200 times the amount from hydro.

Green power

A new energy service has been offered in many parts of the world where an electricity customer can choose to have a percentage of their electricity supplied from a renewable energy source, such as a wind farm or landfill gas project. These 'green choice' programmes are maturing and proving to be a powerful stimulus for growth in renewable energy supply, according to the National Renewable Energy Laboratory (NREL), which operates the Green Power Network (www.eere.energy.gov/greenpower). Today, more than 50 per cent of all US consumers, for example, have an option to purchase some type of green power product from a retail electricity provider, with 600 utilities offering green power programmes to customers in 40 states.

These programmes allow customers to purchase some portion of their power supply as renewable energy or to contribute funds for the utility to invest in renewable energy development. The term 'green pricing' is typically used to refer to such utility programmes and is often offered at a premium price.

Utility green pricing programmes are one segment of a larger green power marketing industry that counts many government agencies, colleges and universities and Fortune 500 companies among its customers, and helps support more than 2,200 megawatts (the power required for about one million average US homes) of renewable electricity generation capacity. More than 200 universities purchase green power for some (or in a few cases all) of their electricity requirements. The clean-E standard (www.eugenestandard.org/) provides guidance for green power in Europe.

Generate your own

Of course, you can also build your own lower emission energy system, including solar power and lower carbon technologies such as generators powered by natural gas. One of the first companies to do this was beauty products retailer, The Body Shop, buying a 25 per cent stake in a large modern wind generator to provide renewable energy for its UK operations. Other companies installing their own renewable energy plant include 3M, DuPont, General Motors, IBM, Johnson & Johnson and Staples.

US telecom company, Verizon, is using a combination of energy efficient Hypalon roofs, microturbines and fuel cells for some of its facilities. The natural gas mircoturbines generate about 50 per cent of the electricity for two offices in California, reducing carbon dioxide emissions by approximately 6.6 thousand tonnes per year, or the equivalent of taking 1,400 passenger cars off the road for one year. At its switching centre and office building in Garden City, New York, Verizon operates seven fuel cells of 200 kilowatts each, reclaiming the heat and water the cells produce to help heat and cool the building. In addition to environmental benefits, Verizon is saving more than US$600,000 by not having to obtain power from the commercial power grid.

© Tesco

UK retailer, Tesco, has joined the growing number of companies running hybrid delivery vehicles.

State and federal tax breaks and incentives means solar photovoltaic systems and other renewable energy technologies can often be profitably built and used when combined with long term electric contracts. Roofs are being transformed into a source of savings through the installation of solar electric panels that provide energy over time, reduce electricity costs and provide a buffer against price fluctuations.

Solar roofs are also good public relations with a growing number of companies combining government incentives and subsidies to create 'green' brand value. The largest organic and natural food supermarket in the US, Whole Foods, installed a 125 kilowatt solar photovoltaic system at its Edgewater, New Jersey store. The system covers 1,400 square metres of the store's roof and meets more than 20 per cent of the its electricity needs. The one megawatt Green Garden Solar Project in Shenzhen is located in the grounds of

© Martin Bond/Still Pictures

Shenzhen's International Garden and Flower Expo Park and is spread across the roofs of four iconic buildings, each with solar arrays ranging in capacity.

Cleaner fuels and vehicles

As a sector, transport is responsible for 25 per cent of total energy consumption and greenhouse gas emissions, mainly from burning petrol and diesel. Various options for climate action exist in the field of vehicles fleet management and logistics. Take for example fuels. Vehicles can run on a range of fuels, including compressed natural gas (CNG), liquefied petroleum gas (LPG), liquefied natural gas (LNG) and biofuels. Depending on location, these fuels can offer both cost and environmental benefits, although they also often require an additional investment that take some time to payback. Hybrid engines that combine electricity and conventional petrol or diesel engines can also offer substantial fuel savings while reducing emissions.

Biodiesel and bioethanol are biofuels made from crops, such as wheat, soy, corn and sugar cane. They are often blended with petrol or diesel, and almost all vehicles can run on blends up to 10 per cent without modification. Specially enabled biofuel cars, such as the Ford Focus flex-fuel or the Saab Biopower, can run on higher blends, such as a mix of 85 per cent bioethanol and 15 per cent petrol.

© Holger Buse/Fotolia

IT'S THE BILL, NOT THE PRICE

In the short term, many green power programmes carry a slightly higher price. So why should a business pay more for such energy? They shouldn't, and in the medium term, the cost may be cheaper when carbon is priced into the market. However, it's worth remembering that the higher price is in terms of a rate, whereas your business pays a bill for energy: the rate times the number of units used. By becoming more efficient, you can still lower the bill, even though the rate is slightly higher. It's efficiency that enables a switch to cleaner energy that might have a higher price, but a lower bill because you have used less.

Even when blended in small amounts, biofuels can substantially reduce pollution, particularly greenhouse gases as the carbon dioxide released by burning the fuel is absorbed by the crops when they are grown. Depending on how the crops are grown and converted, the greenhouse benefit can be a 30-90 per cent net reduction.

In many parts of the world, biofuels are becoming more popular and easier to find commercially and in various blends. Particularly for companies with automotive fleets, biofuels can offer an attractive alternative, as their cost can be cheaper than conventional fuels. In the UK, for example, Tesco, the country's largest supermarket chain, will run three-quarters of its delivery fleet (2,000 trucks and vans) on a 50 per cent biodiesel mix to cut the greenhouse gases from January, 2008.

Modes

Switching the mode of your transport can also reduce emissions. DHL Ltd, an international express mail carrier, sought to cut costs and reduce waste and also found unexpected improvements in productivity and quality. Instead of using separate delivery vehicles, the company implemented a single 'Team Bus' to transport couriers to a central London location. Once in London, the couriers travelled on foot to their delivery destinations. The company reports that this system significantly reduced fuel and maintenance costs but also improved delivery times and productivity because couriers spent less time in traffic.

Sometimes simple actions can produce a shift. Secure bicycle storage and change/shower facilities, for example, are often inexpensive compared to other parking structures but create a strong incentive for those who can commute by bicycle. In larger cities with adequate public transport, a monthly or yearly pass can be offered instead of parking facilities. Paris and Vienna, for example, offer all its inhabitants a public bicycle system in an action that reduces greenhouse gas emissions and traffic congestion.

7 Invest in offsets and cleaner alternatives

There is a limit to how much efficiency you can squeeze from your operations or how much renewable energy you can employ. The only choice for those that wish to compensate for their remaining emissions is to find them elsewhere in the form of some activity by another party that reduces emissions. This is commonly called a carbon offset or carbon credit. The term carbon neutral refers to the idea of neutralising emissions through carbon savings elsewhere.

Carbon offsets

Offsets are offered in two types of markets: regulated (also called mandatory or compliance), and voluntary. Regulated markets include the Kyoto Protocol and the associated Clean Development Mechanism, and the European Union's Emissions Trading Scheme that began in 2005. In the US, there is currently no federal mandate to reduce greenhouse gas emissions, although substantial effort continues at the state level, with California attempting to create the first regulated carbon market along with several other states. Some large companies are under mandatory programmes to offset their carbon emissions, including those European companies subject to the EU carbon trading scheme. More information can be obtained from the International Emissions Trading Association (www.ieta.org), an independent, nonprofit organisation dedicated to the establishment of effective systems for trading in greenhouse gas emissions by businesses.

To purchase offsets, individuals or businesses pay an offset company to implement and manage projects that avoid, reduce or absorb greenhouse gases. Climate change is a global problem, so carbon reductions will have the same impact no matter where they are implemented. Carbon credits are generated in a number of ways, including:

- Emission free energy generation including renewable energy.
- Reduction of the demand for energy including energy efficiency.
- Sequestration in the form of underground and forestry storage.
- Preservation.
- Chemical conversion.

The first two categories avoid emissions, and are generally considered a higher quality and less risky offset than sequestration offsets. Sequestration projects aim to absorb emissions that have already occurred. They are the most controversial as there are currently no commercial underground sequestration projects and it is unlikely any will be built for a decade. Forestry offsets

© Hazlan Abdul/iStockphoto

"To purchase offsets, individuals or businesses pay an offset company to implement and manage projects that avoid, reduce or absorb greenhouse gases."

are considered risky due to uncertainty of calculating absorbed carbon and the threat of drought or disease that can literally kill a project.

Preservations offsets pay to protect areas that absorb carbon, such as tropical rainforests. The rationale is that by preserving the carbon sink, carbon can be taken from the atmosphere, or not released from activities that reduce the area, such as logging.

According to one report, the highest quality offsets are generated from the flaring of methane from landfills. This is due to the fact that methane has a global warming potential 24 times greater than carbon dioxide over 100 years and 61 times greater in the first 25 years. Greengas International (www.greengas.net) is one company that generates carbon credits by converting waste gas to energy through joint partnerships with mines, landfills and biogas producers. Worldwide benefits of such projects include 125 megawatts (MW) of power, saving four million tonnes of CO_2.

Although sometimes criticised, voluntary markets have the capability to create a much larger change by pushing regulated markets to cleaner outcomes, becoming the tipping point that creates rapid and permanent emission reductions and even a new energy economy based on sustainable energy. Aside from the feel good factor, this is a substantial and concrete reason to participate.

Many companies are finding a marketing value in offsetting some or all of their emissions from a variety of sources directly or indirectly related to their business activities. Others recognise that participating in voluntary carbon markets is excellent preparation for future life under a mandatory cap and trade scheme. Computer maker, Dell, for example, has recently entered into an agreement with a carbon offset company to offset carbon from shipping computer equipment. Some large travel agencies have even started giving their clients the option to purchase offsets for their travel. Non-governmental organisations (NGOs) and corporations have also started to offset their employees' emissions from travelling.

Currently, there are more than 35 companies and organisations active in the voluntary offset markets targeting a range of customers. Eleven companies address the effects of air travel specifically. There are also more than 600 providers of green power. The average price for carbon offsets is US$15/tonne, but costs range from US$5-50 tonne.

Because it is not regulated, the voluntary carbon market is almost anything someone wants it to be. The current market is a case of buyer beware. However, some emerging standards are helping companies choose quality offsets, including Green-e (www.green-e.org), and the CDM Gold Standard (www.cdmgoldstandard.org).

Emissions trading

Whether voluntary or mandatory, emissions trading is an increasingly popular approach to emissions reduction. It is a market based mechanism that aims to achieve environmental objectives at least cost. A central authority, usually a government agency, sets a limit or cap on the amount of greenhouse gases that can be emitted. Companies or other groups that emit CO_2 are given credits or allowances that represent the right to emit a specific amount. The total amount of credits cannot exceed the cap, limiting total emissions to that level. Companies that pollute beyond their allowances must buy credits from those who pollute less than their allowances. This transfer is referred to as a trade.

The Chicago Climate Exchange (www.chicagoclimateexchange.com), for example, is North America's only, and the world's first global marketplace for integrating voluntary legally binding emissions reductions with emissions trading and offsets for all six greenhouse gases.

The European Union Emission Trading Scheme (EU ETS) is the largest multinational, greenhouse gas emissions trading scheme in the world where large emitters of carbon dioxide within the EU must monitor and annually report their CO_2 emissions. They are obliged every year to surrender (give back) an amount of emission allowances to the government that is equivalent to their CO_2 emissions in that year (www.euets.com).

8 Get materials efficient

Any serious business leader who examines the flow of materials for their business knows that a linear system that relies on external inputs produces supply risks. But it can also produce disposal risks as well from toxic materials used in the manufacture of products.

Looking at your business through a carbon neutral lens can also help your business in other ways by increasing the efficiency of resource use, producing cleaner, avoiding and reducing waste and ultimately improving your overall performance and reputation. Economists are often fond of saying that there are no banknotes lying on the pavement because someone would have already picked them up. In climate change, however, there are plenty of banknotes just waiting to be picked up. After all, carbon is generally the waste product of producing energy, and reducing waste, becoming more efficient, is always good for business.

Integrating the 3R approach, reduce, reuse and recycle, into a company's waste management and material efficiency processes can lead to multiple benefits. As with energy efficiency, this requires a 'whole of system' approach. A multiple industry approach that promotes a more closed loop, rather than a linear, system can produce even greater savings. In Denmark, for example, a number of companies established operations near each other to 'feed' off the waste streams. In the Netherlands, Shell is selling its carbon emissions to nearby horticultural businesses for their greenhouses. Shell receives direct payments for the excess CO_2, tax breaks to cover the infrastructure costs, as well as credits under the European trading scheme.

WHAT A WASTE

- The amount of waste generated to make a semiconductor chip is over 100,000 times its weight, a laptop 4,000 times its weight. Two litres of petrol and 1,000 litres of water are required to make a litre of orange juice in the Southern part of the US.
- Eighty per cent of products are discarded after a single use.
- Business leader Paul Hawken estimates that 99 per cent of the original materials used in the production of, or contained within, the goods made within the US become waste within six weeks.
- It is estimated that industry moves, mines, extracts, burns, wastes, pumps and disposes of more than one million kilograms (2.2 million pounds) of material to provide one middle class average family's needs in an OECD country for a year.
- Total wastes in the US, excluding wastewater, now exceed 22 trillion kilograms (50 trillion pounds) per year. This includes:
  - 3.3 trillion pounds of CO_2
  - 19 billion pounds of polystyrene 'peanuts'
  - 710 billion pounds of hazardous waste

© Hazlan Abdul/iStockphoto

9 Offer low carbon products and services

The market for climate friendly products and services is growing rapidly, from energy efficient products to new renewable energy systems. To offer such products, however, it's important to begin at the design stage where 70 per cent of the costs of product development and manufacture are invested, and significant impact on end of life management for a product is determined. Actions as simple as adding energy efficient specifications into the design process, for example, can produce a design that minimises energy consumption during its use and saves customers the time and energy from making adjustments to a product after a purchase, such as wrapping water heaters with insulation blankets.

A more thorough and systematic approach comes from the field of 'design for sustainability', which includes life cycle design and environmentally conscious design and manufacturing. This new approach to design considers environmental aspects at all stages of the product development process to create products with the lowest environmental impact throughout the product life cycle. This can reduce waste, generating financial benefits on resource use and waste disposal costs. Ecodesign is an important strategy for small and medium sized companies both in developed and developing countries to improve the environmental performance of their products and at the same time improve their competitive position on the market.

Product service systems (PSS) that enable new, climate friendly models of efficiency can also be introduced. These systems shift the business focus from designing and selling physical products only, to selling a system of products and services which are jointly capable of fulfilling specific client demands. What the company or an alliance of companies conceive, produce, and deliver, is not simply material products, but in fact, a more integrated solution to a customer demand, producing a satisfactory utilitarian result. Perhaps the most famous example is that of carpet supplier Interface that decided to offer its products in a modular tile system as a 'floor covering service' where the units are eventually returned to the company for re-manufacture into new units.

“The market for climate friendly products and services is growing rapidly, from energy efficient products to new renewable energy systems.”

© Interface Americas, Inc.

Interface carpets are sold in units which are eventually returned to the company for re-manufacture into new units.

© Interface Americas, Inc.

10 Talk to consumers

The market for green products and services is growing rapidly but remains underdeveloped because people may find it difficult to locate these products and trust their environmental claims. According to a recent *Business Week* article, "As green products proliferate, innovative marketing will be key to attracting consumers' attention."

You may want to revisit your marketing strategy, aiming to listen more carefully to the needs of consumers, your customers, as well as finding ways to help them consume in more climate friendly ways. This can take many forms, from the online click for carbon offsetting on a tourism booking website to the label on a product at the local retail store.

Close attention to consumer loyalty and behaviour needs to consider carefully renewed consumer interest in climate matters. In many countries consumer surveys report that growing numbers of consumers are willing to buy green products if given the choice. Innovative product design and presentation as well as responsible marketing and communications can help ensure that this consumer interest translates into consistent purchasing behaviour at the point of sale.

The Natural Marketing Institute (NMI) has identified sales of "health and sustainability" products at an estimated US$550 billion worldwide in 2006. These customers are characterised by strong environmental and social values and purchase decisions based on those values. They are less price sensitive and the earliest adopters of new and innovative products, and they actively try to influence others to purchase green products and services.

Getting these customers to buy requires a company to meet both green and conventional product expectations, be good corporate citizens, and communicate substantive and credible information on all parts of the company's product and business. Your own customers can also be a good source for new products and services that reflect their values. Your own product labelling can help communicate your company's commitment to better environmental practices, and help customers make more informed decisions and change their user behaviour.

"Innovative product design and presentation as well as responsible marketing and communications can help ensure that this consumer interest translates into consistent purchasing behaviour at the point of sale."

PRODUCT LABELS SHOWING GREEN ATTRIBUTES

THE 10 COMMUNICATIONS ELEMENTS

# of Successful Corporate Sustainability

How can you distinguish a corporate sustainability program that's destined for success from one that's headed for the recycling bin?

One way is to examine how firms communicate. In 2007, Fleishman-Hillard's Sustainability Communications specialists researched communications indicators associated with successful sustainability programs in companies on both sides of the Atlantic. What we found was that whether the companies were in Toronto or Torino, Dublin or Detroit, the hallmarks of sustainability success were the same.

Sustainability communications should be aspirational.

At this point in the evolution of sustainability communications, companies are best served when they articulate what they want to be, as opposed to what they are. Framing environmental initiatives as goals and aspirations shows long-term commitment to environmentally sustainable corporate conduct. But the alternative — promoting only those things you've already done, calling yourself green, or engaging in hyperbole about your environmentalism — is viewed as disingenuous, and sets you up for charges of "greenwash."

Let others sing your green praises.

It's far better and more credible to have your environmental initiatives praised by a third party than to do it yourself. Third parties, such as NGOs and academics, have more impact when lauding an organization's green efforts, and their independent status makes them more credible.

Consumers expect companies to act responsibly simply because it's the right thing to do.

Consumers take it for granted that companies are incorporating environmentally responsible programs into their operations. This is true for all kinds of companies, from manufacturing to service firms. Companies without environmental policies and proactive sustainability programs are vulnerable to reputational damage, loss of market share, and even loss of shareholder value.

It's okay to profit from your sustainability initiatives.

A growing number of consumers and environmentalists realize that dealing with climate change requires a cooperative effort between government, industry, and the scientific community. What's more, consumers and environmentalists recognize that companies need to make profits to stay in business. After all, companies need to be financially sustainable to maintain environmentally sustainable programs.

Corporate sustainability programs need to be integrated into the firm's strategic plan. Occasional environmental tactics are no longer adequate.

One of the best ways to be stamped with a "greenwash" label is to characterize your company as environmentally active, simply because it built a bike path or landscaped its headquarters with indigenous plants. To be credible, sustainability programs need to be comprehensive and built into the company's business model; one example would be a shipping firm that searches relentlessly for transportation energy efficiencies.

6 Internalize your sustainability strategy.

If an organization's sustainability program is to succeed, environmental responsibility must be part of its corporate culture. It's difficult, if not impossible, to build an external environmental corporate identity if that identity isn't first embraced internally by employees.

7 Be aware of downstream consequences of your corporate actions.

Inventory everything your organization does that has an environmental impact on others, from water use to waste disposal, carbon dioxide production, and nonrenewable energy consumption. In the months and years to come, companies will be held accountable for their entire environmental footprint.

8 It's all about solutions.

Consumers value companies that offer solutions to environmental problems. Even companies whose primary function has nothing to do with ecology can frame themselves as problem solvers, simply by having a functioning sustainability program.

9 Leverage your influence.

Companies, nonprofits, and universities are not self-sufficient islands. To stay in business, they need the services of many other firms, from power companies and food vendors to paper distributors and computer manufacturers. Increasingly, companies are expected to leverage their influence with their supply chains, encouraging — or even requiring — them to meet certain environmental standards as a condition of doing business.

10 Maintain transparency about your policies and their implementation.

An organization should be willing to share information about its sustainability programs. Communicating how you built employee enthusiasm or reduced your carbon emissions can provide others with ideas, and position your institution as a leader. Conversely, firms that are unwilling to be part of the public dialogue about their corporate environmental behavior risk damage to their reputations, market share, and shareholder value. One way to share environmental action information is by participating in a voluntary climate-change information program, such as the Carbon Disclosure Project (CDP). By 2007, more than 2,400 of the world's major corporations participated in the CDP, providing data on their carbon emissions reductions and other environmental efforts. The CDP makes this information available to the public.

These 10 elements can be used by anyone — from financial analysts trying to identify firms with winning corporate sustainability initiatives, to CEOs designing leading-edge environmental strategies. And our work isn't limited to Europe and North America; we've got experts in Mumbai, Beijing, Hong Kong, Tokyo, Mexico, Johannesburg, and a dozen cities in between.

To learn more about the rapidly evolving world of sustainability communications, or to find out how to design a meaningful sustainability approach, talk to us. We'll identify optimal sustainability practices for tomorrow, so you can lay the groundwork for them today.

For more information, contact Malin Jennings, Fleishman-Hillard Global Sustainability Communications in Washington, D.C., at **202-828-5062 or malin.jennings@fleishman.com**

11 Team up

Addressing climate change presents opportunities to profit from working with other organisations inside and outside the business sector. Many companies are increasingly working with non-governmental organisations, cities or governments to identify and implement best practice solutions to profitably reduce emissions. The Carbon Disclosure Project (www.cdproject.net), for example is an independent nonprofit organisation providing information for institutional investors with a combined US$41 trillion of assets under management. On their behalf, CDP seeks information on the business risks and opportunities presented by climate change and greenhouse gas emissions data from more than 2,000 of the world's largest companies.

Similarly, governments at all levels are seeking out opportunities to partner with business on delivering low carbon solutions for their citizens. In countries, such as Canada, government institutions and power utilities supported the setting up of Energy Service Companies (ESCos). In the US, the federal Environmental Protection Agency (EPA) started the Energy Star program (www.energystar.gov) in 1992 as a voluntary partnership to reduce greenhouse gas emissions through increased energy efficiency. In 2006, American businesses and consumers with the help of Energy Star saved US$14 billion on their energy bills and reduced greenhouse gas emissions equal to those of 25 million vehicles annually. Sometimes programmes have financial incentives that help to break the finance barrier. Programmes or initiatives range from simple information to finance incentives with low or no interest loans to make changes that produce environmental benefits, including those related to climate change. Under the Clinton Climate Initiative, US president, Bill Clinton has recently negotiated an agreement with private companies to help 1,100 cities buy energy efficient products at volume discounts (www.clintonfoundation.org)

Collective industry initiatives and your local business association can also help, among others, to share experience and know how. Many business and climate initiatives have been launched in recent years. A special contribution can also be made through UN business partner initiatives, such as the climate working group under the UNEP Finance Initiative and the Caring for Climate initiative of the UN Global Compact, UNEP and the World Business Council for Sustainable Development. Partner initiatives with UN agencies can help facilitate work with stakeholders and regulators to raise the playing field and use recognised global standards in all regions of the world.

© claudiobaba/iStockphoto

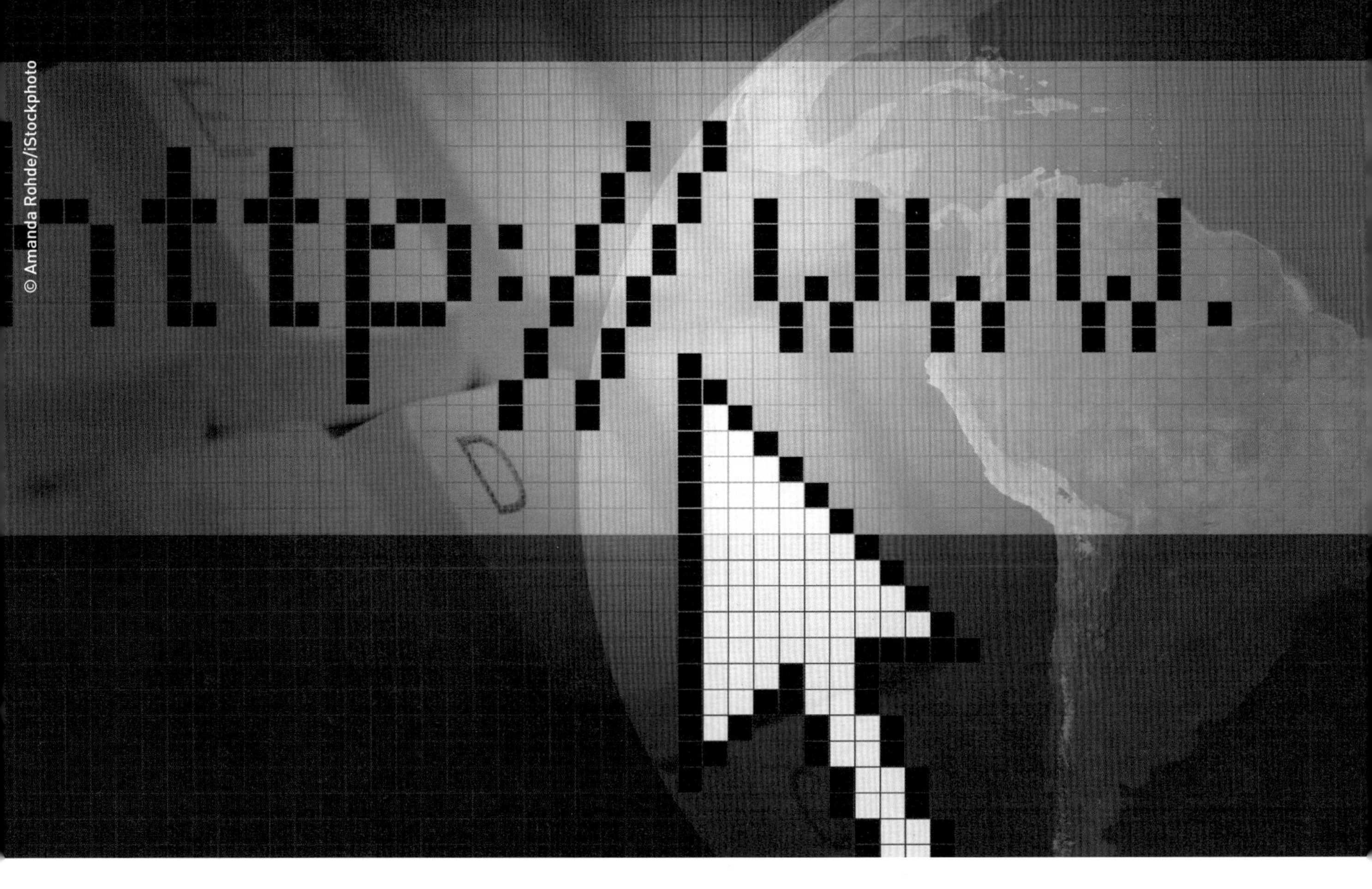

12 Communicate and report results

The increasing importance of climate change means that companies and organisations will need to effectively communicate their responses in various ways and through various channels, both internally and externally. As with assessments of a company's greenhouse gas emissions, transparency is critical. The internet and other new media mean that companies cannot hide behind the greenwash thin veneer. This is where tools for verification and reporting guidelines with recognised indicators are critical, an obvious example being the Global Reporting Initiative (GRI) (www.globalreporting.org). The G3 version of the GRI Guidelines includes special guidance on boundary questions and protocols on for example energy and emissions-related indicators that can be used by your company to measure, track, benchmark and communicate progress.

> **"Internal communications via intranets and company publications can report progress and acknowledge contributions by individual staff or teams."**

Internal communications via intranets and company publications can report progress and acknowledge contributions by individual staff or teams. It's also important to let your shareholders know. Reducing emissions, particularly by improving efficiency is a win-win situation that can also enhance a company's reputation. Many climate change related shareholder resolutions are now underway by institutional shareholders such as state and city pension funds, foundations, religious institutional investors and socially responsible investment firms. These resolutions are requesting information on a company's response to risks and opportunities related to climate change such as climate change policy, greenhouse gas emissions reports, and emission reduction plans.

Naturally, industries that have a direct impact on emissions are more targeted, such as fossil fuel, forestry and the electric utility sectors. Past resolutions have resulted in changes in corporate governance, accounting practices, and now to a growing extent, climate change policies and non-financial reporting. In response to a shareholder resolution, for example, Ford Motor Company began an open dialogue with the resolution's proponents on greenhouse gas reporting and future emissions reductions.

As mentioned earlier, reducing carbon emissions can serve as a driver for reducing, or eliminating, other environmental impacts of your business and its products and services. Communicating this to other parts of the company and encouraging staff to look for opportunities is an important element to improve the way your company does business. This underlines the potential role of reporting as a management tool and helping your organisation find deeper solutions. It brings us back to the change in culture and business model highlighted earlier.

ADVERTISER	PAGE	WEBSITE
Arctic Paper	94	www.arcticpaper.com
BARD Engineering	146	www.bard-offshore.de
Basque Government (IHOBE)	59	www.euskadi.net / www.ingurumena.net
Dalum Papir	272	www.cyclus.dk
EDF	141	www.edfenergy.com
Fleishman Hillard	288 & 289	www.fleishman.com
France Telecom	278 & 279	www.francetelecom.com
Green Gas	157	www.greengas.net
Grundfos	284	www.energyproject.com
InterCall	14 & 82	www.greenconferencing.com
Johnson Controls	264 & IBC	www.johnsoncontrols.com
KPMG	2	www.kpmg.co.uk
Megaman	275	www.megaman.cc
Novo Nordisk	66	www.novonordisk.com
Regione Abruzzo	246	www.araen.org
Ricoh	93	www.ricoh.com
Sappi	18	www.sappi.com
SGS	74	www.sgs.com
Siemens	32	www.siemens.com
U.S. Environmental Protection Agency	65	www.epa.gov
Verizon	26 & 27	www.verizonbusiness.com
Vestas	8	www.vestas.com
Webex	214	www.webex.com
WestLB	112 & 113	www.westlbmarkets.com
Yara	16 & 17	www.yara.com